THE EXPERIMENTAL BASIS OF MODERN BIOLOGY

TO

JAMES GRAY

THE EXPERIMENTAL BASIS OF MODERN BIOLOGY

BY

J. A. RAMSAY

Fellow of Queens' College and
Professor of Comparative Physiology in the
University of Cambridge

CAMBRIDGE
AT THE UNIVERSITY PRESS
1970

Published by the Syndics of the Cambridge University Press
Bentley House, 200 Euston Road, London N.W. 1
American Branch: 32 East 57th Street, New York, N.Y. 10022

Standard Book Numbers:
521 06027 3 clothbound
521 09575 1 paperback

First published 1965
Reprinted 1966 1970

Printed in Great Britain
at the University Printing House, Cambridge
(Brooke Crutchley, University Printer)

CONTENTS

PART III. INHERITANCE

PREFACE

Of all the features of this book there is probably none which so openly invites criticism as does the title. A glance through the chapter headings can hardly fail to provoke the comment: is this, then, the sum total of modern biology? is it implied that physiology, ecology, the study of evolution, are worked-out subjects, of no further interest, shouldered aside in the press of new ideas? I confess, lamely, that the title is not a good one, but it is the best I can think of. I might, with stricter attention to accuracy, have called the book 'The Observational and Experimental Basis of Certain Aspects of Modern Biology', or better still '... of Those Aspects of Modern Biology which are Currently Written Up in the Popular Scientific Press'. One can be too pernickety about these things. I do not think that the precise title matters very much; no title could be concise and at the same time convey what I would wish it to convey. But I think it does matter that I should try to explain how the book came to be written and what it hopes to achieve.

Many students, with time and interest to spare, are encouraged to engage in fieldwork to supplement the formal courses given at school and university. But the interests of others lie elsewhere, in the direction of cell biology, and for them a literature exists which presents the results of recent advances in a very appealing way. Such boys find themselves in a situation where on the one hand they learn in school a traditional biology which is still preoccupied with the names of the arteries of the frog, while on the other hand the colourful pages of the *Scientific American* invite them to participate in discovering 'the secret of life'.

I was for many years an examiner for Cambridge college entrance scholarships. It has often been made out that the examinations for college entrance scholarships require a much wider factual knowledge than is specified in the syllabuses and are therefore an important contributory factor in the intensive specialisation at schools, which is so loudly deplored. In view of this it was our policy, in the group of colleges for which I examined, to ensure that the questions we set were fully covered by the syllabuses. Having in mind the large number of hours allotted to the anatomy and physiology of the mammal, I once set a question:

'Describe how a horse converts a meal of oats into work'. I was taken aback, when I came to read the answers, to find that a majority of the candidates, after a few perfunctory remarks about digestion, plunged into the details of anaerobic glycolysis and the Krebs cycle, covering pages with the structural formulae of glyceraldehyde 3-phosphate, *cis*-aconitic acid and what have you. In cases where I was able to follow this up I asked these candidates if they could tell me anything about how the Krebs cycle was discovered, or anything about the methods which were used to investigate these matters. Not a single one of them had any idea. All the elaborate chemical relationships, so painstakingly memorised, had for them no observational basis whatsoever. Since that time I have made a point of putting the same sort of questions to freshmen who have claimed an interest in modern biology. It is always the same story. Everybody knows that the information coded in DNA is passed to RNA in the ribosomes where it acts as a template, etc. But if you ask what evidence there is that DNA has anything to do with inheritance they are at a loss and probably end up by saying that it is present on the chromosomes or something equally inadequate.

I find this all rather disturbing. In the education of scientists we rightly insist that the principles of science rest upon observation and experiment. It is not enough merely to know that there is a mechanical equivalent of heat; we expect our students to know how this equivalence was measured and established. Popular science does not set out to be educative in the true scientific sense; it is written for the information and entertainment of those who want to know what science is about. To be popular it must be easy to read; basic facts, generalisations and working hypotheses, even the latest outrageous speculations, all have to be streamlined into a convincing whole. This is especially true of what is now being written about recent advances in biology. There is much that has been established upon an experimental basis and amply confirmed. There is also a very great deal which is as yet no more than plausible speculation. But from reading some articles it is often hard to know where the one leaves off and the other begins.

What is it that we require of a scientific education that is not provided by a course of popular science? First and foremost we require that it will develop in the student a critical attitude to his subject. We expect the student, as he matures, to appreciate that many of the so-called facts of science are really inferences based upon measurements made with a particular kind of apparatus under specified conditions, that the measurements are liable to error and that the inferences may be compromised by other observations in other fields. As he approaches the frontiers of knowledge he finds the issues are seldom clear cut and the evidence is incomplete and often contradictory; he has to learn to weigh each piece of evidence against the others, and to judge its weight he has to take account of the circumstances under which it was obtained.

I seem to be getting myself into the absurd position of saying that all popular science is pernicious and that you can only approach modern biology by reading the original papers. To discourage the reading of popular science would be unthinkable. Modern biology is wildly exciting and lively young minds are just not going to be told that they must work steadily through traditional biology and put off modern biology to their third year at university. And they are quite right. There is nothing about modern biology which makes it unsuitable for discussion at an elementary level and nothing about its presentation as popular science which cannot be made wholesome with a pinch of salt. But we cannot inculcate a proper scientific attitude of mind merely by flitting from one bright idea to the next. There must be some discipline —some recourse to the original observations and how they were made, some discussion of the arguments, some appraisal of the interpretation.

What I have set out to do in this book is to examine some of the recent advances in biology and to test the strength of the evidence upon which our present conceptions rest. It has also been my aim to provide something of a bridge between school science and popular science. I have assumed that the book is to be read by students who already have a basic knowledge of physics, chemistry and biology, such as is required for the British General Certificate of Education, Scholarship Level, and similar examinations elsewhere. I have not presumed knowledge of any principles not included in the 'S' level syllabus and, where it is necessary to go further in describing modern research methods, as in discussing the resolution of microscopes, oxidation-reduction potentials and other topics, I have tried to introduce these as extensions of the syllabus. If the student has already encountered the diffraction of light by a grating it should be possible, without having to write at excessive length, to convey to him the manner in which the same principles apply in X-ray diffraction, the nature of the information so obtained and how this information can be combined with information from other sources in working out the structure of things like DNA. In pursuit of this policy my problem has been to tell the truth and nothing but the truth but not the whole truth—and at the same time to avoid statements which are dangerously misleading.†

I must also make it clear that I have not tried to write a textbook of cell biology. I have not tried to give a balanced treatment of what should be included in this field, but instead I have chosen certain parts of the field, those which seemed best to lend themselves to my purposes, for discussion at greater length than could be attempted in a balanced textbook. In making this selection I have been guided by my own estimate of the educative content of the subject matter and the possibility of keeping the treatment at an elementary level; for example, I decided not to include nerve and muscle because it did not seem that they could be usefully discussed except at a biophysical level which was more difficult than I thought desirable. I also decided to keep the details of organic chemistry out of the text as far as possible, and to refer to compounds by name, providing their structural formulae in an appendix.

Being myself no more than an elementary student of the subjects with which this book is concerned, I have had perforce to seek the help of various friends and colleagues who can speak with authority. The book has been read in whole or in part by Dr J. W. L. Beament, Dr M. J. P. Canny, Dr R. R. A. Coombs, Dr A. V. Grimstone, Dr B. S. Hartley, Dr R. Hill, Dr J. C. Kendrew, Dr D. G. I. Kingston, Dr K. E. Machin, Dr E. A. C. MacRobbie, Dr J. A. Pateman, Dr L. E. R. Picken, Dr G. Salt and Dr A. O. W. Stretton, to whom my gratitude is due. With their kind help many glaring errors were removed from earlier drafts. I also solicited and received from them many suggestions as to approach and treatment. Most of these suggestions I was only too willing to accept, but there were some issues on which I was resolved to persist in the line I had chosen to follow. I have to say this so as to make it clear that those whose generous help I have just acknowledged do not necessarily endorse everything I have written.

J. A. R.

Queens' College, Cambridge
May, 1963

† The following quotations are examples of what I would describe as dangerously misleading:

'The second law states that there are two forms of energy: "free", or useful, energy; and entropy, or useless or degraded energy.'

'Fibers of these macromolecules can be prepared and an X-ray picture of them can be obtained in a manner based on principles similar to those used to obtain an X-ray picture of the human hand.'

ACKNOWLEDGEMENTS OF ILLUSTRATIONS

I am very grateful to the following, who have given me permission to reproduce their photographs and many of whom have kindly supplied prints:

Dr T. Alderson (Figs. 2.3 and 25.7), Dr T. F. Anderson (Fig. 8.9 and 25.4), Prof. J. T. Bonner (Fig. 1.5), Prof. D. W. Fawcett (Fig. 7.10), Prof. J. G. Gall (Fig. 6.11), Dr S. A. Henderson (Fig. 6.7), Prof. R. M. Herriott (Fig. 8.8), Dr R. W. Horne (Figs. 8.5, 8.6 and 8.7), Prof. B. P. Kaufmann (Fig. 6.9), Dr E. Kellenberger (Fig. 8.3), Dr J. C. Kendrew (Figs. 3.10 and 3.11), Dr L. F. La Cour (Fig. 6.4), Dr H. Latta (Fig. 7.15), Prof. K. Mühlethaler (Figs. 7.22 and 7.23), Dr Ö. Ouchterlony (Fig. 3.12), Dr G. E. Palade (Figs. 7.11, 7.12, 7.13 and 7.24), Dr D. C. Pease (Figs. 7.17 and 7.18), Dr D. Peters (Fig. 8.2), Dr K. R. Porter (Figs. 7.2 and 7.8), Dr K. R. Raper (Figs. 1.3 and 1.4), Prof. J. D. Robertson (Fig. 7.4), Prof. C. E. Schwerdt (Fig. 2.5), Dr B. M. Shaffer (Fig. 1.2), Prof. F. S. Sjöstrand (Fig. 7.7), Dr W. Stoeckenius (Fig. 7.5), Dr A. Vatter (Fig. 7.25), Prof. M. H. F. Wilkins (Figs. 23.3 and 23.6).

I also gratefully acknowledge the permission to reproduce published illustrations granted by the following holders of copyright:

Academic Press Inc. (Figs. 7.3, 7.5, 7.7, 7.15, 7.19, 8.5, 8.6 and 8.7), Messrs Addison-Wesley Publishing Co. (Fig. 4.4), Messrs Allen and Unwin (Fig. 6.4), The American Genetic Association (*J. Heredity*) (Figs. 6.9 and 6.10), The Anatomical Society of Great Britain and Ireland (*Electron Microscopy in Anatomy*) (Figs. 7.11, 7.12 and 7.13), The Editors of *Annales de l'Institut Pasteur* (Figs. 8.9 and 25.4), Brookhaven National Laboratory (Fig. 6.11), The Clarendon Press (Figs. 7.22 and 7.23), The Managing Editor of *Cytologia* (Fig. 26.1), The General Biological Supply House, Inc., Chicago (Fig. 6.2), Messrs Ginn and Company, Boston, Massachusetts (Hilliard, *A Text book of Bacteriology and its Applications* (Fig. 2.1), The Histochemical Society, Baltimore (Figs. 7.17, 7.18 and 7.24), Dr W. Junk (Krebs and Johnson, *Enzymologia*, 1937, 4, 148–56) (Fig. 17.3), S. Karger, Basel/New York (*Progress in Allergy*, 1958, v, 1–78) (Fig. 3.12), McGraw-Hill, Inc. (*Biological Effects of Radiation*, Vol. 2) (Fig. 26.1), Messrs Macmillan and Co. (Figs. 4.10 and 5.1), The Editorial Board of *Proceedings of the National Academy of Sciences*, Washington (Figs. 2.5*a*, 3.9, 23.8, 25.8 and 25.9), D. Van Nostrand Company, Inc., Princeton, N.J. (Meyer and Anderson, *Plant Physiology*, Copyright Van Nostrand, 1952) (Fig. 7.21), Pergamon Press Ltd (Robertson, *Progress in Biophysics*, 1960, Vol. 10) (Fig. 7.4), Prentice-Hall Inc. (Stanier, Doudoroff and Adelberg, *The Micobial World*, 2nd edition) (Fig. 2.4), Princeton University Press (Bonner, *The Cellular Slime Moulds*, 1959) (Figs. 1.3, 1.4 and 1.5), The Rockefeller Institute Press (*J. Biochem. Biophys. Cytol.* 1955, 1; *J. Gen. Physiol.* 1957, 40; *J. Exp. Med.* 1944, 79) (Figs. 7.2, 7.8, 7.26, 7.27, 8.8 and 22.1), Messrs W. B. Saunders, (Figs. 6.1, 6.3, 6.5 and 6.6), Scientific American, Inc. (copyright 1961) (Figs. 7.10 and 7.25), The Society for Experimental Biology and Medicine (Fig. 2.5*b*), Verlag Zeitschrift für Naturforschung (Fig. 8.2), John Wiley and Sons Ltd (Figs. 8.3 and 26.1).

INTRODUCTION

1

ORGANISMS AND CELLS

We recognise that living matter differs from non-living matter in its capacity for growth and reproduction. A crystal can grow, but only by adding to itself more of the ready-made units of which its structure is built up. The growth of living matter differs significantly from the growth of crystals in that living matter grows by taking into itself materials of one kind and fashioning them into another kind which it then adds to its organised structure. There is yet another distinction. Given suitable conditions there is no theoretical limit to the size to which a crystal can grow; but living matter grows only to a size which is characteristic of the type of organism to which it belongs, after which the organism reproduces itself and thereby growth results in an increase in the number of organisms. The idea of an organism is bound up with this characteristic natural discontinuity of living matter, which exists not as an indefinite continuum but as lumps of various sizes and shapes, belonging to recognisable categories, such as those which in higher organisms we call species.

Let us first consider the size of organisms. Fig. 1.1 sets out the size range of organisms with a few physical points of reference at the lower end of the scale. The range is in fact so very great that we have to use a logarithmic scale to display it. The largest known animal, the blue whale, is about 30 metres long, while one of the smallest known organisms, the poliomyelitis virus, is about 300 Å† in diameter. The blue whale is about one thousand million times—9 orders of magnitude—longer than the virus, and organisms of all sizes lie between these two extremes.

We may next ask ourselves whether we can point to any factors which may set upper and lower limits to the possible size of an organism. It seems likely that the upper limit is set by the strengths of the materials of which organisms are made. For bodies of the same shape the area is proportional to the square of the length and the weight is proportional to the cube of the length. If we scale up a horse by a factor of 2 in length we scale up its weight by a factor of 8 and we scale up the area

† $1\ \mu = 10^{-3}$ mm, $1\ \text{m}\mu = 10^{-3}\ \mu$, $1\ \text{Å} = 10^{-4}\ \mu$.

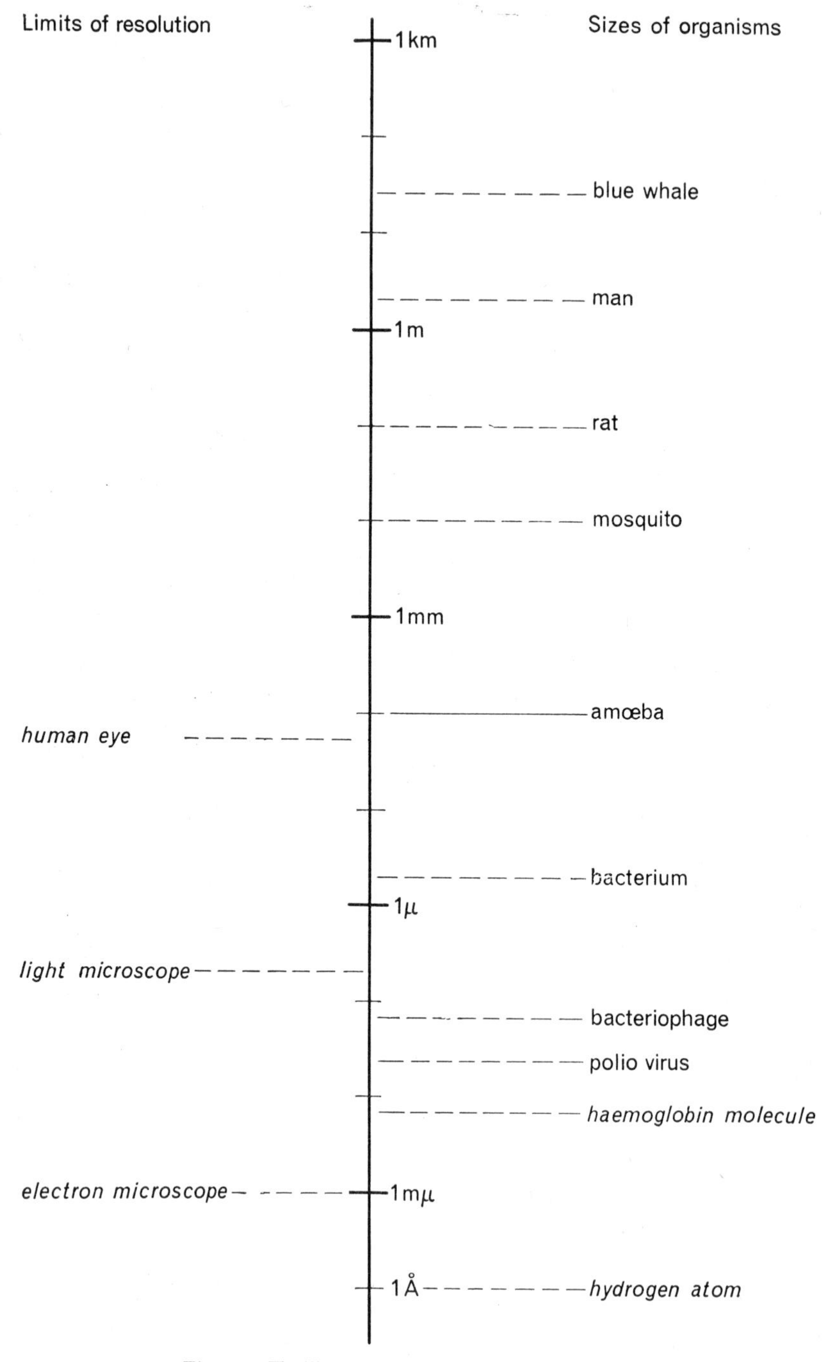

Fig. 1.1. To illustrate the size range of organisms.

of its hoof by a factor of 4; thus the weight borne per unit area of hoof is scaled up by a factor of 2. But we are not able to scale up the strength of the materials of which the hoof is made. Obviously, if we continue to scale up the length of the horse there will come a time when the hoof is unable to bear the weight applied to it. It is not possible to state in any precise way how these considerations may affect the form of large terrestrial animals, but in view of the susceptibility of horses to injury when galloped on hard surfaces we are at liberty to suppose that Nature is working to fairly narrow margins of safety. Any marked increase in size would almost certainly entail some sacrifice of speed and a modification of the form of the body in the direction of that of the elephant. In the case of the whale, although it is less affected by gravity by being water-borne, one can see that there must be some limit to size here too, imposed by mechanical considerations.

At the lower end of the scale we observe that the size of the smallest organisms approaches the size of protein molecules. We know that the chemical processes which go on in living organisms require enzymes and that all enzymes are proteins. It is therefore easy to appreciate that a small virus can accommodate only a limited number of protein molecules and that on this account the complexity of its metabolic processes may be drastically restricted. Indeed this seems to be the case. Viruses cannot grow and reproduce in isolation but only within the cells of other organisms, and as we shall see later they are only able to grow and reproduce by taking over the metabolic machinery of the host cell and directing its operations to their own ends.

By contrast with the great size range of organisms, the size range of cells is very small. The largest single cell produced in the animal body is the yolk of a bird's egg, say 5 cm in diameter in the case of the ostrich, and the smallest we may take as the spermatozoon, say 5 μ excluding the tail. This represents only 4 orders of magnitude, but this size range is quite unrepresentative of non-reproductive cells. Typical cells from the bodies of plants and animals are all of the same order of magnitude, about 20 μ across.

Cells were so named by Robert Hooke who built the first compound microscope. In 1685 he described how he had cut a thin slice of cork with a sharp pen knife and how, placing it under his microscope, he had observed that it was divided up into minute compartments which he called cells. Little more was heard of cells for about 150 years, no doubt because in those days one had to make one's own microscope and it was not every investigator who had the necessary skill and enthusiasm. By the beginning of the nineteenth century one could have a microscope made by a professional instrument maker, and as these early instruments became available Hooke's observations were soon extended. Cells were seen and described in tissues other than cork, and it was realised that they were not empty but contained a viscid substance to which Purkinje gave the name 'protoplasm'. It was also observed that within each cell there was generally a visibly differentiated body, the nucleus. In retrospect it is clear that in the early years of the nineteenth century the significance of cellular organisation was beginning to impress itself on the minds of a great many microscopists; and in 1839 Schleiden and Schwann, independently, proclaimed the generalisation that the bodies of plants and animals were composed of cells. To this was later added the idea of cell lineage by Virchow in 1858: *omnis cellula e cellula*—cells arise only from pre-existing cells.

In this way there emerged one of the most important generalisations in biological science, particularly important in its implication that all the cells of multicellular organisms are basically the same, however much they may be specialised for different functions. All modern work has confirmed and extended this conception of the essential similarity of the structural and functional endowment of unspecialised cells from multicellular organisms of all types.

Unfortunately, there were those who were not content with these substantial gains, and in their hands the cell theory, or 'Cell Doctrine' as it was sometimes unhappily called, gathered unto itself a certain mystique, and from it there grew the conception of the cell as the unit of life.

There is a certain amusement to be had from contemplation of the difficulties which biologists have got themselves into by their preoccupation with units and definitions, and in the idea of the cell as the unit of life we have an excellent example of man being trapped in a pit of his own digging. Once the cell concept had been raised to this higher level and extended to organisms other than multicellular plants and animals all sorts of awkward questions presented themselves to be answered. What are the essential attributes of a cell—does it have to have a nucleus, for example? If so, bacteria are not cells, yet they would appear to be units of life. What is a unit of life? Many would reply that the organism—the rabbit, the oak-tree, the amoeba—is the unit of life. Yet even so simple a conception as this runs into difficulties.

Among the lower organisms there is a group known as the slime moulds, of which the genus *Dictyostelium* is an example. This organism is propagated by asexual spores. When the spores find themselves in a suitable medium they germinate and from each spore there emerges a single small amoeba. The amoebae feed on bacteria and reproduce asexually by binary fission. When the population density of the amoebae

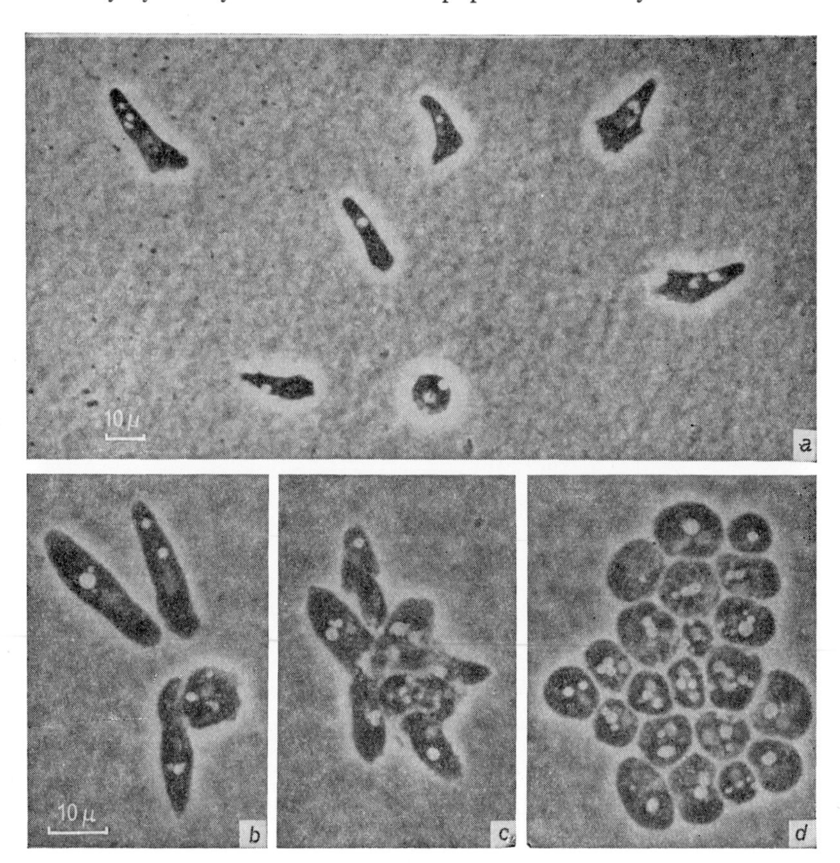

Fig. 1.2. *Polysphondylium*, a relative of *Dictyostelium*. The beginning of an aggregation can be traced to a single amoeba which withdraws its pseudopodia and rounds up, and towards which other amoebae make their way.

reaches a certain level, and if at the same time the supply of food is becoming exhausted, their behaviour changes. Quite suddenly a few amoebae will gather together into a clump and almost at once the amoebae at some distance from it are seen all to be moving so as to join the clump (Fig. 1.2). As a result of this activity the whole population, previously widely dispersed, becomes aggregated into a limited number of closely packed masses. Each mass now begins to move over the substratum like a slug (Fig. 1.4). After some little time the mass settles down and grows into the fruiting body, some of the cells forming the stalk and others forming the spores (Fig. 1.5). The process of aggregation, at least in its early stages, is reversible. If the streaming amoebae

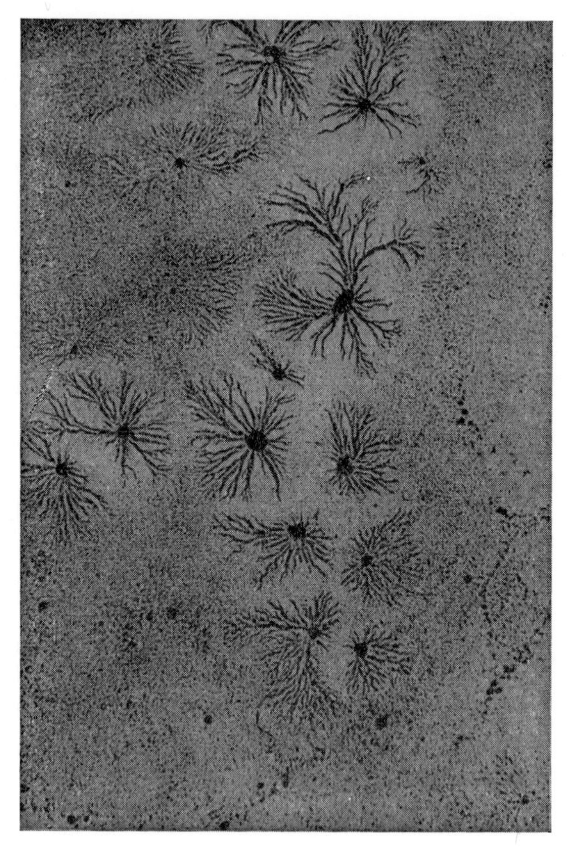

Fig. 1.3. *Dictyostelium.* Aggregation well under way. Root-like organisations of aggregated amoebae spread through the culture and soon all the amoebae are in aggregates.

are mechanically dispersed again, and if more bacteria are provided for them to feed on, they revert to their independent existence.

Suppose, then, that a biologist is introduced to *Dictyostelium* in its feeding phase and that he is unaware of its reproductive behaviour. He sees a population of unicellular amoebae feeding, growing and reproducing in complete independence one of another. If he is willing to accord the status of an independent organism to the familiar *Amoeba proteus* on what grounds would he deny it to these amoebae of *Dictyostelium*? If on the other hand he were to meet with *Dictyostelium* during its reproductive phase he would see a multicellular organism whose component cells were obviously co-operating in the activities of the organism as a whole. Why should he hesitate to pronounce the fruiting body an independent organism?

These few examples will suffice to show how hopeless it is to try to set clear limits to the concept of the cell and the concept of the organism,

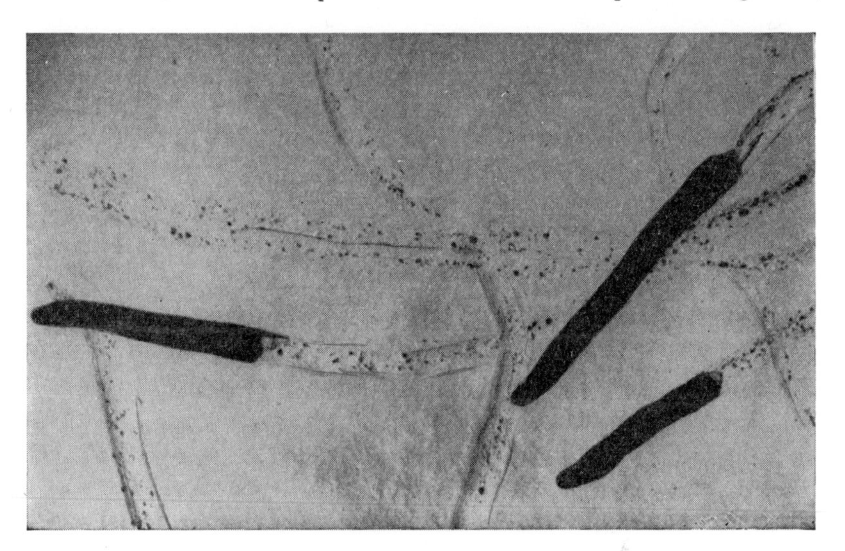

Fig. 1.4. *Dictyostelium*. Aggregates in movement, leaving slime trails.

and they serve us with notice of the sort of difficulty which attends upon the definition of the unit of life.

There are two kinds of unit with which physical science has to deal. First, we have the units of length, mass and time—the centimetre, gram and second, or c.g.s. system of units. These units are arbitrarily defined for our own convenience; they do not correspond to any patterns of discontinuity in nature. Secondly, we have discovered entities, such as the proton and the electron, and certain physical constants such as the velocity of light in space. These are natural 'units'; we are forced to recognise their existence as a result of our investigations of the natural world, and better understanding of the natural world follows upon this recognition. But are we forced to recognise that there is a unique natural unit of life? Or is biological science at an impasse from which it can advance only if we arbitrarily define a unit of life? Or is it that in taking the cell as a unit of life we are trying to impose our own preconceptions upon Nature and to ascribe fundamental significance to certain compartments which from Nature's point of view may be merely matters of convenience?

Fig. 1.5. *Dictyostelium.* Formation of the fruiting body.

2

MICRO-ORGANISMS

Micro-organisms are here, arbitrarily, taken to mean all organisms which are not multicellular. The distinction is not necessarily one of size; some of the larger Protozoa are larger than some of the smaller Metazoa. Like nearly all attempts to separate organisms into clearly defined categories it runs into the usual difficulties over intermediate types. Nevertheless, the distinction is a useful one, which is its justification. For our present purposes micro-organisms will include unicellular plants and animals, bacteria and viruses.

1. PROTISTA

The higher multicellular organisms are either animals or plants—either Metazoa or Metaphyta—and we do not find among them any which we have difficulty in referring to one kingdom or the other. We were brought up to believe that plants have cellulose cell walls, can carry out photosynthesis and have no powers of locomotion, whereas animals lack cellulose cell walls, lack the ability to carry out photosynthesis but have powers of locomotion—and, strangely, this is true; by these criteria we can always distinguish a multicellular plant from a multicellular animal. When we come to unicellular organisms these same criteria fail to distinguish between plants and animals. *Euglena* has chlorophyll and can carry out photosynthesis; but it has no cellulose cell wall and it can move by means of its flagella. *Chlamydomonas* is another green cell, and it has a pair of flagella, and a cellulose cell wall as well. *Polytoma* is a close relative of *Chlamydomonas* but it has no chlorophyll. In recognition of these difficult cases many authorities do not attempt to define the Protozoa and Protophyta as separate taxonomic groups but assign all unicellular organisms to a single group, the Protista.

The Protista have this in common with the Metazoa and the Metaphyta that they have clearly distinguishable nuclei and that these nuclei divide in a very characteristic way. The nuclear arrangements of the Protista are highly diversified, but there are certain features which undoubtedly relate to the process which takes place with such remarkable uniformity throughout the higher organisms. The appearance of the chromosomes prior to cell division, their assembly upon the equator of the mitotic spindle and their rapid simultaneous passage to its poles make it clear that all these organisms share the same basic replicative mechanisms. This is in complete contrast to what we find in the bacteria.

2. BACTERIA

It was a long time before anything corresponding to a nucleus was discovered in a bacterium and the bacterial nucleus is not separated from the rest of the cytoplasm by any nuclear membrane. There are no visible chromosomes and no spindle is formed when the bacterial cell divides. These important differences make possible a valid taxonomic distinction between the Protista and the bacteria.

Although they are relatively small the Protista exhibit a sufficient diversity of structure to make it possible to classify them using their structural features in accordance with the same taxonomic principles as are used in the classification of higher animals and plants. This cannot be said of the bacteria. The bacteria present a wide range of growth habit, whereby some major groups may be defined. Some form separate spherical cells, others have rod-shaped cells; sometimes these cells are motile, being provided with flagella. In some bacteria the cells remain attached together after cell division, forming filaments; in others the cellular organisation breaks down into a filamentous mycelial growth like that of a fungus. Among the bacteria are included the spirochaetes, each of which is a coiled flexible filament. Fig. 2.1 shows some of these bacterial types. But within these major groups so defined there is seldom sufficient variation in observable structure to make possible division into families, genera and species. In these circumstances bacteriologists have found it necessary to make use of other features such as reaction to stains and nutritional requirements, though as we shall see later (Chapter 25) the nutritional requirements of micro-organisms are liable to spontaneous change and for this reason may be taxonomically unreliable.

This reference to the nutritional requirements of bacteria in the taxonomic context affords an opportunity for a necessary digression into the nutrition of micro-organisms and indeed of organisms in general. It is commonly stated in textbooks that in their modes of nutrition organisms are either (1) holophytic, building up complex

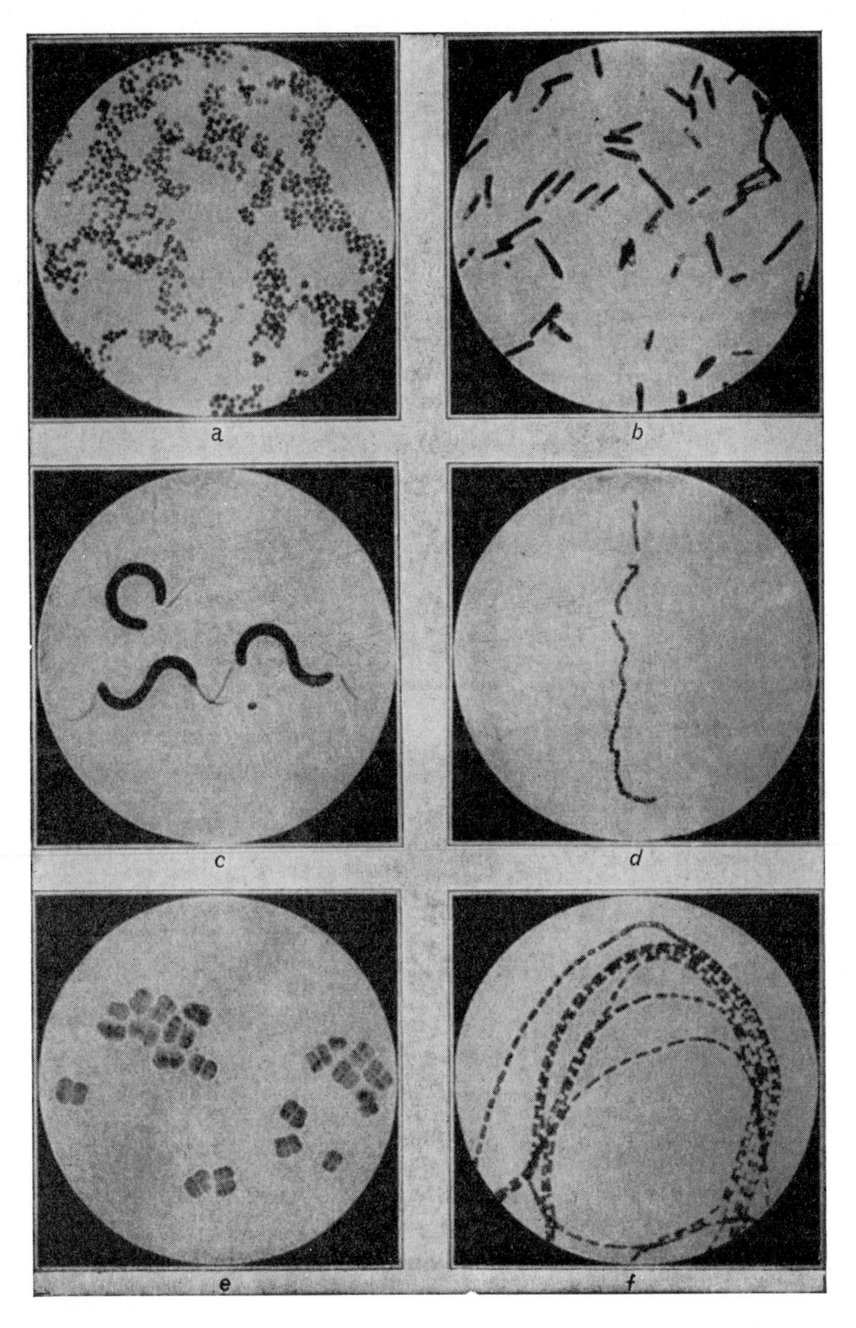

Fig. 2.1. Various types of bacteria: (*a*) *Staphylococcus*, (*b*) *Clostridium botulinum*, (*c*) *Spirillum undulans*, (*d*) *Streptococcus*, (*e*) *Sarcina agilis*, (*f*) *Bacillus anthracis*.

organic substances from simple inorganic ones, using the energy of sunlight, or (2) holozoic, swallowing complex organic substances in solid form, or (3) saprophytic, absorbing relatively complex organic substances from solutions. These distinctions are not very helpful. Not all micro-organisms which make use of the energy of sunlight are able to make do with simple inorganic compounds; conversely, some are able to make do with simple inorganic compounds but obtain their energy from sources other than the sun. And the distinction between swallowing whole and absorbing from solution is hardly a fundamental one. In fact, these terms have outlived their usefulness and now only distract our attention from the real issues. It is more useful to begin by recognising that organisms have a material requirement and an energy requirement. Plants derive their energy from the sun and for this we retain the not very appropriate adjective 'photosynthetic'. Animals derive their energy from the breakdown—mostly oxidative—of complex organic compounds, and—even more inappropriately—we call this the 'chemosynthetic' mode of obtaining energy. Among bacteria we find that both these modes are made use of, and that in addition there are some bacteria which can exploit a variety of energy-yielding inorganic reactions:

nitrifying bacteria	$NH_3 \longrightarrow NO_2$
sulphur bacteria	$S \longrightarrow H_2SO_4$
iron bacteria	$FeCO_3 \longrightarrow Fe(OH)_3$
hydrogen bacteria	$H_2 \longrightarrow H_2O$

In principle the energy requirement is quite distinct from the material requirement, though in many organisms the distinction is not altogether easy to establish. In plants the distinction is quite clear; the energy comes ultimately from the sun and the material requirement is for water, carbon dioxide, nitrate, and for traces of various elements with which we shall not concern ourselves here. The important point is that from these simple materials the plant can synthesise all the organic compounds it requires; to describe such organisms we use the word 'prototrophic'. Organisms which cannot supply their requirements by synthesis have to have them provided and such organisms we call 'auxotrophic'. All animals, and also many micro-organisms, are auxotrophic, being 'exacting' towards a great many components of their normal diet; we may think of vitamins as molecules which form essential parts of their metabolic machinery which animals are unable to synthesise for themselves. In the higher animals it is not always a simple matter to ascertain the minimal nutritional requirements for the reason that these animals commonly harbour an extensive bacterial flora within their bodies. A cow, for example, is provided with many vitamins by the bacteria which multiply in the rumen and subsequently pass to the intestine, there to be digested. But it is usually not difficult to grow micro-organisms in pure culture, and their nutritional requirements can be ascertained by growing them on artificial media of accurately known composition.

In pursuing this matter we shall avoid discussing the cultivation of the pathogenic bacteria with which medical bacteriologists are concerned; these organisms are usually highly exacting in their requirements and very elaborate media have to be provided for them. Instead, we will take as an example a bacterium which we shall have occasion to mention many times in other contexts—*Escherichia coli*, found as a harmless commensal in the human intestine, a prototrophic chemosynthetic organism from which various exacting strains have been bred. *E. coli* is 1·5 μ long and 0·5 μ in diameter. It will grow well on a medium which contains glucose (0·5 %) as a source of energy and a source of carbon, ammonium chloride (0·1 %) as a source of nitrogen, and various other elements in inorganic form. A small amount of human faeces incubated in this medium would soon produce a thriving culture of *E. coli*. But this culture would also be likely to contain micro-organisms of other kinds. Our first aim, then, would be to obtain *E. coli* in pure culture.

A culture of bacteria growing in suspension in a fluid medium, in a flask or in a test tube, is called a 'broth culture'. An alternative method of growing bacteria is a 'plate culture'. For plate cultures we make use of a material called agar, a polysaccharide from seaweed which dissolves in hot water and sets to a jelly when cold, and which few bacteria can decompose. A small amount of agar is added to fluid medium which is then heated to 100° C and poured into a Petri dish, a shallow glass dish about 4 inches in diameter with a glass lid (Fig. 2.2). After the agar has set a suspension of bacteria is poured over it to form a thin layer. One of the advantages of this method is that it affords a means of ascertaining the number of bacteria present in a broth culture. For this purpose we first dilute the broth culture by a known amount (usually a very great amount if it is a healthy culture) so that the bacteria are widely dispersed. A measured volume of the dilute suspension is then poured over an agar plate. Each bacterium after settling upon the surface of the agar grows into a dense colony which is readily visible to the naked eye (Fig. 2.3) and these colonies can be counted. If the

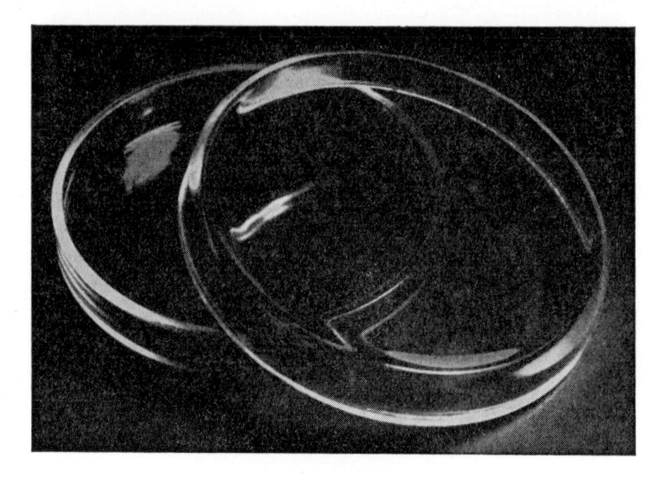

Fig. 2.2. Petri dish.

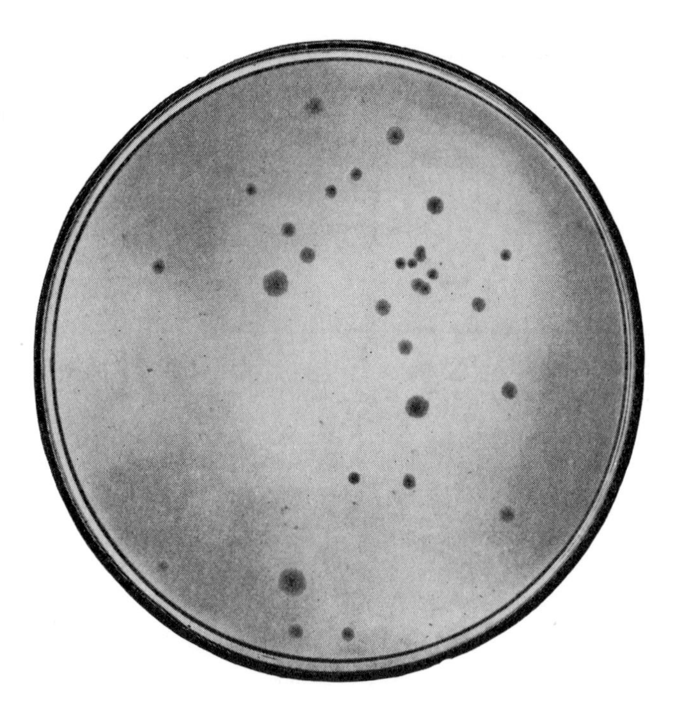

Fig. 2.3. Bacterial colonies growing on the surface of a nutrient culture medium set in agar.

A diluted bacterial suspension is spread over the surface of the agar and is incubated overnight. Each visible colony is a clone, being the offspring derived by repeated fission from a single bacterium.

culture has been sufficiently diluted so that the colonies are well separated we can assume that each colony has arisen from a single bacterium. We thus find the number of bacteria in the known volume of suspension which was poured over the plate, and so calculate the number of bacteria originally present in the broth culture.

The same procedure can be used as a means of obtaining a pure culture. An isolated colony, produced in the way described above, should be a 'clone', that is, a population derived by asexual reproduction from a single parent individual. But just in case the colony we picked from happened to be a mixed one, having arisen from two bacteria of different types, we would probably take the precaution of repeating the process; we would pick bacteria from an isolated colony and grow them in a broth culture, replate at high dilution and pick again from an isolated colony. In critical experiments where it is necessary to know that one is dealing with an absolutely pure clone it may be necessary to isolate a single bacterium. This is difficult, since it has to be done under a very high power of the microscope. A small drop of culture is placed on the underside of a supported coverslip and a single bacterium is pulled out with a fine glass hook. This requires the use of a micromanipulator by means of which the movements of the glass hook can be controlled.

It was for long believed that bacteria reproduced themselves exclusively by asexual methods, and as far as increase in number of organisms is concerned this is true. The discovery of sexual processes in bacteria is described in Chapter 25.

3. VIRUSES

What we now know as the viruses were known to exist as early as the beginning of the present century, although their ability to reproduce was not established until much later. Attempts to isolate the causative organisms of certain plant diseases led to the discovery that something which could pass through an ordinary bacterial filter was able to cause the disease. This agent was referred to as a 'filterable virus', meaning no more than that it was a poison; but it was suspected of being alive since it appeared to increase within the plant, more virus being recoverable from the infected plant than had been used to infect it. Parallel observations were also made on diseases of man and of animals, and it was soon found that bacteria too were susceptible to the attacks of filterable virus to which the name 'bacteriophage' (nowadays abbreviated to 'phage') was given. With the discovery of phage it became possible to show that filterable virus was particulate and to measure the concentration of particles in the filtrate. This is done very simply, making use of the same principle as is used for counting bacteria. A dilute suspension of phage is poured over a continuous culture—a 'lawn'—of bacteria growing on an agar plate. Wherever a phage particle settles it proceeds to infect a bacterium and starts a centre of infection

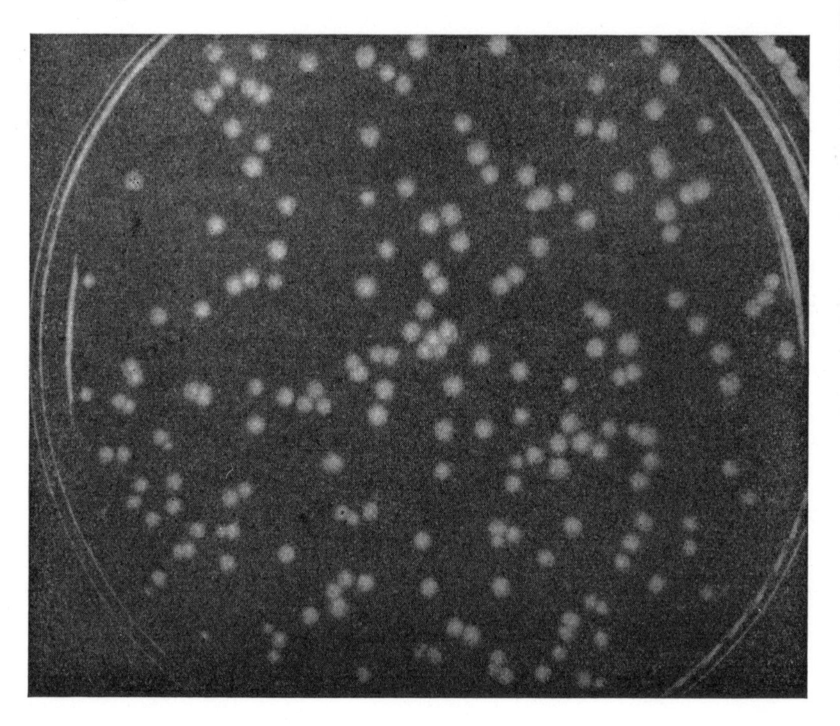

Fig. 2.4. Plaques formed by bacteriophage on a continuous culture ('lawn') of bacteria growing on an agar plate. The plaques are clear areas where the bacteria have undergone lysis.

from which the phage spreads out to attack other bacteria. The areas of infection are very easily seen since the infected bacteria undergo lysis—dissolution—and show up as clear areas, known as plaques. By counting the plaques and knowing the volume of suspension poured over the plate we find the number of phage particles per unit volume of suspension (Fig. 2.4).

Some idea of the sizes of virus particles was obtained from the pore size of membranes which retained them. It was established that they were composed of protein and nucleic acid.† But their structure remained unknown until the electron microscope came into use.

All viruses are parasites and they are all intracellular parasites. They are classified into three groups according to their hosts: animal viruses, plant viruses and bacterial viruses (bacteriophage).

Among animal viruses the virus of psittacosis, a rare disease contracted by kissing parrots, is one of the largest known, being about 3000 Å in diameter, and is mentioned here for that reason alone. The

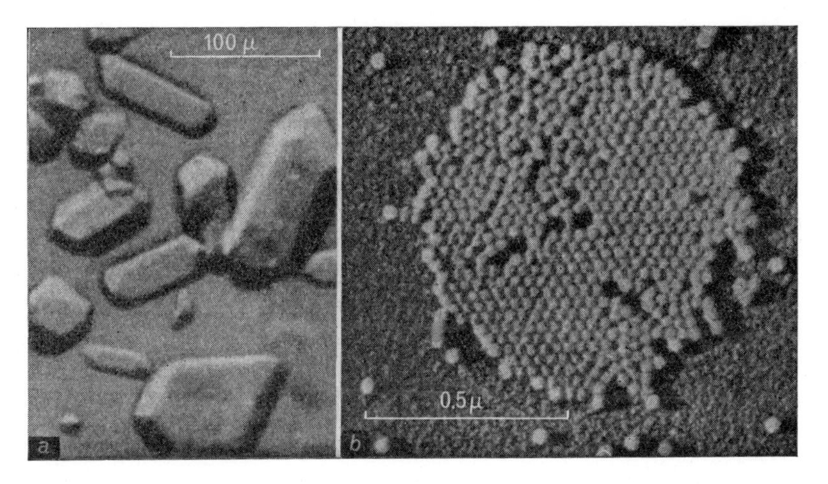

Fig. 2.5. (*a*) Crystals of purified poliomyelitis virus. (*b*) Electronmicrograph (shadowed) of poliomyelitis virus particles showing the way in which they pack together in orderly array.

influenza virus is of intermediate size, being 1000 Å in diameter, and is one of the most intensively studied of the animal viruses. The same may be said of the virus of poliomyelitis which is one of the smallest, 300 Å in diameter, and has been obtained in crystalline form (Fig. 2.5). In animal viruses the nucleic acid may be either DNA or RNA.

Plant viruses also show a wide range of size, from tobacco mosaic virus, which is in the form of rods 3000 Å long by 150 Å in diameter, to turnip yellow mosaic virus which is spherical and 200 Å in diameter.

† Nucleic acids and their biological significance are discussed in Chapters 22 and 23. For the benefit of any reader who may not have heard of nucleic acids (they are not included in some syllabuses of biology), let it simply be stated that nucleic acids are large molecules which are believed to determine the heritable characters which are transmitted from parent to offspring. There are two kinds of nucleic acid: ribonucleic acid (RNA) and deoxyribonucleic acid (DNA); these nucleic acids are attacked by the enzymes, ribonuclease (RNase) and deoxyribonuclease (DNase) respectively.

All the plant viruses can be obtained as crystals and all contain nucleic acid in the form of RNA.

It is not surprising that we know more about phage than about other forms of virus, for undoubtedly phage is the most convenient for study, mainly on account of the ease with which the hosts can be reared. When a bacterial culture has been heavily infected with phage events take the following course. For the first 10 minutes or so no free phage is detectable (assuming that any phage remaining from the original inoculum has been removed). During this period, which is known as the eclipse period, no free phage can be detected even if the bacteria are disrupted and their contents set free. During the next 10 minutes free phage begins to appear in increasing numbers rising to a maximum some 20 minutes after the original infection was made. The phage has multiplied within the bacteria, and the bacteria, when they eventually undergo lysis, liberate on the average about 150 phage particles each. Thereafter the number of free phage particles does not increase because all the bacteria have now been destroyed. Once liberated, the phage particles remain stable and inert for an indefinite period.

The nucleic acid of phage is exclusively DNA. The structure of phage, which is more elaborate than that of other forms of virus, has been worked out in some detail, but a description of this is reserved for a later chapter (Chapter 8, p. 121).

In general, free virus particles consist of a protein coat or case enclosing the nucleic acid. As will be described in other contexts it is the nucleic acid and not the protein which is the essential infective agent. The free particles of viruses are biologically completely inert. They do not grow or reproduce, they show no signs of respiration or of any other metabolic activity. The question is therefore often asked—are viruses alive? The fact that many viruses can be crystallised may seem to lend point to this question, for we have insisted that the growth of crystals has nothing in common with the growth of living organisms. But this is to misunderstand the position. The ability of particles to crystallise depends upon their regularity of size and shape; there is no reason why regularity of size and shape should be incompatible with life.

The free virus particle may be likened to the dried seed of a plant. We cannot tell by looking at a plant seed whether it is alive or dead; the only test is to place it in conditions which we know will allow of its germination, and see whether it germinates. If it does, then by the standards of everyday life it is alive. In the same way we may place a free virus particle in a situation where it can infect a host cell and thereby determine whether it is infective or non-infective. If it is infective we may, if we so wish, agree to regard it as alive. In order to decide whether something is alive or dead we have to devise some test and agree to abide by its result. The test is not necessarily the same for all organisms. There is therefore a certain arbitrariness about the decision; and in view of this we may well ask ourselves whether the question—are viruses alive?—is really worth answering.

PART I

STRUCTURE

3

THE CHEMICAL BASIS OF LIVING MATTER

We were most of us taught at one time or another that protoplasm was made of proteins, carbohydrates and fats; and although our experience soon outgrew this conveniently simple conception, its recognition of these three classes of materials still has relevance. We shall continue to take this as our basis of treatment.

1. PROTEINS

The central importance of proteins in the structure and activities of living matter has never been doubted. Although the discovery of the role of nucleic acids as the repositories of the information which directs the synthesis of proteins has perhaps elevated the nucleic acids to a superior rank in the biochemical hierarchy, this role is in fact a narrowly restricted one. Proteins are the very stuff of life. As the structural basis of the matrix of bone, as tendon, as the working parts of muscle, as blood pigments and above all as enzymes, they take part in all the manifold activities of organisms. Proteins are at once very large molecules and very unstable molecular configurations. They are liable to suffer irreversible changes as a consequence of exposure to temperatures above 60° C or encounter with the ions of heavy metals, and their properties are profoundly altered by small changes of pH. To all these agencies living organisms are likewise extremely sensitive, and with the very best of reasons we ascribe this sensitivity to the participation of proteins in every aspect of vital function.

By treatment with strong acids or alkalis proteins may be broken down into the units of which they are built up. These units are known as amino acids, of which some 20 are found in proteins. By biochemical standards they are very simple compounds; glycine is the simplest, and the most complex is tryptophane (see page 26).

All the naturally occurring amino acids have in common the feature that they are α-amino acids; the amino group is always attached to the carbon atom next to the carboxyl group.† That is not to say that there

† The amino acid proline is an exception (see Appendix, p. 316).

cannot be other amino groups and other carboxyl groups elsewhere in the molecule; but it is to say that we can always represent any naturally occurring amino acid as

$$R\text{—CHNH}_2\text{—COOH},$$

where R stands for any of the 20 different amino acid residues.†

```
                                     H               H    H
H       H       O                    |                \  /
 \      |      //                    C            H    N        O
  N-----C-----C                    //  \          |    |       //
 /      |      \              H--C      C-----C---C----C-----C
H       H       OH               |      ||    ||  |    |      \
                              H--C      C     C   H    H       OH
                                  \\   / \   / \
                                    C      N     H
                                    |      |
                                    H      H
     glycine                                tryptophane
```

The outstanding significance of the fact that all naturally occurring amino acids are α-amino acids will be seen from Fig. 3.1. By condensation of the α-amino group of one amino acid with the carboxyl group of another, water being eliminated, it is possible for α-amino acids to be joined together into molecular chains of indefinite length. This is how protein molecules are built up; and the important thing is that no matter what amino acids go to build them, the backbone of the molecule always takes the form

$$\text{—C—C—N—C—C—N—C—C—N—}$$

which it would not do if the amino acids were not α-amino acids.

This linkage between two amino acids, which is of the same nature as in an amide, is called the peptide bond. Compounds made up of two amino acids linked in this way are called dipeptides, compounds made up of several amino acids are called polypeptides, and when a polypeptide is of such a size as to comprise some 50 amino acids we may dignify it with the name of protein.

Such a protein, which might have a molecular weight of 5000, would be reckoned very small indeed; a very large protein molecule, such as the blood pigment haemocyanin, found in crustacea and molluscs, has a molecular weight of the order of 5,000,000. But even in a small protein molecule, if there are 50 'places' for amino acids and if each 'place' may be occupied by any of the 20 naturally occurring amino

† The formulae of the 20 amino acids are given in the Appendix, p. 314.

acids, it is clear that the possible number of different protein molecules of this size is so vast as not to be worth calculating.

Suppose, then, that we want to find out as much as can be known about some protein molecule, which we will assume to be provided for us in a pure state. We would want to know, first, which of the 20 amino acids enter into its composition and, second, the relative amounts of these amino acids. We would begin by heating the protein for about 24 hours with strong acid which would result in the

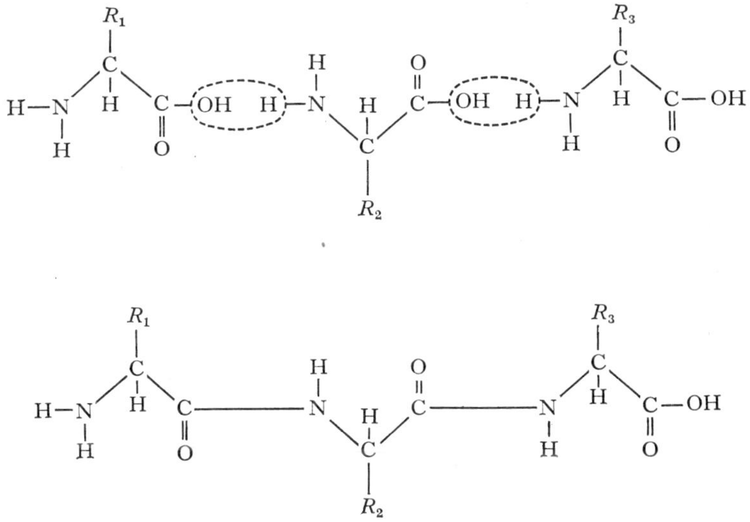

Fig. 3.1. To show how α-amino acids can be linked together by peptide bonds.

hydrolysis of the peptide bonds and we would then have a mixture of up to 20 different amino acids in solution. Before the Second World War the task of separating each of these 20 amino acids in a sufficient state of purity would have been possible but infinitely tedious. By the end of the Second World War it had become a matter of simple routine. This revolution—it was no less—in biochemistry followed upon the introduction of the method of chromatography, developed by Martin and Synge.

The word 'chromatography', which unfortunately is here to stay, is most misleading. It was first used in relation to an early application of the principle, by the botanist Tswett, for purposes of separating pigments, and fixed attention upon the unessential feature of the process, namely that the substances were coloured. Chromatography is first and foremost a means of separating substances, whether they are coloured or not. In principle, that variant of the method most widely used, paper chromatography, is as follows. A drop of solution containing the mixture of substances to be separated is placed near one side of a sheet

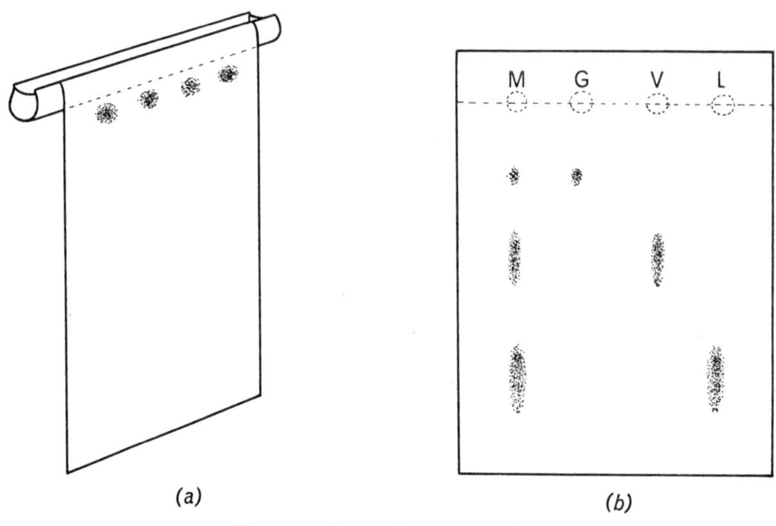

Fig. 3.2. Paper chromatography.

This diagram illustrates the separation of three amino acids, glycine (G), valine (V), and leucine (L), from a mixture (M). The mixture is placed on the filter paper to the left of the starting line and the pure compounds in order towards the right. The paper is hung over the edge of the solvent trough, as in *a*, and is then set up in a tank which is saturated with the vapour of the solvent. When equilibrium has been established the solvent is poured into the trough and allowed to percolate through and down the paper. After a suitable period (hours or days) the paper is removed and dried. The amino acid spots can be made visible by staining with the reagent ninhydrin, as in *b*. If the aim was to separate one of the components, say valine, from the mixture one would stain up the right-hand side of the chromatogram only. Then, having located the level of the valine spot, one would cut out a piece of the paper at the same level on the left (mixture) line and wash the valine out of it.

of filter paper and dried down to a spot of about 1 cm diameter. The paper is then hung over the edge of a glass trough as shown in Fig. 3.2. A suitable solvent, e.g. propyl alcohol, is placed in the trough and allowed to flow slowly down the paper under the combined actions of capillarity and gravity. As it flows past the spot the solvent washes the material down the paper with it. If a compound has a higher affinity†

† The word 'affinity' as used here is not susceptible of any precise explanation. Just as molecules of water pass freely back and forth across the interface between

for the solvent than for the paper it will be washed down more easily and so will travel farther than a compound which has a higher affinity for the paper. Different substances therefore travel at different rates in accordance with differences in affinity, and by a suitable choice of solvent, or by a combination of different solvents, the various components of a mixture can be separated in this way. The method of paper chromatography requires no elaborate apparatus and can be used with very small quantities (about 0·001 mg). The same principles can be used, in more elaborate apparatus, for separation in bulk.

Granted, then, that we can readily discover what amino acids are present, and in what amounts, the next task is to find out in what order they are linked up in the protein molecule; this is what we call the primary structure of the molecule. To establish the primary structure of a protein is a work of immense labour, even with the most modern methods. In 1954 Sanger described the primary structure of insulin. As proteins go, insulin is a very simple one, having a molecular weight of only 5700, but it took Sanger and others 10 years to work it out. This is a landmark in the history of biochemistry and the way in which the primary structure of insulin was discovered deserves some description here.

Prolonged acid hydrolysis will separate all the amino acids in the protein molecule, but by shorter treatments it is possible to break the molecule up into various peptides containing small numbers of amino acids (oligopeptides). Within such an oligopeptide it is possible to establish the order of the amino acids. The method used by Sanger depends upon the fact that the backbone of a peptide has an amino group at one end—the N-terminal—and a carboxyl group at the other—the C-terminal. The reagent dinitrofluorobenzene can be made to react with the free amino group whereby a dinitrophenyl group is substituted for one of the hydrogen atoms, and this combination is very stable, surviving the complete hydrolytic dissociation of the backbone. Suppose, then, that we have a tripeptide comprising the amino acids A, B and C in an unknown order. We treat the tripeptide with dinitrofluorobenzene and then hydrolyse it completely. We separate the products by chromatography. One of the amino acids set free will have the

liquid and vapour so also do molecules of solute pass freely between the state in which they are adsorbed upon the cellulose fibres of the filter paper and the state in which they are in free solution in the solvent. The average time which a solute molecule spends in free solution, relative to the average time which it spends adsorbed upon the cellulose, measures its relative affinity for these two components of the system.

dinitrophenyl group attached to it, and can be identified by its yellow colour. Suppose that it is amino acid C; then we know that C is the N-terminal. We then submit the tripeptide to partial hydrolysis so as to separate it, if possible, into a dipeptide and a single amino acid. If the single amino acid is A, then the order is CBA; if the single amino acid is B, then the order is CAB. If the single amino acid is C, then we have to repeat the dinitrofluorobenzene treatment upon the dipeptide AB to discover which amino acid is its N-terminal, and this will be the one to which C attaches. With patience, then, it is possible to establish the order of the amino acids in the small pieces of the molecule which result from partial acid hydrolysis.

(*a*) Original portion of the molecule, on complete hydrolysis, yields the following amino acids:

asp cys glu gly his leu phe val;

dinitrophenyl group attached to phe; therefore N-terminal is phe.

(*b*) Original portion of the molecule, on partial hydrolysis, yields the following oligopeptides (N-terminal to the left in all cases):

asp—glu, glu—his—leu—cys, his—leu—cys—gly,
phe—val—asp, val—asp—glu.

(*c*) Solution:

```
phe—val—asp
    val—asp—glu
        asp—glu
            glu—his—leu—cys
                his—leu—cys—gly
-----------------------------------
phe—val—asp—glu—his—leu—cys—gly
```

Fig. 3.3. For explanation see text. The contractions used for the amino acids may be identified in the Appendix, p. 314.

The strategy of the attack is first to break up the protein molecule into fairly large pieces. This can be done in various ways, notably by using mammalian digestive enzymes which attack protein molecules only at certain points (e.g. pepsin which attacks next to an aromatic amino acid). These large pieces are then separated and purified by chromatographic methods. A sample of one of the large pieces is first treated with dinitrofluorobenzene and then completely hydrolysed; this enables us to ascertain the molar quantities of the amino acids which it contains and also to identify the amino acid occupying the N-terminal (Fig. 3.3*a*). Further samples are then partially hydrolysed under a variety of conditions so as to yield a variety of oligopeptides. The amino acid sequence of each oligopeptide is then worked out by the method described above (Fig. 3.3*b*). The final step is to set out the oligopeptides as in Fig. 3.3*c* so that from their overlaps the complete sequence can be established. This final step is simply a matter of trial and error, like solving a jig-saw puzzle.

The foregoing is only to describe the principle of the method; the many difficulties and problems at the detailed level have not been touched upon. The primary structure of insulin as worked out in this

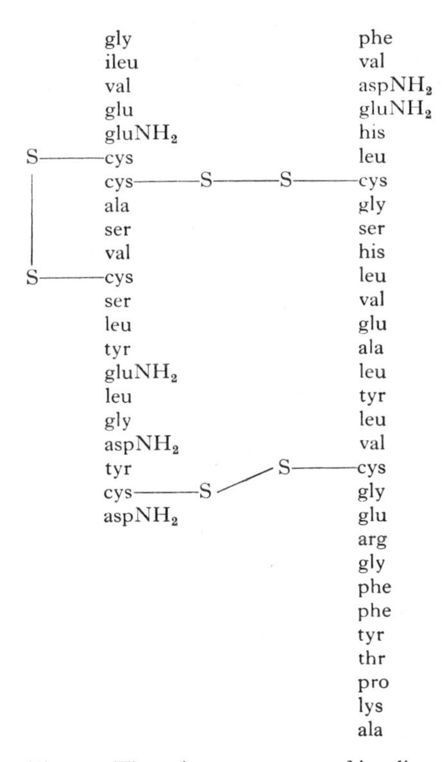

Fig. 3.4. The primary structure of insulin.
The contractions used for the amino acid residues may be identified in the Appendix, p. 314

way is shown in Fig. 3.4 and is seen to comprise two polypeptide chains. Since 1954, when this was published, the primary structures have been worked out for four other proteins, of which the largest is haemoglobin, having a molecular weight of 68,000, and four polypeptide chains.

This, however, is only the end of the beginning. We have found the order in which the amino acids are linked together, but this tells us nothing about the configuration of the molecule. A protein molecule could conceivably have its backbone stretched out in the form of a long thread; or it could be rolled up into a ball; or it could be folded in some special way. How are we to approach the problem of the configuration of a protein molecule?

Broadly speaking, there are two kinds of proteins, which differ in respect of their configuration: fibrous proteins which are thread-like, and globular proteins which are more compact. Fibrous proteins can be obtained from such structures as hair and tendon. We can tell in a general sort of way that they are extended because when a solution of fibrous protein is made to flow through a narrow tube it can polarise a

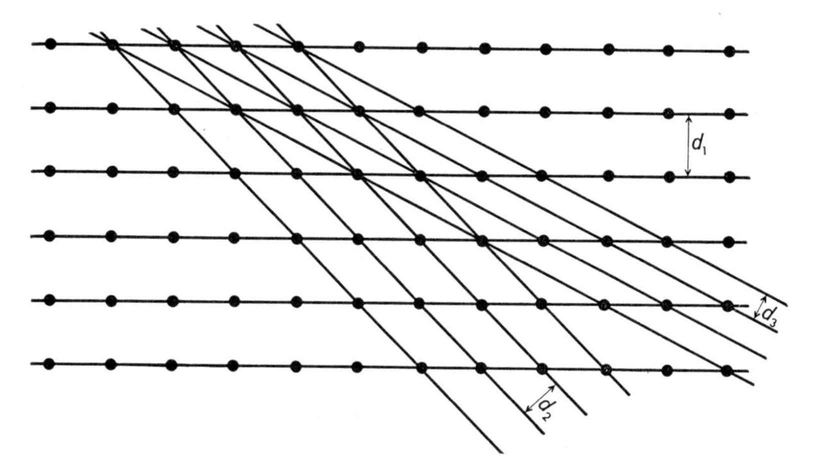

Fig. 3.5. To illustrate some of the imaginary series of planes which can be drawn through a crystal lattice.

beam of light.† The fact that solutions of fibrous proteins show flow birefringence, and show it very markedly, indicates that the ratio of length to diameter in such proteins must be very large. On the other hand, blood plasma, though it contains much protein of various molecular species, does not show flow birefringence, from which we may conclude that in these proteins the backbone is extensively folded. These, however, are merely qualitative indications of the configuration of protein molecules. For accurate information we must have recourse to the methods of X-ray diffraction analysis.

The methods of X-ray diffraction analysis have long been used by

† It is known (1) that elongate bodies suspended in a viscous fluid align themselves parallel to the direction of shear and (2) that parallel arrays of long molecules show birefringence, i.e. can polarise light.

crystallographers for investigating the arrangement of atoms in crystals and they can only be applied to crystalline material, or at least only to material which shows some regularity of structure. The use of X-rays is related to the fact that the wavelength of X-rays, about 1 Å, is comparable with the interatomic distances in crystals. X-rays are scattered by the electron clouds surrounding atoms and they can be diffracted by the lattice structure of a crystal in the same way as light is diffracted by a grating. In studying the diffraction of X-rays by a crystal lattice

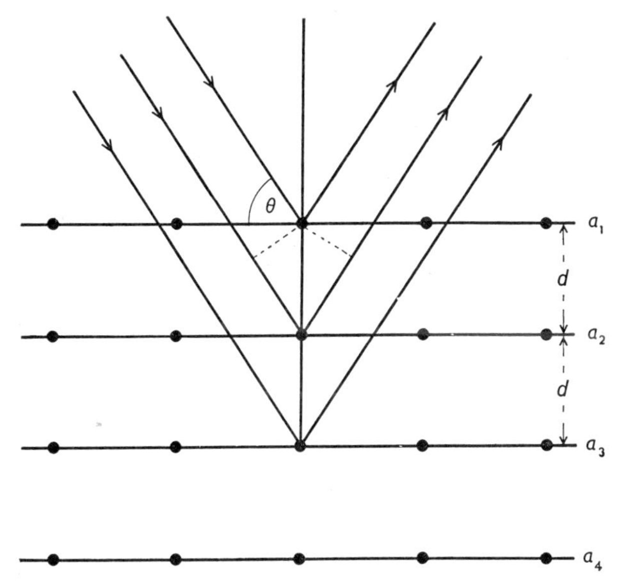

Fig. 3.6. The path of the wave reflected from plane a_2 is longer than the path of the wave reflected from plane a_1, by $2d \sin \theta$. When θ is such that the path difference is a multiple of the wavelength the waves reflected from all the planes are in phase and combine to form a beam which is reflected as though by a mirror.

it is permissible—and very much simpler—to regard the crystal as being made up of various series of parallel planes in which the atoms lie. Fig. 3.5 shows, in two dimensions only, how such imaginary planes can be drawn, and obviously there are a great many different series of such parallel planes, each series having a characteristic separation. If we know the wavelength of the X-rays we can find the distance separating the planes. How this is done is indicated in Fig. 3.6. A wave reflected from plane a_2 has to traverse a longer path than a wave reflected from plane a_1. If d is the distance between the planes and θ is the angle between the incident ray and the planes, then the path difference is given by $2d\sin\theta$. If the path difference is equal to the wavelength, λ, then the wave reflected from plane a_2 will be in phase with the wave reflected from plane a_1 and also with the waves reflected from all the other planes of the series. There is therefore a critical value (or values) of θ, for which $2d\sin\theta = n\lambda$, where n is a whole number, and at this critical angle the X-rays will be reflected from the crystal as though from the surface of a mirror. Knowing λ and the critical value of θ we obtain a value for d.

The principle of a method commonly used by crystallographers is illustrated in Fig. 3.7. Monochromatic X-rays (i.e. rays of uniform

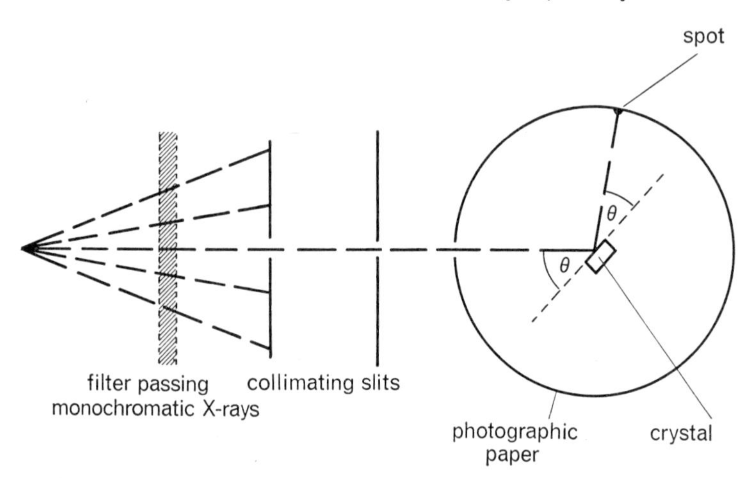

Fig. 3.7. Apparatus for studying the diffraction of X-rays by a crystal. The crystal is rotated about an axis which is perpendicular to the plane of the paper and is surrounded by a co-axial cylinder of photographic paper. When the angle of incidence of the beam of X-rays upon the crystal passes through a critical value the beam is reflected and produces a dark spot upon the paper.

wavelength) are obtained by the use of suitable filters, and a parallel beam is produced by means of pinholes in lead sheets. The X-ray beam then falls upon the crystal which is so mounted that it can be rotated about an axis perpendicular to the plane of the paper. When the angle of incidence passes through a critical value the beam is reflected and produces a spot upon a sheet of photographic paper which forms a cylinder co-axial with the axis of rotation of the crystal. From the positions of the spots we obtain the critical angles, and from the darkness of the spots we obtain the intensities of the reflected beams. Then follows an immensely complicated series of calculations from which we finally arrive at a 3-dimensional map showing regions from which the X-rays are more strongly scattered than from other regions. These regions of high scattering power indicate the positions of atoms. What sorts of atoms these are, the X-ray diffraction method can seldom tell us. In order to interpret the X-ray data we have to know something about the chemical composition of the material and we have to try to reconcile chemical composition with spatial configuration.

To apply the X-ray diffraction method to a protein it is clearly necessary to have the protein in the crystalline state. This is not always possible but some proteins crystallise very much more readily than others. Crystals of relatively small globular proteins, such as haemoglobin, were first prepared many years ago. Fibrous proteins, on the other hand, do not crystallise so readily, and we often have to make do with natural or artificial fibres in which the molecules probably form small crystallites with their long axes parallel but otherwise disposed at random. The information which X-ray diffraction provides about such paracrystalline preparations is correspondingly less precise.

Among the earliest studies of fibrous proteins were those made by Astbury on the protein keratin which is the main constituent of wool. When a wool fibre, mainly composed of the protein keratin, is heated under tension it 'gives' and can be stretched out to a greater length. This change in length is associated with a change in the X-ray diffraction pattern, from what is called the α-pattern to the β-pattern. A silk fibre, mainly composed of the protein fibroin, cannot be extended in this way and shows only the β-pattern. The β-pattern is shown by a backbone which is nearly fully extended. The α-pattern is seen in many other fibrous proteins, and since what all proteins have in common is their backbone structure, these observations suggest that the α-pattern indicates some partially folded state which the backbone normally assumes.

In 1937, Pauling and Corey set on foot a very thorough investigation of protein structure. They began by studying the configuration of simple amino acids, then of dipeptides and so on, working out the positions of the atoms to an accuracy of 0·02 Å. This provided a secure basis from which to interpret the rather vague information from the X-ray diffraction studies of fibrous proteins. By 1951 they had come to the conclusion that in proteins showing the α-pattern the backbone was twisted into a helix and held in that configuration by hydrogen bonds between neighbouring parts of the backbone.

Because of the very great importance of hydrogen bonding in maintaining the configuration of proteins and other molecules of biological interest some description of its nature is called for. The elements nitrogen, oxygen and fluorine, which in co-valently bonded compounds have respectively 1, 2 and 3 unshared electron pairs, are capable of sharing a hydrogen atom between them, and this act of sharing holds the atoms together. According to the usual convention, a hydrogen atom shared between a hydroxyl group and a carbonyl group might be represented as follows:

$$\rangle C::\ddot{\underset{\cdot\cdot}{O}}: \qquad H:\ddot{\underset{\cdot\cdot}{O}}:\underset{|}{\overset{|}{C}}-$$

$$\rangle C::\ddot{\underset{\cdot\cdot}{O}}:H:\ddot{\underset{\cdot\cdot}{O}}:\underset{|}{\overset{|}{C}}-$$

Alternatively, we may think in terms of the electric fields surrounding the atoms. Because the oxygen atom of a carbonyl group has a positively charged atomic nucleus (carbon) to one side only, the group is electrically out of balance; it is a dipole, the carbon end being positive relative to the oxygen end. Similarly, the hydroxyl group is a dipole because of the exposed position of the hydrogen nucleus. Consequently, there is an attraction between the two groups which brings them together until the force of attraction is balanced by the repulsion due to interpenetration of their electron clouds.

$$\rangle\overset{+}{C}{=}\overset{-}{O} \qquad \overset{+}{H}{-}\overset{-}{O}{-}\underset{|}{\overset{|}{C}}-$$

$$\rangle\overset{+}{C}{=}\overset{-}{O}\text{---}\overset{+}{H}{-}\overset{-}{O}{-}\underset{|}{\overset{|}{C}}-$$

As already indicated, hydrogen bonds can be formed between two electrically negative atoms, which in the biological context means oxygen and nitrogen. Thus in addition to the above we can also have

$$\rangle N{-}H\text{---}O{=}C\langle$$

and

$$\rangle\!\!\!/\, N\text{---}H{-}O{-}\underset{|}{\overset{|}{C}}-$$

Hydrogen bonds are relatively weak, having bond energies (the energy required to pull the atoms apart) of 5000–10,000 cal./mole as compared with about 10 times that amount in the case of co-valent bonds.

Returning now to the work of Pauling and Corey on the protein backbone, let us look at Fig. 3.8. This is an idealised diagram showing the protein backbone wound as a helix upon the outside of a cylinder. If the diameter of the cylinder is appropriately chosen the –NH group of a peptide linkage on one turn can form a hydrogen bond with the –CO group on a neighbouring turn, and by the establishment of a large

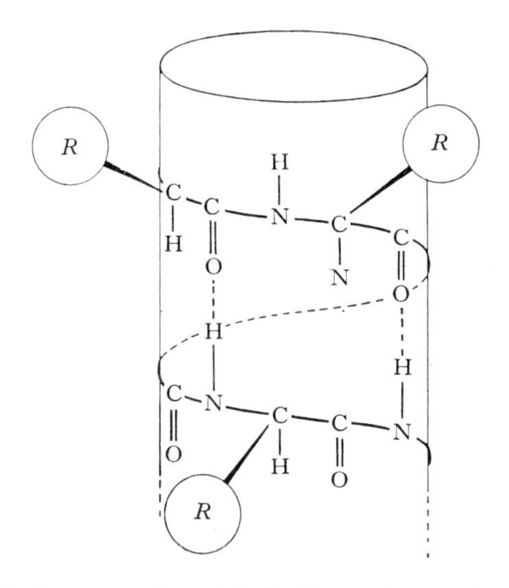

Fig. 3.8. Idealised representation of the backbone of a protein molecule wound as a helix upon a cylinder, with amino acid residues extending radially. Hydrogen bonds are formed between neighbouring turns of the helix.

number of such hydrogen bonds the helical configuration can be stabilised. In point of fact the true configuration is very much more angular than this, as is shown in Fig. 3.9. This configuration, which is called the α-helix, is found in a great many (though not all) proteins, and is what we mean by the secondary structure of the protein.

Because of the constraints imposed by the hydrogen bonds the α-helix is less flexible than the fully extended backbone of an isolated protein molecule; but the helical configuration is not present over the whole length of the backbone. Where it is not present the backbone can be folded and this allows the protein to assume a compact form in which it is retained by further bonding between the amino acid residues.

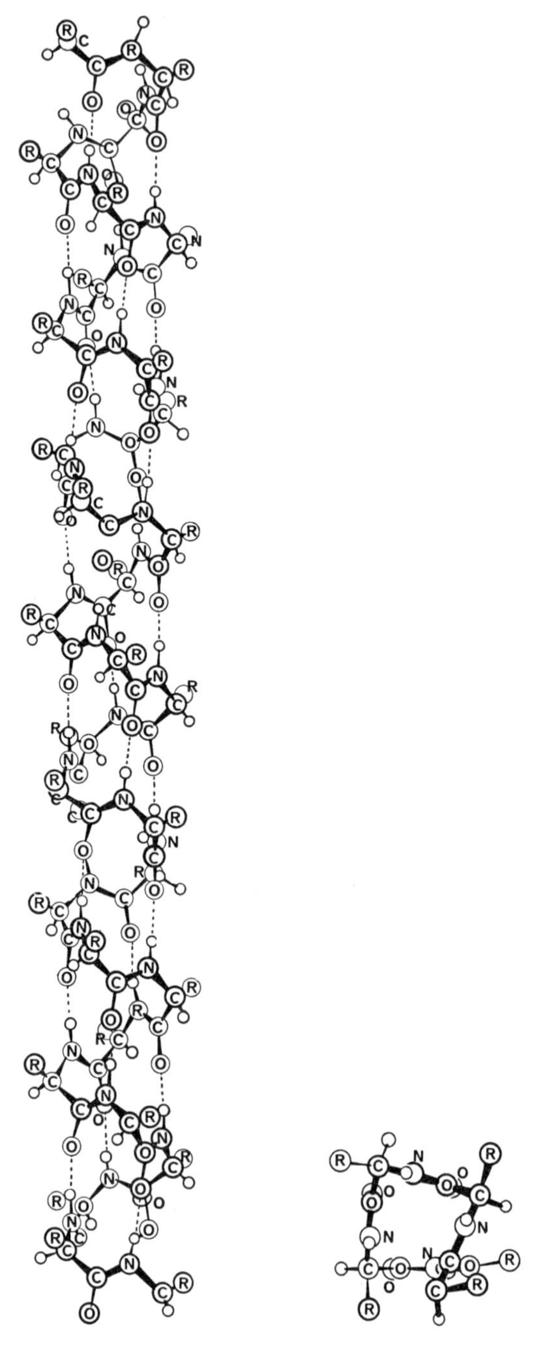

Fig. 3.9. The α-helix configuration of the protein backbone. Left, seen from side. Right, seen from end.

This can be co-valent bonding as between the sulphydryl groups of two cysteine residues (as in insulin, see Fig. 3.4), or hydrogen bonding, or even the still weaker van der Waals forces.† The relative ease with which globular proteins are denatured, with transition towards the extended condition, by comparatively low temperatures or even by spreading upon a surface film, testifies to the weakness of the forces which maintain this, their tertiary structure.

At the time of writing, the complete secondary and tertiary structure has been worked out in the case of one protein only, the respiratory pigment myoglobin from the muscles of the sperm-whale. This is a much larger molecule than insulin, having a molecular weight of 17,000, but its greater size and complexity are offset by the fact that the relatively heavy iron atom, which is part of the molecule, acts as a kind of landmark for the crystallographer. There are also other advantages for which only highly technical reasons can be given. A first account of the structure of myoglobin was published by Kendrew in 1958. This analysis had a resolution of 6 Å, which meant in effect that the distribution of electron density (X-ray scattering power) was sufficiently detailed to indicate the shape of the α-helix and the position of the iron atom, but was insufficient to show the positions of the other atoms. The analysis was subsequently refined to a resolution of 2 Å, at which the whole structure of the molecule is disclosed. Photographs of Kendrew's models are shown in Fig. 3.10, and one of the X-ray diffraction patterns from which the structure was derived is shown in Fig. 3.11. The calculations required to reach the 6 Å resolution took 6 years to work out, even with the use of electronic computers.

From what has already been said it will be abundantly clear that to identify a protein by physico-chemical methods is a particularly arduous undertaking. It is therefore fortunate that we have an alternative method which, although it does not tell us anything about the physico-chemical nature of the protein, is very much easier to carry out. This is the method of serological reaction.

The science of immunology, which deals with antigen-antibody reactions, came into being as a branch of medicine and may be said to date from Jenner's work on vaccination against smallpox, which he began in 1796. The lines of its early development not unnaturally ran in those

† These are the forces of attraction between molecules represented in the a/v^2 term of the van der Waals equation

$$\left(p+\frac{a}{v^2}\right)(v-b) = RT.$$

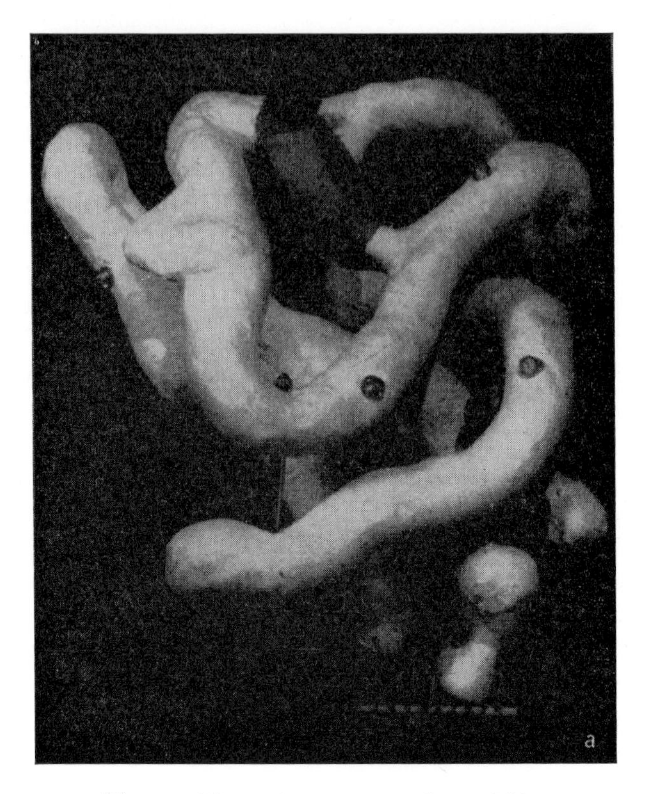

Fig. 3.10. The tertiary structure of myoglobin.
(*a*) The 6 Å model, showing the folding of the α-helix and (darker) the haem group and iron atom.

directions in which it could most usefully contribute to medical practice and only within comparatively recent years has its significance for other branches of biology been appreciated. It is a fact of common observation that a patient who survives an attack of an infectious disease is seldom the victim of a second attack. We say that he has an actively acquired immunity. During the latter part of the nineteenth century it was shown that this immunity could be conveyed to another person by the injection of serum from a convalescent. This we call passively acquired immunity. It was later possible to show that the pathogenic organisms when exposed to immune serum *in vitro* were visibly affected, suffering, for example, agglutination or lysis. The presence of some kind of 'antibody' in the blood of the immune person was therefore inferred. It was next discovered that the 'antigen' which evoked the

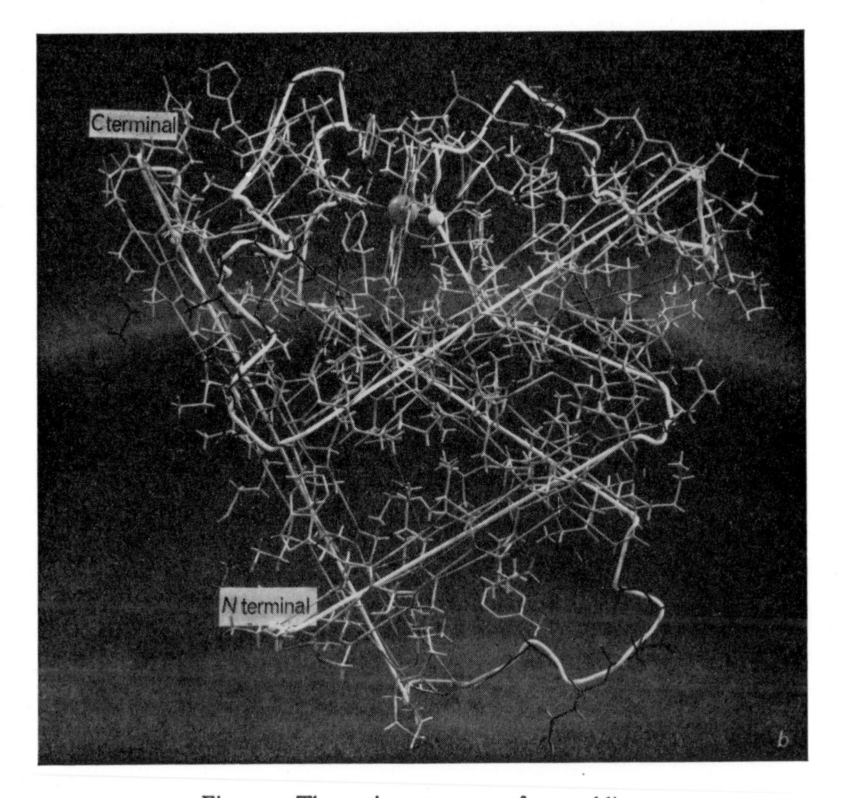

Fig. 3.10. The tertiary structure of myogoblin.
(*b*) The 2 Å model. The position of the iron atom is indicated by a sphere slightly above and to the right of the centre of the model.

production of antibody need not be a living pathogenic organism; antibodies could be formed in response to the injection of the toxins which the pathogenic organisms produced. Eventually it was realised that antigenic properties were possessed by a wide range of materials other than pathogenic organisms and their products.

The first requirement of an antigen is that it must be foreign to the body of the animal into which it is injected. The second requirement is that it must be a large molecule. Soluble substances of molecular weight less than 10,000 are not usually antigenic though they may become antigenic if they are adsorbed upon particles of indifferent material such as carbon or kaolin. Proteins are antigens *par excellence*, though not invariably so; gelatin (chemically a very simple protein) is an exception. Most polysaccharides (p. 48) of high molecular weight are antigenic.

Antibodies are always proteins and they belong to the class of proteins known as globulins, defined by their properties of solubility. The molecular weights of globulins range from 90,000 to 900,000. The globulin fraction of the serum proteins is a complex mixture of different species of protein molecule—probably the majority of them are antibodies to something or other—and it is not generally possible to separate

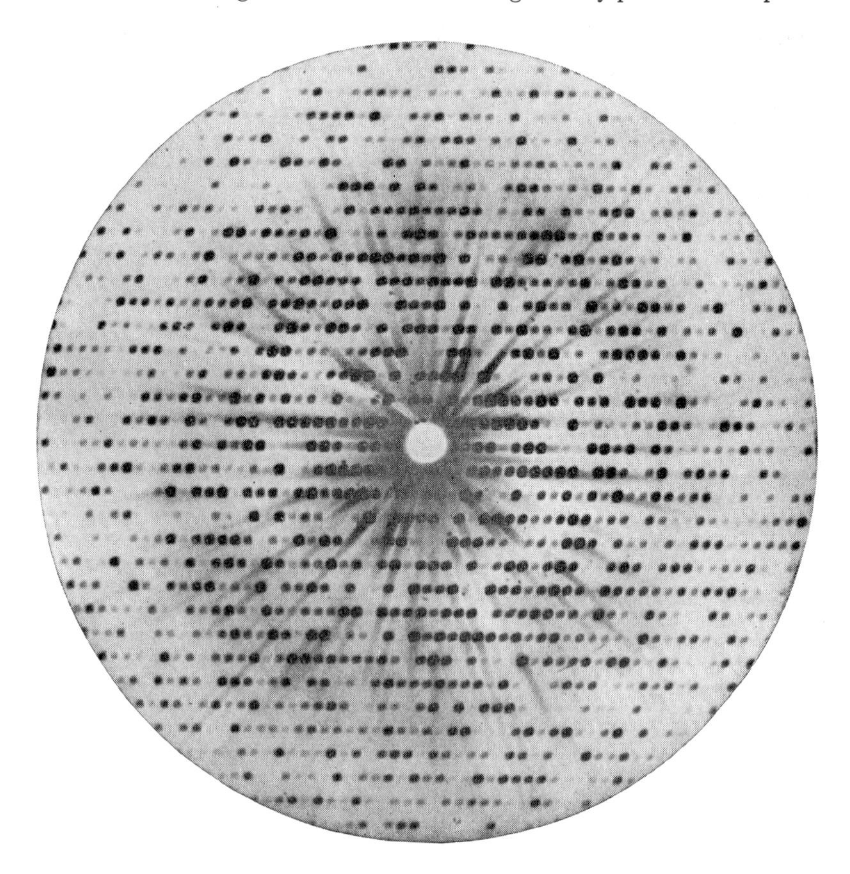

Fig. 3.11. An X-ray diffraction photograph of myoglobin.

one particular antibody from the mixture by any purely physico-chemical method. The production of antibody is evoked by the injection of the antigen into the blood, and the cells responsible for antibody production are probably the cells of the lymph nodes, bone marrow and spleen. A few days after the first injection of the antigen a small amount of antibody can be detected. If a second injection is given within one or two weeks there is a much more rapid response and the amount of antibody present in the blood rises to a high value which declines slowly over a period of months or years. In medical practice it is sometimes necessary to give a third injection, after which the acquired immunity can be of very long duration.

The production of antibodies in response to antigens is best seen in the higher animals. All the classes of vertebrates are known to be able to produce antibodies and among the invertebrates antibody production has been described in arthropods; but it does not seem that the antibodies of arthropods are so highly specific as those of mammals towards the antigens which produce them. In annelids and molluscs no production of antibodies has been found.

The presence of antibody in the blood of an immune animal can be demonstrated by its reaction with the antigen which evoked it. This reaction may be detected in various ways: in the intact animal, by the failure of the pathogenic organism to infect; in serum prepared from the drawn blood, by the agglutination or lysis of the organisms; or, if the antigen is not an organism, by the formation of a precipitate in which antigen and antibody are combined when the antigen is added to the immune serum. This last method of detection, when it can be used, is by far the most convenient.

An antigen-antibody precipitate is not formed in accordance with solubility product as in simple chemical reactions; instead, it follows a principle of optimum proportions. If we place some antiserum in a test-tube and add successive small amounts of antigen no precipitate is formed at first. As we go on adding antigen the antiserum becomes opalescent and if it is allowed to stand for a short time the particles will flocculate and settle out. But if we go on adding more antigen the opalescence passes through a maximum and then declines until the antiserum is clear once more. It has been established that for any antigen-antibody reaction the amount of precipitate is a maximum at a certain proportion of antigen to antibody. Optimum proportions of antigen : antibody range from 1:7 to 1:200 or more depending on the size of the antigen molecule.

The concentration of antibody present in antiserum, as compared with a standard antiserum, can be roughly determined by making successive dilutions and noting the dilution corresponding to the least detectable opalescence. The endpoint is difficult to determine and it is usual to increase the dilution by a factor of 2 at each step, since smaller steps would not give any greater accuracy. In medical work it is of course important to know how much antibody is present, but for general biological purposes we are usually content to know whether antibody is present or not. This being so we can avail ourselves of the following very simple and convenient method, known as the double-diffusion method of Ouchterlony. We allow the antigen and the antibody to diffuse towards each other through a gel; where they meet a precipitate is formed along a line which corresponds to the optimum proportions. The test is usually carried out in a shallow dish with a thick layer of agar jelly at the bottom. Small cup-like depressions are made in the agar into which the antigens and antisera can be placed in such arrangement as the experiment demands. In this way we can allow one antigen to react with various antisera or one antiserum to react with various antigens, and we can record the results photographically (Fig. 3.12).

It would be quite wrong to leave the reader with the impression that immunology is nothing more than the technical handmaiden of medical practice. Immunology is concerned with facts and problems that are basic to biological science. But we are here concerned only with the application of the techniques of immunology as a means of identifying proteins. Suppose that we wish to test for the presence of a particular protein in a mixture. We take the pure protein and by successive injections into a rabbit we produce an antiserum to that protein. We then place the pure protein in one depression, *a*, of the agar gel (Fig. 3.12), the mixture in another, *b*, and the antiserum in the third, *c*. A precipitate will form as a line between *a* and *c*. If a precipitate forms between *b* and *c* there is a *prima facie* case that the protein in question is present in the mixture. But a more exacting criterion can be applied. As diffusion proceeds the lines of precipitate extend in length. If the protein in *b* is serologically identical with that in *a* the two lines will fuse into one; if the proteins are not identical the lines will cross.

It is proper to point out that serological specificity, although it can be very high, is a matter of degree. Interpretation is sometimes difficult, and in any case it is to be supposed that only the outer parts of a globular protein molecule take part in antigen-antibody reactions. Serological identity, therefore, cannot be taken to mean that the proteins are identical in every aspect of their primary, secondary and tertiary structures. Nevertheless, the serological method has, and will continue to have, very great usefulness in the biological, as distinct from the purely chemical, aspects of the study of proteins.

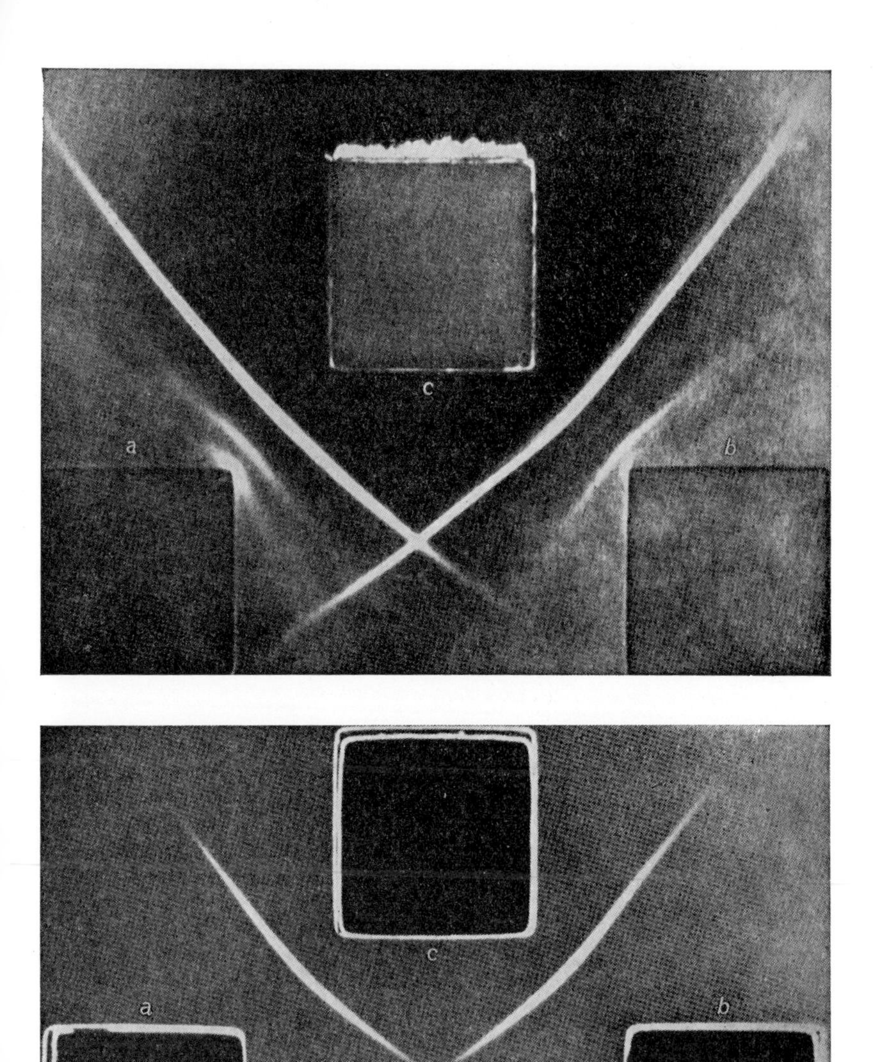

Fig. 3.12. The Ouchterlony double-diffusion method.

a contains the known protein, *b* contains a mixture of proteins, *c* contains an antiserum to the protein in *a*. In the lower picture the two lines of precipitate fuse into one, indicating serological identity. In the upper picture the two lines cross, indicating serological non-identity.

These are photographic prints made directly from the agar gel and are therefore negatives.

2. CARBOHYDRATES

Carbohydrates are much simpler than proteins, both in their chemical form and in their physiological and structural roles. The sugar glucose is found in the blood of the higher animals and also in the sap of plants and it is clear that in these organisms its physiological role is that of a ready-use fuel intermediate between the energy-using mechanism of the cells and the energy reserves of the organism, which are often laid

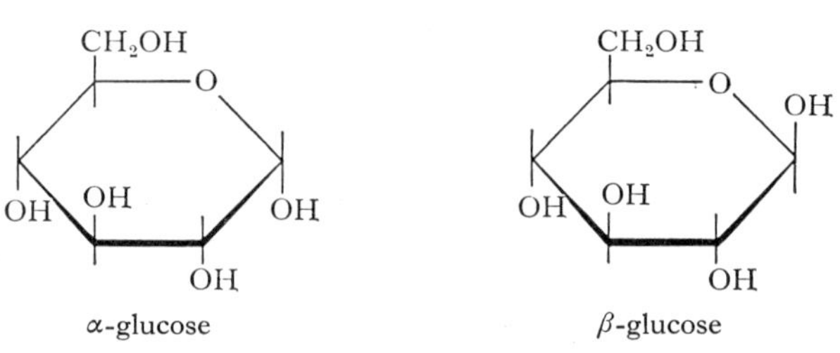

Fig. 3.13. The cyclic form of glucose.

Conventionally, the ring is represented as lying in a horizontal plane, the heavy lines connecting carbon atoms (1), (2), (3), (4) indicating that this side of the ring is nearer to the observer. The hydrogen atoms, the hydroxyl groups and the terminal $-CH_2OH$ group can then be shown as lying above or below the plane of the ring.

Below, a simplified representation omitting the carbon atoms of the ring is used to show the difference between α-glucose and β-glucose. This difference affects the arrangements at carbon atom (1) only. See also Appendix, p. 316.

down in the form of other carbohydrates, such as starch or glycogen. In its structural role, in the form of cellulose, carbohydrate forms the basis of the cell wall of plants.

The unit molecules of carbohydrates, analogous to the amino acids which are the units of protein, are simple sugars—monosaccharides—

whose structure we shall presently examine. Most monosaccharides have the empirical formula $(CH_2O)_n$, where n has the values 5 or 6, occasionally 7. Glucose and fructose are monosaccharides, and having 6 carbon atoms they are also called hexoses. Ribose and xylose are pentoses, having 5 carbon atoms. Monosaccharides can be linked together, with the elimination of water, to give disaccharides such as sucrose, maltose and lactose, and by still further polymerisation they can be built up into polysaccharides such as starch, glycogen and cellulose.

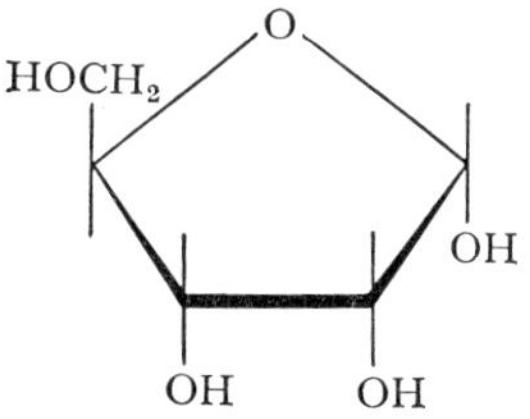

HOCH$_2$ O OH OH OH

ribose

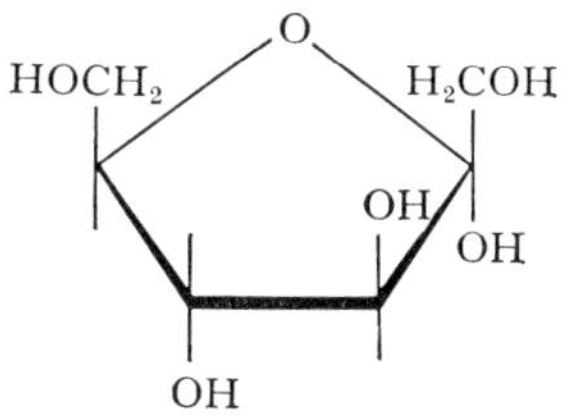

HOCH$_2$ O H$_2$COH OH OH OH

fructose

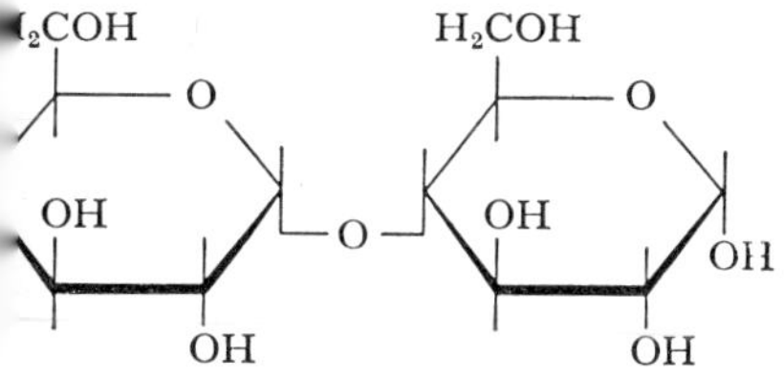

H$_2$COH O OH OH H$_2$COH O OH O OH OH

maltose

H$_2$COH O HO OH OH HOCH$_2$ O H$_2$COH OH O OH

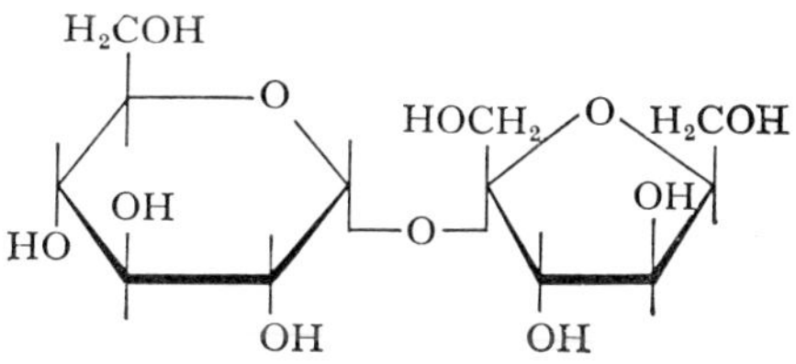

sucrose

Of all the monosaccharides glucose is the one most commonly found in organisms in the free uncombined state. It was originally believed that all monosaccharides were open-chain compounds but towards the end of the nineteenth century Emil Fischer and others showed that they exist mainly in a cyclic—more correctly a heterocyclic—form; the ring is formed out of 4 or 5 carbon atoms and one oxygen atom. The conventional representation of the cyclic form of glucose is shown in Fig. 3.13 and further explained in the legend to that figure. Its relationship with the open-chain form is explained in the Appendix, p. 317. In other monosaccharides, for example ribose and fructose, the ring may contain only 4 carbon atoms.

The hydroxyl group at carbon atom (1) is the most reactive, and will condense with a hydroxyl group on another molecule to give a disaccharide. This linkage is called a glycosidic bond and it is analogous to the peptide bond of proteins. The disaccharide maltose is made up in this way out of two molecules of glucose, and the disaccharide sucrose out of a molecule of glucose and a molecule of fructose.

Proceeding beyond the disaccharides to the consideration of polysaccharides we will take two examples only: amylose and cellulose. Amylose is found as a component of starch grains, which do not show any fibrous structure. Cellulose, by contrast, forms a highly ordered fibre. Both of these polysaccharides are high polymers of glucose, having the form of unbranched chains. In both cases the glycosidic bond is between carbon atom (1) of one glucose unit and carbon atom (4)

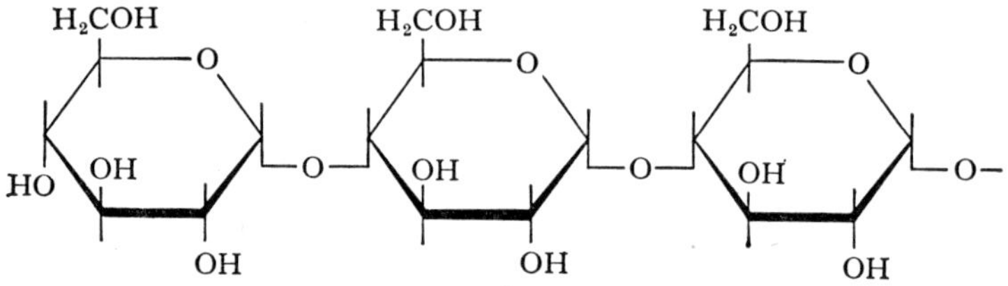

amylose

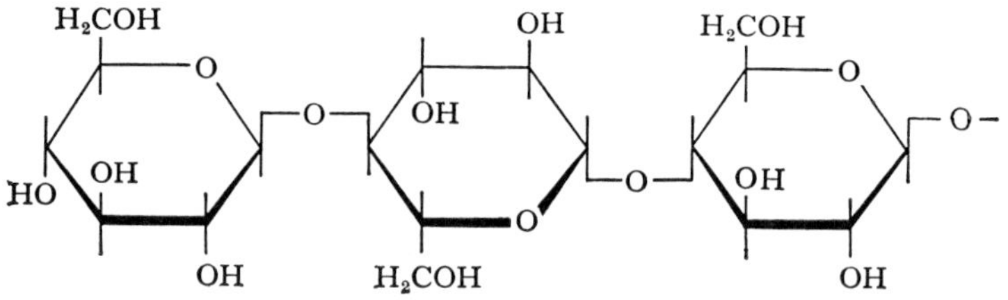

cellulose

of the next. The only difference between them is that amylose is made up of α-glucose whereas cellulose is made up of β-glucose. As a consequence alternate units in the molecular chain of cellulose are inverted.

By this time the reader may well be wondering, in some impatience, why all these details of the structural chemistry of carbohydrates have been inflicted on him. There has, however, been a purpose in this which is now to be revealed.

As a further consequence of the alternate inversion of the units in cellulose extensive hydrogen bonding becomes possible between molecular chains which are aligned in parallel, and in this way long crystalline aggregates, known as micelles, can be built up of any length and thickness. This example has been chosen in order to illustrate the way in which a seemingly trivial difference in chemical structure can have the most far-reaching biological implications. α-glucose and β-glucose differ only in the position of the hydroxyl group at carbon atom (1); they are interchangeable, being in equilibrium with the open-chain form of glucose. Yet while amylose is no more than a convenient form of energy reserve cellulose lends itself to the formation of a strong fibre. To relate biological role to chemical structure, as has been done for cellulose, is one of the general aims of molecular biology.

3. FATS

The word 'fat' is nowadays taken to mean neutral fat. The word used to describe fatty substances in general is 'lipid'. Lipids have in common the property of being insoluble in water but soluble in such solvents as alcohol, acetone and ether. There are three main groups of lipids: neutral fats, phospholipids and steroids.

Neutral fats are the glycerol esters of long-chain aliphatic acids. A typical neutral fat is glycerol tri-stearate.

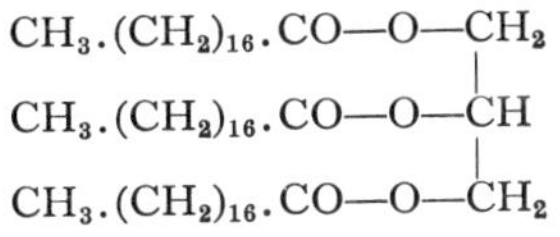

$$
\begin{array}{l}
CH_3.(CH_2)_{16}.CO{-}O{-}CH_2 \\
\qquad\qquad\qquad\qquad\quad | \\
CH_3.(CH_2)_{16}.CO{-}O{-}CH \\
\qquad\qquad\qquad\qquad\quad | \\
CH_3.(CH_2)_{16}.CO{-}O{-}CH_2
\end{array}
$$

glycerol tristearate

In the higher animals neutral fats form an important part of the diet and provide an important part of the energy reserves. In other organisms neutral fats may be conspicuously absent. It does not appear that neutral fats have any special role in the molecular architecture of organisms or any essential biochemical function other than provision of energy.

Phospholipids have a general resemblance to neutral fats, in fact they are neutral fats in which one of the fatty acid residues is replaced by phosphoric acid and an organic base. In lecithins this organic base is choline.

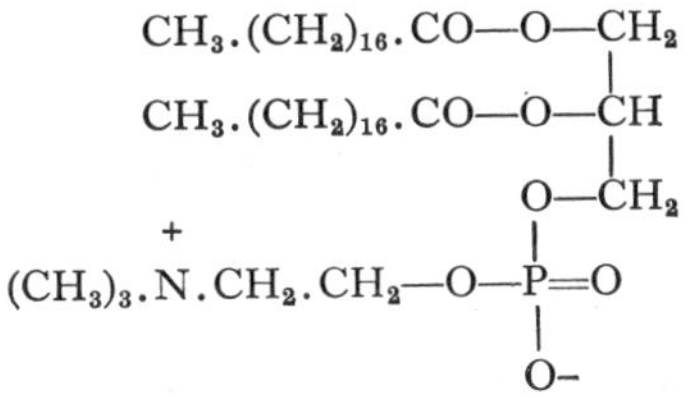

$$
\begin{array}{l}
\qquad\qquad\qquad CH_3.(CH_2)_{16}.CO{-}O{-}CH_2 \\
\qquad\qquad\qquad\qquad\qquad\qquad\qquad\quad | \\
\qquad\qquad\qquad CH_3.(CH_2)_{16}.CO{-}O{-}CH \\
\qquad\qquad\qquad\qquad\qquad\qquad\qquad\quad | \\
\qquad\qquad\qquad\qquad\qquad\qquad\qquad O{-}CH_2 \\
\qquad\quad + \qquad\qquad\qquad\qquad\qquad\quad | \\
(CH_3)_3.N.CH_2.CH_2{-}O{-}P{=}O \\
\qquad\qquad\qquad\qquad\qquad\qquad\quad | \\
\qquad\qquad\qquad\qquad\qquad\qquad\quad O^-
\end{array}
$$

a typical lecithin

Phospholipids are obtained in quantity from nervous tissue, but they are also present, though in much smaller amount, in nearly all cells. We have some reason to believe that they may be an important constituent of cell membranes, and this point will be taken up again presently. But we will first dispose of the steroids.

Of the steroids we shall have very little to say because although they are of the very greatest physiological importance it so happens that we shall not have occasion to mention them again in this book, and they are included here only for the sake of completeness. The steroids are mostly very complex monohydric alcohols, having one –OH group attached to a characteristic cyclic framework.

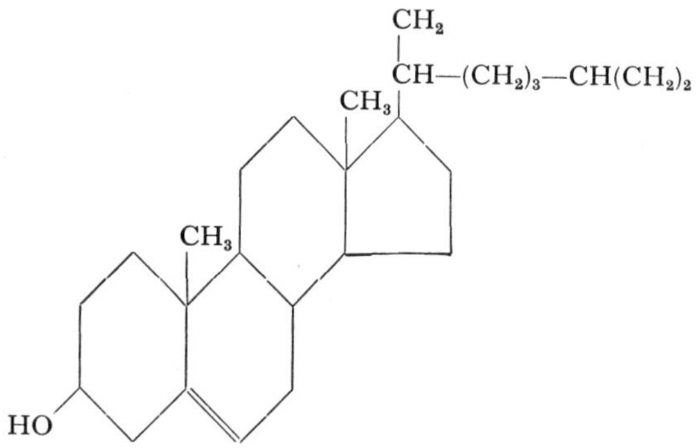

cholesterol

Among biologically important steroids we may mention in passing vitamin D, corticosterone (from the adrenal cortex) and the various sex hormones.

The following are some of the reasons why we believe that lipids are an important constituent of cell membranes. First, the penetration of various substances into cells is correlated with their lipid-solubility; for example, the internal pH of a sea-urchin egg can be changed more readily by butyric acid, a weak acid but one which is lipid-soluble, than by hydrochloric acid. Secondly, small amounts of lipid solvents have profound effects upon the permeability of cell membranes and especially upon the functions of the nervous system in which we know that the cell surface is primarily concerned; the usefulness of ether and chloroform as anaesthetics illustrates this point. Thirdly, some lipids have physicochemical properties which peculiarly fit them for activity at surfaces.

The lipids which are important in surface phenomena are pre-eminently those which incorporate a group which has affinity for water, in contrast to the rest of the molecule which has little if any affinity for water. To this class belong the phospholipids, such as lecithin. Choline phosphate by itself is water-soluble, whereas glyceryl tristearate is water-insoluble. Suppose that we take a very small quantity of lecithin, dissolve it in ether and allow a drop of this solution to spread upon a clean surface of water. The ether presently evaporates leaving the

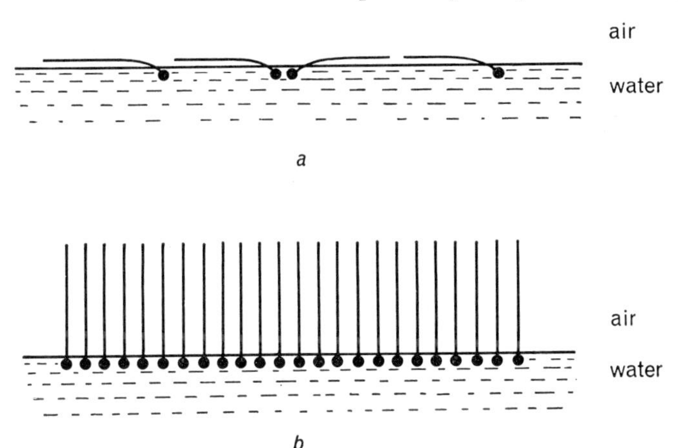

Fig. 3.14. Lipid monolayers.

Polar lipids tend to spread as films upon a surface of water, the water-soluble ends of the molecules entering the water while the water-insoluble parts remain outside.

In the absence of restraint the water-insoluble parts tend to lie parallel to the surface, as in *a*, giving an expanded monolayer. When the surface is restricted the water-insoluble parts are forced into an orientation more or less normal to the surface, as in *b*, giving a compressed monolayer. A monolayer of this latter type effectively obstructs the free passage of water molecules across the interface.

lecithin as a thin film upon the surface. Under these conditions the choline phosphate part of the molecule, being water-soluble, enters the water, whereas the rest of the molecule, being water-insoluble, remains outside the surface and lies flat along it. We may represent this situation as in Fig. 3.14*a*.

If the area of the water surface is decreased, or if we add more lecithin to it, the molecules become crowded together and their water-insoluble ends are forced into a vertical orientation. In this way we obtain a surface of water covered by a film of oriented lecithin molecules in the form of a monomolecular film, or monolayer. The presence of a monolayer of this type upon a free water surface has very important physical effects, among which the most significant is that it greatly retards the rate of evaporation of water molecules from the surface.

Now although one does not always realise it, the surfaces of cells are remarkably impermeable to water. The following considerations will perhaps serve to bring this into perspective.

Many of the Protozoa, *Amoeba* among them, possess an organelle known as the contractile vacuole. The surface membrane being, to some extent at least, permeable to water and the protoplasm containing inorganic ions at greater concentration than in the external medium, there is a tendency for water to enter the cell by osmosis. If this process were to go on uncompensated the cell would swell up and disintegrate; but as fast as the water enters it is removed by the activity of the contractile vacuole. If we measure the diameter of the vacuole just before it empties and count the number of times it empties per minute we can calculate the rate at which water is pumped out by the contractile vacuole, which is the same as the rate at which water enters through the surface. The necessary figures were obtained by Kitching for the protozoon *Zoothamnium*, and are as follows. The volume of *Zoothamnium* is 16,800 μ^3 and its surface area is 3500 μ^2. The maximum diameter of the vacuole is 7 μ, from which its volume is 180 μ^3. It empties on the average 3 times per minute, so it eliminates water at a rate of 540 μ^3/min. At this rate it will evacuate a volume of water equal to the whole volume of the protozoon in about half an hour, so it would appear that water must enter the cell rather rapidly. But let us calculate the rate of entry of water per sq.cm of surface. The rate of entry is 540 μ^3 per 3500 μ^2/min. $1\ \mu \equiv 10^{-4}$ cm. The rate of entry of water per minute is then

$$\frac{540\ \mu^3}{3500\ \mu^2} = \frac{540 \times 10^{-12}\ \text{cm}^3}{3500 \times 10^{-8}\ \text{cm}^2} = 1{\cdot}55 \times 10^{-5}\ \text{ml/cm}^2.$$

This does not look very much, but it is difficult to visualise quantities such as 10^{-5} ml; so let us try it the other way round.

1 ml passes through 1 cm^2 in

$$\frac{1}{1{\cdot}55 \times 10^{-5}}\ \text{min.} = 6{\cdot}45 \times 10^4\ \text{min.}$$

There are $4{\cdot}03 \times 10^4$ minutes in a month. Therefore 1 ml passes through 1 cm^2 of cell membrane in 6·45/4·03 months, that is, in about 6 weeks.

Our original impression of a rapid entry of water turns out to be illusory; and when we further consider that the difference in osmotic pressure on the two sides corresponds to a column of water over 30 feet high it is clear that the cell membrane is about as good a waterproof as we could wish to buy. The explanation of the illusion is that when we look at a small organism through a microscope we instinctively think of it as scaled up to macroscopic size. But whereas the volume of an object varies as the cube of its linear dimension, l, its surface area varies only as the square. Thus the ratio of surface to volume varies as l^2/l^3, i.e. as $1/l$. Consequently the smaller the object the greater is its surface/volume ratio, with the result that although water may enter at what seems to be a rapid rate in relation to the volume of the cell, the rate of entry reckoned per unit of surface area can be very low.

This, however, is by the way. For our present purposes the object of the calculation is to show that the cell membrane is very impermeable to water. This is what one would expect if the entry of water is opposed by a barrier of oriented molecules of lipid, and this observation, taken in conjunction with the other evidence mentioned earlier, reinforces our belief that lipids have a most important role in determining the physiological properties of the cell membrane. This is a subject to which we shall return later (p. 96).

Of course, not all the biologically important compounds can be referred to protein, carbohydrate or fat. The reader will have noted the omission of any reference to nucleic acids. These are to be considered later as the material basis of inheritance (Chapter 22). Other organic compounds of biological importance will be described in appropriate contexts.

4

METHODS OF STUDYING CELL STUCTURE

The successful use of the method of dissection to display the anatomy of the dogfish depends mainly upon the manual skill of the dissecter. Once the overlying tissue has been removed, a nerve or a blood vessel is plain to see; or, if it is very small, a low-power dissecting microscope will enable us to see it. In establishing the factual details of the anatomy of the larger animals we do not as a rule encounter problems of interpretation. If the anatomical feature is present, we can see it; or, if we cannot see it after careful dissection, we are entitled to conclude that it is not present. And it does not matter greatly whether we take a living animal or a dead animal for dissection. During the relatively short time required for dissection post-mortem changes are unlikely to make appreciable progress, and in any case they can be arrested if necessary by the use of preservatives.

These freedoms do not extend to the investigation of fine structure at the cellular and subcellular levels. The drastic changes which can follow upon the application of preservatives are unmistakable, and as our techniques of observation improve we are increasingly able to detect changes which set in as soon as the cell finds itself in an unfavourable environment. Yet in order to be able to see anything at all of the fine structure of a cell it is usually necessary to prepare it in some manner which involves its death. The prepared cell which we then examine is different from the cell as it was in life—how different we have no direct means of discovering. Our conception of the living cell must therefore be inferred from observations made mainly upon dead cells, and it is at this stage that the investigator's knowledge, experience and judgement are called in to interpret the observations. Even the observations themselves, the structural features which we seem to discover in the dead cell, have to come under searching scrutiny. In studying fine structure we are very often working at the extreme limits of the microscope's performance, and under these conditions we begin to encounter phenomena which have their origin partly in the material to be examined and partly in the method of examination. Not only must we proceed with the greatest caution in drawing inference from what we observe; we must also constantly ask ourselves how far that which we see in the microscope is representative of the actual structure. We have always to bear in mind that certain features, which on other grounds we believe to be present, may not, in theory, be observable with the means at our disposal.

The need to maintain constant vigilance against the dangers of these pitfalls is the justification for devoting so much space to the methods currently used in the study of the structural organisation of cells.

1. MICROSCOPES

(i) The light microscope

The oldest and still the most widely used instrument for studying cells is the ordinary light microscope. It is surprising how many students of biology regard the microscope as no more than a magnifying device and believe that its quality can be assessed in terms of magnifying power. This is to disregard the very much more significant attribute of resolving power. Magnification, *per se*, presents no very serious problems; there is no theoretical limit to the degree of magnification that can be attained. But as every amateur photographer knows, in the enlarging of a negative there comes a stage at which the imperfections of the negative—the grain, for example—begin to appear in the enlargement; and when this stage is reached further enlargement gives a bigger picture but reveals no more detail. In the case of the microscope magnification is only useful up to the stage at which the imperfections of resolving power become apparent.

The magnifying property of the microscope is adequately described by the methods of geometrical optics, which treat light as rays travelling in straight lines. The reader will already be familiar with the type of diagram reproduced in Fig. 4.1. To describe the resolution of the microscope it is necessary to use the methods of physical optics, which treat light as wave motion. Let us first take the case of a beam of parallel light which is brought to a focus by a simple lens as shown in Fig. 4.2. According to geometrical optics the rays should all converge upon the principal focus, F; the point F should therefore be bright and the rest of the screen should be dark. But the methods of physical optics lead to a different conclusion. According to physical optics the parallel beam of light is envisaged as being a plane wave front which after passing

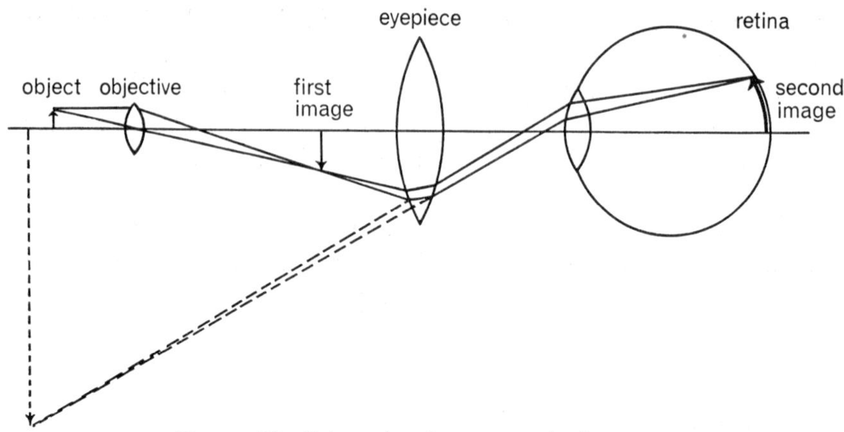

Fig. 4.1. The light paths of a compound microscope.

A real inverted magnified image of the object is produced by the objective lens. This is brought to a second real erect image upon the retina by the lens of the eyepiece and the lens of the eye. What the eye sees is a virtual inverted image.

through the lens becomes a concave wave front centred upon F. Making use of Huygens' principle, let us consider two wavelets originating at the points A and B on the concave wave front as it emerges from the lens. Being equidistant from F the two wavelets will arrive at F in phase; they will interfere constructively, their amplitudes will be added and the screen at F will be bright. Now take a point P on the screen such that the distance AP is x and the distance BP is $x+\lambda/2$, where λ is the wavelength of the light. When the two wavelets arrive at P they will be in antiphase, they will interfere destructively, the one cancelling the other, and the screen at P will be dark. Beyond P where the path difference becomes one whole wavelength the wavelets will again arrive in phase and the screen will be bright. Clearly there will not be a single bright spot on the screen but a pattern of bright and dark areas. To discover the form of this pattern it is necessary to take account of all the wavelets arising at all points on the concave wave front and to find their resultant at F; and to repeat this process for all points on the screen. This can be done by the methods of the integral calculus and a mathematical expression is obtained which describes the relation between brightness and distance from F. What this tells us is that a beam of parallel light does not produce a single bright point at F upon an otherwise dark screen—as geometrical optics would have it—but that centred upon F there is a bright disc surrounded by alternate bright and dark rings, as in Fig. 4.3.

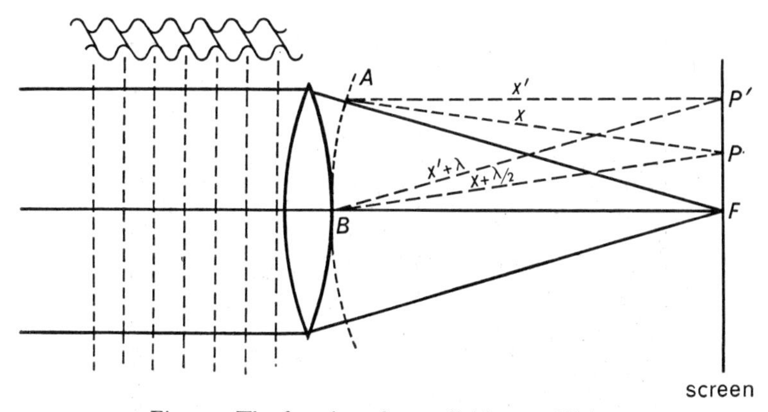

Fig. 4.2. The focusing of a parallel beam of light.

According to geometrical optics all the rays pass through F, the principal focus, and the rest of the screen is dark. According to physical optics wavelets arising at A and B will arrive in phase at F, which will be bright; they will arrive in anti-phase at P, which will be dark, and in phase again at P', which will be bright. As the focal length of the lens is decreased—and hence the angle at which the light converges is increased—the distance FP becomes smaller.

The dimensions of this concentric diffraction pattern depend in part upon those parameters of the lens which determine the angle of the cone of light as it converges upon the focus. A lens of short focal length and large diameter causes the light to converge more strongly than a lens of long focal length and small diameter; and for this reason it produces a smaller diffraction pattern. But however much we increase the convergence P will never coincide with F. There is a definite lower limit to the dimensions of the concentric diffraction pattern; and if d is the diameter of the bright central disc (or equally the diameter of the first dark ring, see Fig. 4.3) then d will never be less than λ.

The limitations set by wavelength upon the resolving power of a miscroscope may be understood in similar terms. There is one difference: in the case described above, the size of the diffraction pattern was governed by the angle of the cone of light *leaving* the lens and converging on the image, while in a microscope the size of the observed pattern depends mainly on the angle of the cone of light coming from the object and *entering* the objective lens. Let us consider a microscope objective lens forming an image of a minute hole in an opaque diaphragm. No matter how small this hole is made, its image will take the form of the concentric diffraction pattern just described. If we make two holes close together in the diaphragm two overlapping patterns

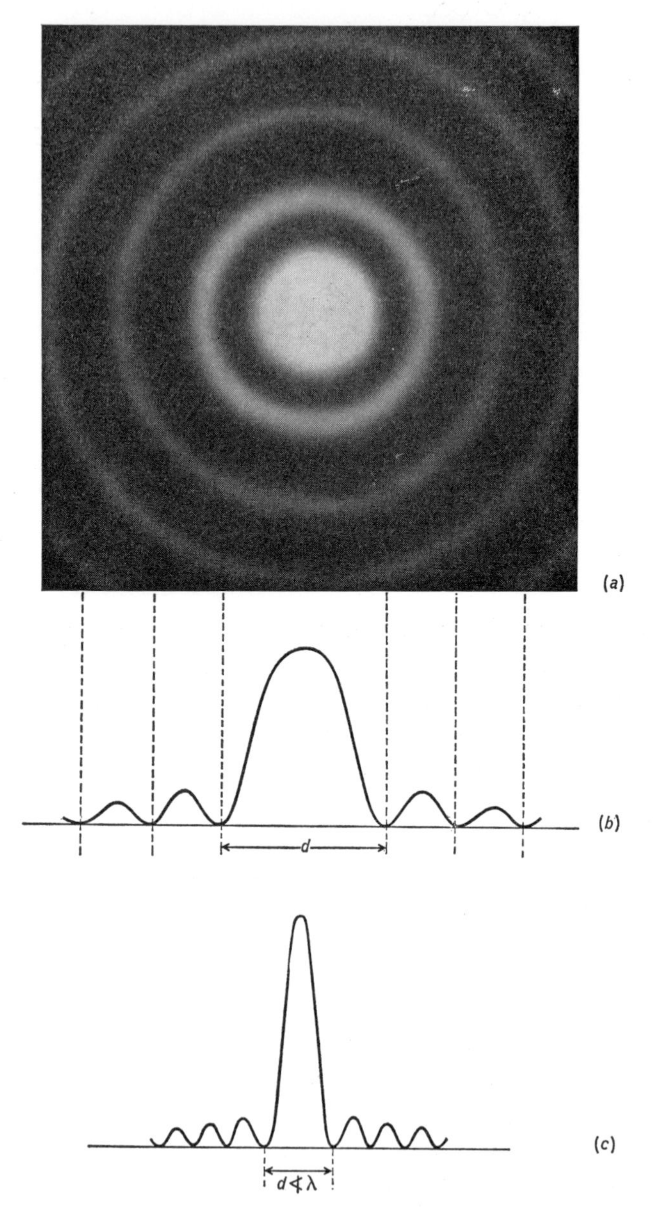

Fig. 4.3. (*a*) Pattern such as would be produced when a beam of parallel light is brought to a focus on a screen. (*b*) the same, brightness plotted against distance. (*c*) The same, for maximum convergence of the light, i.e. shortest possible focal length. The width of the central bright disc cannot be less than the wavelength of the light.

will be produced. Our ability to decide whether there are two holes or one will depend upon our ability to decide whether there are two diffraction patterns or one. If we screen off the outer parts of the lens by means of an iris diaphragm, thereby cutting off the highly convergent outer rays, the resolution will deteriorate, the patterns will expand and we will see only a single, apparently circular, disc of light. As we open the iris diaphragm the patterns will contract, resolution will improve and we will be able to see that the bright disc is double (Fig. 4.4). Whether the holes can or cannot be resolved depends to some extent on the observer. To avoid this difficulty of subjective assessment Rayleigh proposed the following objective criterion: two objects are deemed to be resolved if the centre of the bright disc of one falls upon the first dark ring of the other (Fig. 4.5). The diameter of the first dark ring in the image plane corresponds to a distance of about one wavelength in the object plane, from which it follows that using Rayleigh's criterion we cannot decide that there are two holes and not one if the holes are less than half a wavelength apart. This is an approximation which has the merit of being easy to remember. For green light, of wavelength 5000 Å, resolution is possible down to 2500 Å, that is, down to 0·25 μ. This is the maximum possible resolution, set by the wavelength of the light used, and implying that the lens is perfectly free from aberration.

From the foregoing it will be apparent that the aim of the designers of microscopes is to achieve the greatest possible convergence of light at the object. The converging properties of a microscope objective are specified in terms of its numerical aperture, which is explained in Fig. 4.6. To make effective use of the full numerical aperture of an objective it is necessary to use it in conjunction with a condenser of comparable converging power (Fig. 4.7). It may also be noted that the numerical aperture is proportional to the refractive index of the medium between the objective and the object. With objectives of the highest power it is customary to fill the space between the objective and the coverslip with oil of high refractive index; such objectives are known as oil-immersion objectives. The use of oil increases the maximum possible convergence of light, as explained in Fig. 4.8.

Light microscopes have been developed to a stage at which further significant improvement of the optical system can hardly be looked for as we are up against the fundamental limitation set by wavelength. Efforts are therefore made to develop microscopes working at wavelengths shorter than those of visible light. One obvious direction to be explored is the use of ultraviolet (UV) light. The lower limit of visibility is about 4000 Å and this is not far from the lower limit of wavelength which is transmitted by glass, about 3500 Å. The lenses of a UV microscope have to be made of quartz, which transmits wavelengths down to 1800 Å, and thus offers an improvement over visible light by a factor

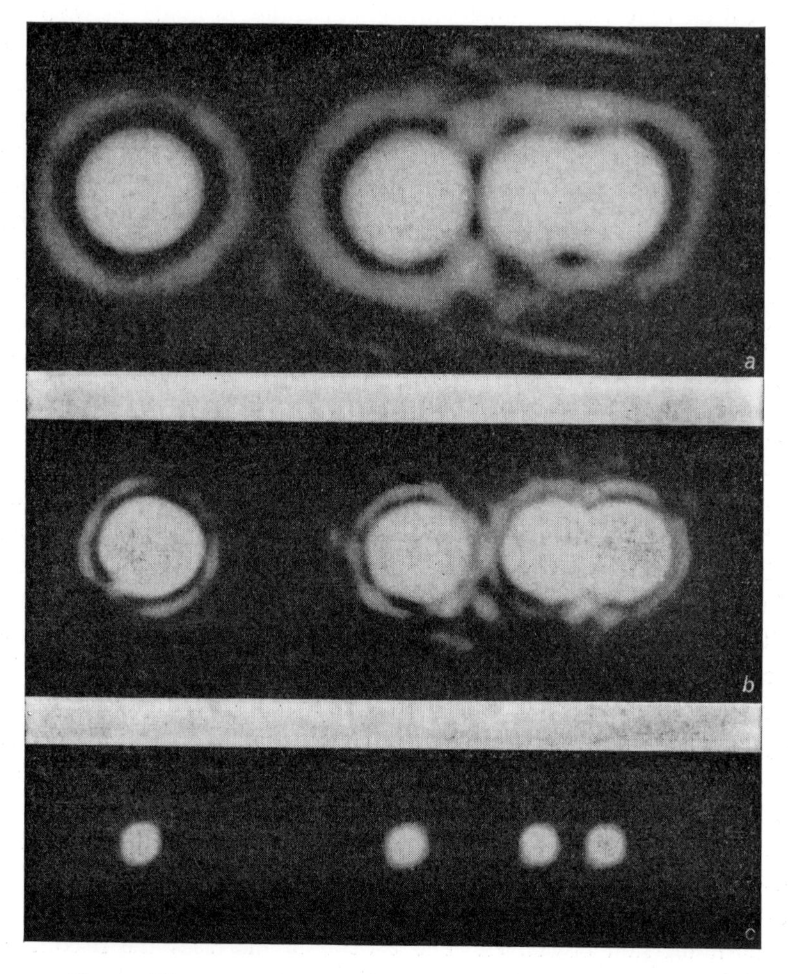

Fig. 4.4. The concentric diffraction patterns from four point sources with a circular aperture of variable diameter in front of the lens.

(*a*) The aperture is small, the resolution is poor and the two right-hand sources only just satisfy Rayleigh's criterion. (*b*) The aperture is larger, the diameter of the central bright disc is smaller, but the outer bright rings are still very conspicuous. (*c*) The aperture is larger still and the brightness of the central disc is very great in comparison with the brightness of the rings which do not show up in this photograph.

of 3. But there are also disadvantages, the principal disadvantage being that quartz lenses are very expensive. This means that the UV microscope is never likely to be anything but a specialist's instrument. It

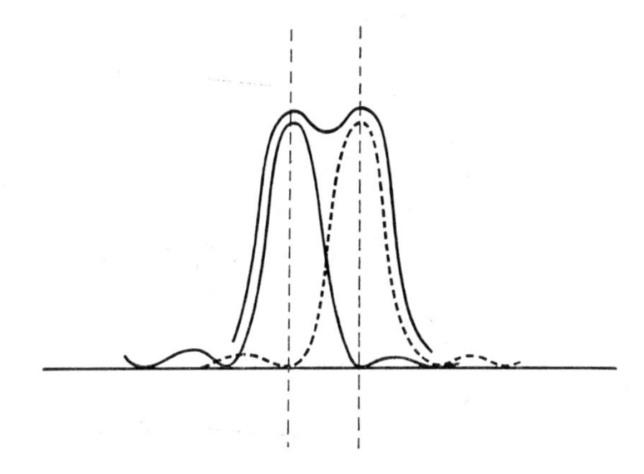

Fig. 4.5. Rayleigh's criterion of resolution.
The central maximum of the concentric diffraction pattern from one source falls on the first minimum of the pattern from the other source.

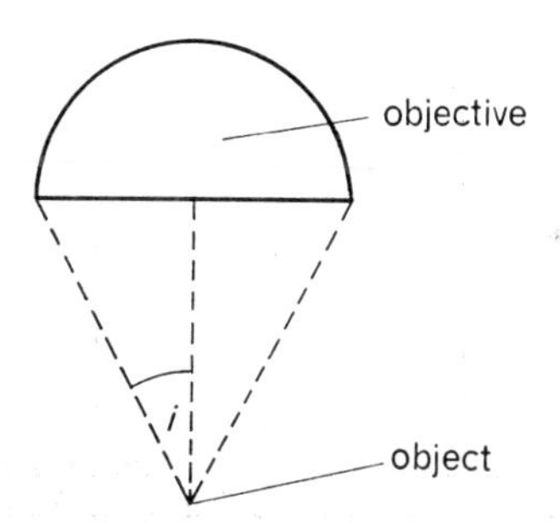

Fig. 4.6. The numerical aperture (N.A.) of an objective.
N.A. $= 2\,\mu \sin i$, where i is the half-angle of convergence, μ is the refractive index of the medium between object and objective.

finds its most important use in cytochemical investigations—for example, in revealing the distribution of nucleic acids which absorb strongly at about 2600 Å. X-ray microscopes have also been constructed. X-rays have wavelengths from 10 Å downwards, which offers prospect of very substantial improvement in resolution. But X-rays are not refracted by workable materials and therefore one cannot make an X-ray microscope of conventional design. In any case, most of the advantages of short wavelength are to be had with the electron microscope which is easier to make and to use.

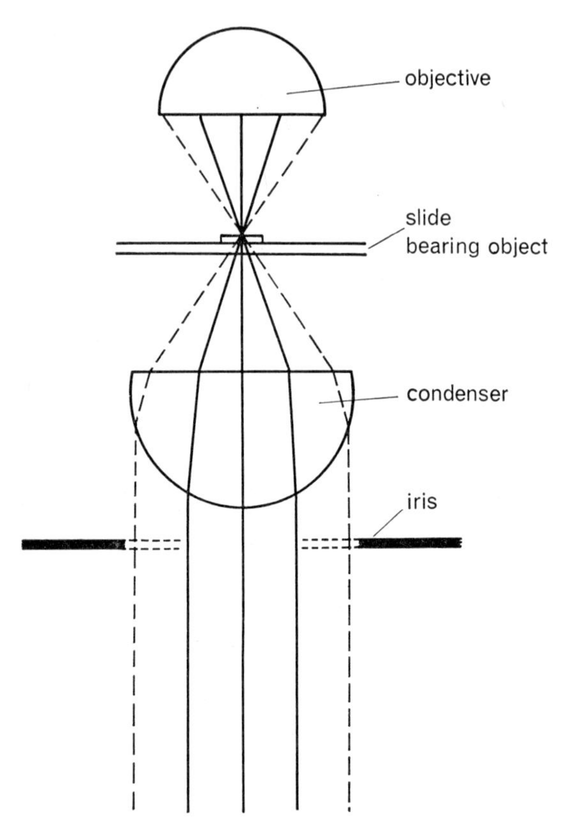

Fig. 4.7. To illustrate the importance of using a condenser whose numerical aperture is not less than that of the objective if the full resolving power is to be attained. If the iris diaphragm is used to cut down the light, the outer strongly convergent rays are cut off and the effective numerical aperture of the objective is reduced.

(ii) The electron microscope (EM)

Electrons not only behave as though they are charged particles; they also behave as though they are waves. Because electrons behave as charged particles a beam of electrons can be deflected, and therefore focused, by electrostatic and/or magnetic fields. Because electrons behave as waves the resolution of an electron-optical system depends upon wavelength.

An electron beam is produced in the EM in the same way as in a cathode-ray tube: electrons are emitted by a hot cathode and accelerated towards an anode which is provided with a hole allowing a narrow beam to pass through. Most EMs use magnetic focusing. An electron entering a uniform magnetic field is not deflected if its path is parallel to the

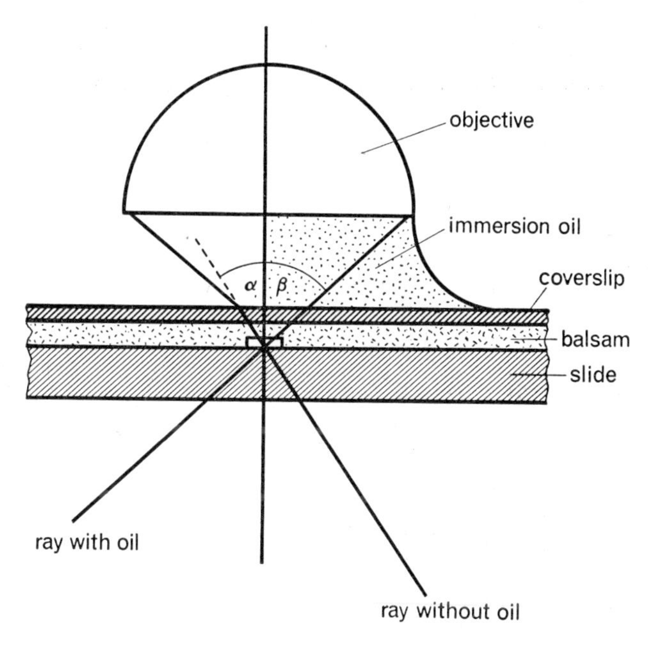

Fig. 4.8. To illustrate the advantage of using oil of high refractive index between the coverslip and the objective.

Starting from the right-hand bottom corner a ray is drawn passing through the object and being refracted on emerging into the air space between the coverslip and the objective; this ray makes an angle α with the axis. Starting from the left-hand bottom corner a ray is drawn passing through the object and, since the space between the coverslip and the objective is filled with oil of high refractive index, this ray is not refracted at the surface of the coverslip; it makes an angle β with the axis. The angle β being greater than the angle α, the use of immersion oil is seen to increase the convergence of light at the object.

lines of force. If its entering path makes a small angle with the lines of force the path of the electron through the field is helical. If the field is non-uniform the electrons entering as a parallel beam are constrained to follow converging helical paths and in this way are brought to a focus. Diagrams of a magnetic lens and of the paths of electrons through it are given in Fig. 4.9. Geometrically, the EM follows the same principles as the light microscope; it has a condenser and an objective lens and the principal difference is that the lenses of the eyepiece and the eye are replaced by a projection lens which forms the second image upon a fluorescent screen or photographic plate.

The greater the speed of the electrons the shorter is their wavelength. With an accelerating voltage of 1000 V the wavelength is 0·4 Å; at 60,000 V it is 0·05 Å. The distances between atoms in molecules being of the order of 1–5 Å, a wavelength of 0·05 Å should give us all the resolution we could want. But unfortunately in the EM resolution is

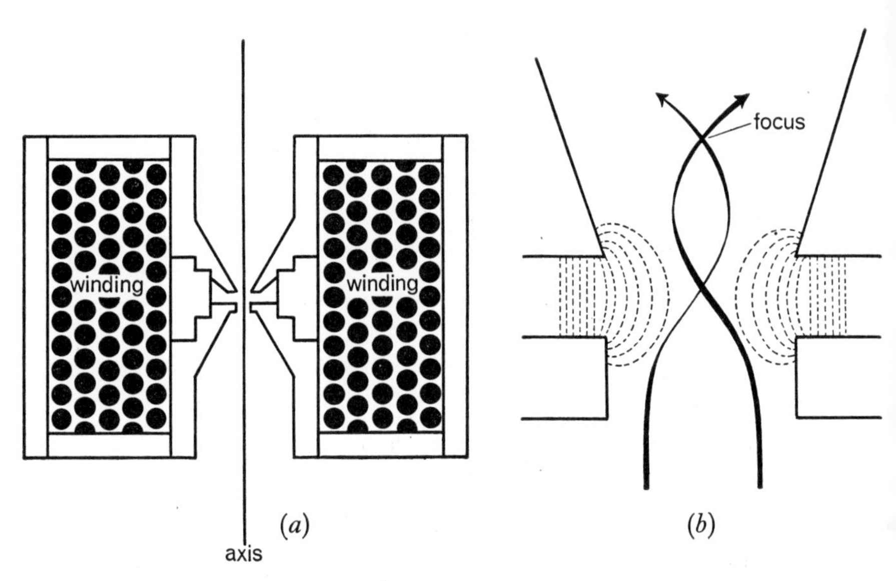

Fig. 4.9. (*a*) Section to show the construction of a magnetic lens; it is in the form of a cylinder with the electron beam passing along the axis. (*b*) To show the way in which electrons, encountering the non-uniform magnetic field in the region of the pole pieces, follow converging helical paths to a focus.

limited not by wavelength but by the shortcomings of electron lenses. Whereas a good light-microscope objective can have a numerical aperture of 1·5, the numerical aperture of an electron objective is not better than 0·01. This limit is set by the properties of materials. By increasing the current in the windings of a magnetic lens we can increase the magnetic field, but only up to a certain point, because there is a limit to the magnetic flux density which can be produced in the pole pieces—the iron becomes 'saturated'. If we use electrostatic lenses we cannot increase the potential difference beyond a certain point because the insulation breaks down. As a consequence, electron lenses are of relatively long focal length, and their angles of convergence are low. The maximum resolution obtainable in practice is only about one-hundredth of that which we might hope for from considerations of wavelength alone. For materials of high contrast, such as those studied by metallurgists, the resolution is 10 Å or even less; for biological materials it is 15–20 Å, i.e. 0·0015–0·002 μ as compared with 0·25 μ for the light microscope. So there is a net improvement in resolution by a factor of about 100.

Against this we must set certain disadvantages. First, the EM operates in a high vacuum; this means that we cannot use it to examine living material. Secondly, as a consequence of the low numerical aperture the depth of focus is very great. With a good light microscope we can focus sharply and thus separate structures lying at different depths. In the electron microscope this is impossible; all structures at all depths appear as though they were superimposed, and for this reason the material to be examined must be very thin. Thirdly, an electron beam of high energy, which may be needed to penetrate electron-opaque material, can produce excessive heating which may destroy the specimen.

Other modifications and refinements of the light microscope are seen in the interference microscope, the phase-contrast microscope and the polarising microscope, but these will not be described since we shall not require to refer to them elsewhere in this book.

2. PREPARATION OF MATERIAL

Hooke's original observations of the cells in cork were made on sections of cork cut freehand with a pen-knife. Fresh plant tissues, being supported by their cellulose cell walls, also lend themselves to the freehand cutting of useful sections. But fresh animal tissues are so soft that they offer too little resistance to the knife. The early microscopists found that in order to cut sections of animal tissue they had first to harden it; and it was the necessity for hardening which set in train the whole massive research effort which has been devoted to the study of fixed cells. For some very delicate tissues hardening by itself is inadequate, and the requirement for further support led on to the technique of wax embedding and the use of microtomes for section cutting.

With these improvements in preparative methods, and with the continuing improvement in the quality of microscopes, the amount of detail visible in fixed cells steadily increased. Cell organelles such as mitochondria and chloroplasts were observed and described, and by the end of the nineteenth century interest had moved to the 'ground substance' of the cytoplasm in which the organelles were suspended. Some believed that it was granular in nature, others that it was reticular and others again that it was alveolar, being a sort of foam. The minutest details were painstakingly described and often heatedly discussed. Into this lively activity, in 1899, there crashed A. Fischer and W. B. Hardy. Independently, they both pointed out that all the diverse

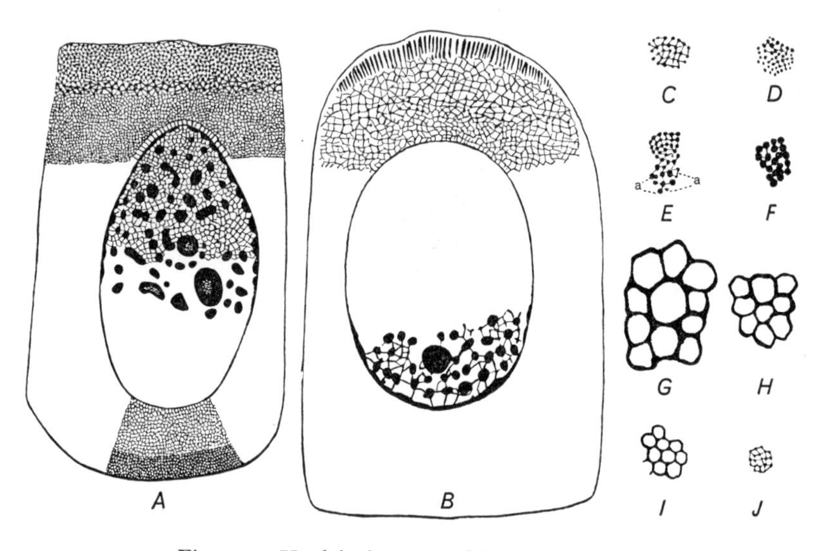

Fig. 4.10. Hardy's drawings of fixation artefacts.
A and *B* are epithelial cells from the intestine of the woodlouse *Oniscus*. *A* is fixed with osmium tetroxide and shows a finer 'grain' than *B* which is fixed with mercuric chloride. *C–F* are egg-white and *G–J* are gelatine, treated with various fixatives and showing structures such as are also to be seen in fixed cells.

appearances of fixed cytoplasm, upon which so much controversy centred, could be reproduced at will upon inanimate matter such as ordinary egg-white by the appropriate use of fixatives—they were nothing more than fixation artefacts.

The effect of this announcement was far reaching. Some dampening of enthusiasm was perhaps in order, and so also was a return to the study of the living cell, for so long neglected. But as so often happens, the pendulum swung far to the other side. It became fashionable in some circles to dismiss everything not visible in the living cell as fixation artefact, and the work of two generations of cytologists came under a cloud of deep suspicion. Nowadays we have regained a more balanced view. It is admitted that a structure which can be seen in cells prepared according to a variety of methods probably has some counterpart in the living cell; but if only one method will reveal it, judgement will be suspended.

(i) Fixation

Fixation has two main purposes: preservation and fixation proper, in the cytologist's sense. The need for preservation against bacterial action is obvious. But even in sterile material post-mortem changes are of rapid onset. Cells contain a vast number of enzymes. While the cell is alive these enzymes are restricted to their proper spheres of activity, but when the cell dies they run amok. Thereupon ensues the process of autolysis, which is what goes on when meat is hung. In order to prevent autolysis these enzymes have to be inactivated with the least possible delay. The EM reveals that fixation is always better in the outer layers of a piece of tissue than in the inner portions where autolysis can proceed some way in the few minutes taken by the fixative to diffuse in.

It used to be thought that the aim of fixation was to preserve the material in its natural state. Disillusioned, we now realise that the words 'fixation' and 'artefact' are almost synonymous; the cytologist works by producing artefacts which he then interprets. But by appropriate choice of fixative we may avoid artefacts of the grosser kind and so have reason to hope that what we see in the fixed material is not wholly without relation to what was there in life. We may also choose one fixative in preference to another having an eye to the staining procedures which we may thereafter adopt, these being notably affected by the nature of the fixative.

Fixation especially affects the picture which we formed earlier (Chapter 3) of a protein molecule: a long backbone coiled into an α-helix, with the α-helix itself folded in no very regular way to give a compact molecule, held in this shape by hydrogen bonds. We must now add further detail to this picture in order to see how the structure is affected by the fixative used. Although all the amino acids are α-amino acids, having an $-NH_2$ group attached to the same carbon atom as the $-COOH$ group, this does not exclude the possibility that at the other end they may have a second $-NH_2$ group (as in lysine) or a second $-COOH$ group (as in aspartic acid). While the α-amino and

α-carboxyl groups form the peptide bonds which link the amino acids together in a protein molecule, these additional groups, projecting freely from the backbone, are able to become ionised.

\ C /
OH^- ^+H_3N—C
\ N /
C
\ C—COO^- H^+
/ N \

Molecules of water, being electrical dipoles, become oriented by the electric fields in the proximity of the ionised groups. The protein molecule thus becomes highly hydrated, and we may add to our picture a layer or layers of oriented water molecules surrounding each protein molecule like a shell and serving to keep the protein molecules from coming into contact. Many fixatives act by removing this shell of water molecules, thus allowing the protein molecules to come together, whereupon they form aggregates and are precipitated from solution. Alcohol is a fixative by reason of its greater affinity for water, and an outstandingly bad fixative it is, producing excessive distortion and displacement of the cell contents. More suitable fixatives do not cause the general collapse of an extended protein system by removing the water of hydration, but stabilise the system by forming cross-links between the protein molecules. Formaldehyde is one of these.

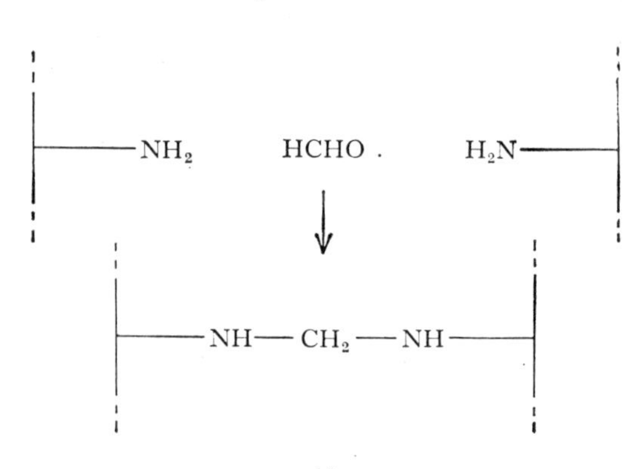

Mercuric chloride is a good fixative from the point of view of the brilliance of subsequent staining, but it causes rather severe contraction. Acetic acid causes swelling. An earlier generation often used a mixture of mercuric chloride and acetic acid, the latter preventing the contraction due to the former. There is no such thing as a really good general fixative; a fixative which gives good results on mitochondria may give poor results on chromosomes. It is therefore necessary to select a fixative which is appropriate to the aims of the investigation.

(ii) Staining

Most of our present knowledge of cell structure was gained from the differential staining properties of different parts of the cell. The earliest staining methods made use of differences in affinity for metallic ions, such as iron and silver, which were subsequently precipitated or reduced *in situ*. Better results were obtained after Ehrlich introduced the use of organic dyestuffs which had been developed to meet the needs of the dyeing industry.

It is useful to distinguish between (i) acid stains, in which the anion is coloured and combines with cell structures which are basic in reaction, and (ii) basic stains whose properties are the reverse. Chromosomes received their name because of their affinities for stains. Chromosomes are rich in nucleoprotein, a compound of nucleic acid with basic proteins containing many free $-NH_2$ groups. On fixation the protein is denatured and loses much of its basic character; the acid groups of the nucleic acid are then free to combine with basic stains.

The affinity of cell proteins for stains depends in the first place upon their state of ionisation, and this in turn depends upon the pH of the medium which bathes them. If the medium is made acid the ionisation of the $-COOH$ groups is repressed while the ionisation of the $-NH_2$ groups is encouraged, giving the protein an affinity for acid stains. This is not a very satisfactory basis for staining procedures; the stain is readily taken up but is just as readily lost again. More satisfactory procedures produce an insoluble coloured compound which withstands subsequent treatment. These procedures involve the use of a 'mordant' —a term borrowed from the technology of dyeing. The idea of a mordant is that it combines with the dye and then 'bites into' the fabric. The combination of a dye with a mordant is called a 'lake' (e.g. crimson lake). A lake is not necessarily insoluble, but the fabric-mordant-dye complex is insoluble. Mordants are commonly metals which form complex ions—iron, aluminium, chromium—and this is no doubt the reason why they work.

To take an example. Haematoxylin is a colourless substance of vegetable origin and is used to prepare the dye haematein which is the oxidised form of haematoxylin. Haematein forms a basic lake with iron or aluminium. The dye is made up together with the mordant in a solution made very slightly acid so as to prevent the precipitation of the lake. Made up in this way the stain has the colour of red wine. When fixed cells are exposed to the stain an insoluble compound is formed between the lake and the nucleic acid. After sufficient time has been allowed for the stain to be taken up the excess stain is removed and the preparation is washed with tap water which is sufficiently alkaline to neutralise the acid made up in the stain and to develop the characteristic purple-blue of this staining technique.

Unfortunately very little of the vast wealth of experience which has been gained in making stained preparations for the light microscope has any application to electron microscopy. Contrast in the EM depends upon the degree to which electrons are scattered by one structure as compared with another. Atoms of heavy elements scatter best; but biological material contains mainly the lighter elements—hydrogen, carbon, nitrogen, oxygen—and by itself scatters very little. Untreated sections show little contrast. We can make out structures under the EM only in so far as they differ in affinity for heavy metals. No single reagent has rendered better service in electron microscopy than osmium tetroxide. Not only is it a 'good' fixative in the sense that it produces little obvious distortion of cell structure, but the osmium is preferentially taken up by various cell structures, in particular those containing lipid, and it is an excellent 'electron stain'. Uranium in the form of uranyl acetate and tungsten in the form of phosphotungstic acid have also proved very useful. 'Electron-stain' technology is only now becoming established, and since no dramatic improvements are expected to be made in the instrument itself, progress in electron microscopy is likely to wait upon progress in 'staining' technique.

It is fair to say that the procedures which cytologists use for fixing and staining have some basis in chemistry. But it is a very slender basis; and it cannot be otherwise when we are still only beginning to understand what a protein molecule is like. It pleases us to put a gloss of science upon what is largely empirical. Cytological methods are developed by trial and error and are assessed on their ability to reveal the structures in which we are interested; afterwards, we try to find a chemical explanation of their success. This situation, however, is now on the mend and a number of cytochemical tests for specific substances have been worked out, having a respectable theoretical basis. But for the most part the preparation of material for cytological examination is not so much a science as a useful art.

(iii) Section cutting

It is assumed that the reader is familiar, at least in principle, with the standard methods of preparing sections for the light microscope.

Paraffin-embedded sections as used with the light microscope are generally cut at a thickness of about 10 μ; with care they can be cut as thin as 5 μ, but below this thickness paraffin wax is too friable. With ester wax we can cut down to 0·5 μ. This is quite inadequate for electron microscopy because the great depth of focus of the EM makes it impossible to separate structures at different depths in the section. For electron microscopy sections are cut at about 0·05 μ, or 0·02 μ in special cases.

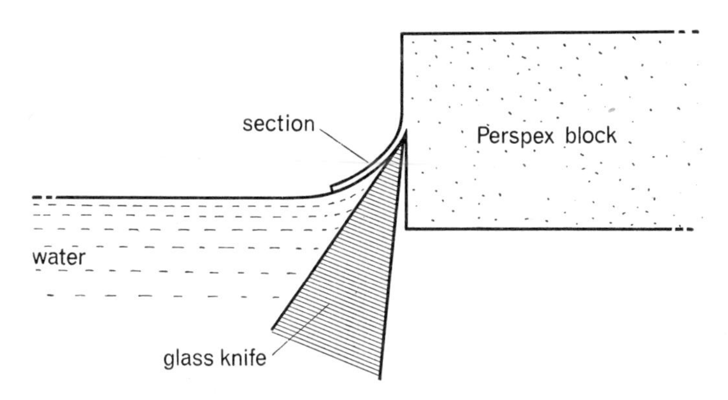

Fig. 4.11. Cutting sections of material embedded in Perspex.
The section is floated off upon the surface of water in a trough behind the knife edge. Observe the very wide angle of the glass knife. This is one of the secrets of success; a knife tapering finely to its edge breaks almost at once.

For EM work we embed the material not in molten waxes but in plastics which harden by polymerisation—Perspex,† for example. Having fixed the tissue in osmium tetroxide we dehydrate it by passage

† Commercial Perspex is a polymer of methyl methacrylate, but for section cutting the polymer of butyl methacrylate is preferred because it is less hard.

through alcohols in the usual way and impregnate it with butyl methacrylate to which a small amount of benzoyl peroxide has been added; this reagent promotes the polymerisation of the methacrylate to form Perspex. To cut the sections we use a glass knife—made by breaking a piece of plate glass—with a small trough of water behind the cutting edge, as shown in Fig. 4.11. As the Perspex block is brought down upon the knife edge the section floats out over the surface of the water. We then pick it up on a grid of very fine wire and remove any adherent water by evaporation. The section is now ready for the EM; it is not necessary to remove the Perspex from it. The design of the microtome, which has to provide for an advance of only 0·02 μ per section, presents formidable problems, but these we shall not enter into.

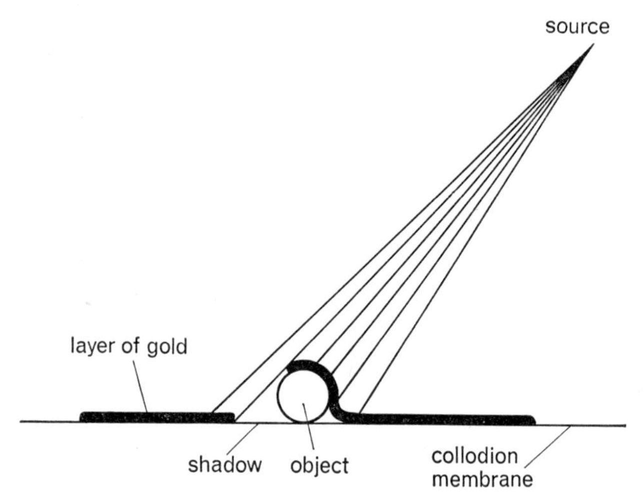

Fig. 4.12. A small object shadowed with gold.

The gold is evaporated from a tungsten filament, and since the process takes place *in vacuo* the atoms travel in straight lines.

(iv) Shadow casting

Another method of preparation for the EM, which is useful for small objects such as viruses, is shadow casting. We spray a suspension of virus particles upon a thin collodion membrane and dry it down. We then set up the membrane in an evacuated vessel with a source from which heavy metal atoms—gold is commonly used—can be deposited upon it (Fig. 4.12). The gold is evaporated from a heated tungsten filament, and being *in vacuo* the atoms travel in straight lines and coat all surfaces directly exposed to them, leaving uncoated shadows. This gives EM pictures such as that reproduced in Fig. 8.8 (p. 127) which not only provide useful information but can be aesthetically most pleasing.

(v) Bulk separation

To describe a variety of structures within the cell and not to be able to discover what they do is too often the frustrating experience of the cytologist. Not many of the methods which physiologists and biochemists are accustomed to use can be brought to bear on systems so very small as those with which the cytologist deals, and there does not seem at present to be any great hope that the necessary refinements can be brought within the range of practical possibility. But progress has been made and we do know something of the functions of at least some intracellular organelles. This progress has been possible because methods have been devised for separating these organelles in bulk, in sufficient quantity to be amenable to biochemical methods.

Making use of suitable tissues, such as mammalian liver, and of carefully graded methods of disruption it is possible to break cells open so as to set free the cell contents without wholesale destruction of intracellular organelles. One would begin, perhaps, by forcing a piece of liver through a fine sieve, thereby removing connective tissue. One might then grind the pulp in a mortar; but more satisfactory methods make use of the hydrodynamic shearing forces developed by the blades of a Waring blendor or between a ground glass cylinder and a rotating piston. Most of the cells now being broken open—homogenised—it becomes possible to separate the nuclei and some of the intracellular organelles by differential centrifugation, taking advantage of differences in particle size and differences in density. The homogenate is suspended in saline or in sucrose solution and is first run in the centrifuge at relatively low speed. This brings down the nuclei and any large cell fragments, leaving smaller particles still suspended in the supernatant fluid. The supernatant is then separated and centrifuged again for longer and at higher speed and this brings down a fraction containing the larger cytoplasmic particles, such as are visible in the light microscope. The process of removing the supernatant and centrifuging it again at still higher speed can be repeated, with the separation of further fractions of decreasing particle size. In some refinements of the method the particles are re-suspended and centrifuged in sucrose solutions of appropriately chosen concentration so that denser particles will settle and less dense particles will rise; or, alternatively, the centrifuge tube is filled with a sucrose solution of graded concentration so that particles of different densities accumulate at different levels. Since the differences in density upon which the method depends are very small, the centrifugal force necessary to effect the separation must be very great, and modern preparatory centrifuges develop forces of up to 300,000 *g*.

Working in this way we can separate a homogenate into a series of fractions and we can then submit these fractions to various biochemical tests. To find out what organelles they contain we take the fraction as a pellet from the bottom of the centrifuge tube and fix, embed and section it for examination with the EM. It must be understood that the fractions thus prepared are far from being pure collections of organelles. All we can achieve is a certain relative separation so that we find, say, most of the nuclei in the first fraction and most of the mitochondria (p. 101) in the second. We then attempt to correlate the distribution of biochemical properties in the various fractions with the distribution of cell organelles. Examples of investigations making use of these methods will be described in Chapter 7.

5

THE CELL IN 1925

The EM was developed during the Second World War and it was not until about 1950 that it came into general use. Being something like one hundred times more powerful than the ordinary light microscope it revealed a whole new intracellular world previously only guessed at. Cytology took on a different look when this new weapon was placed in its hand, and the cell of 1959 was a very different thing from the cell of 1939. But why should we pause to consider what cytologists thought about the cell in 1925?

The reason for choosing this date is that it was the date of the publication of the third and last edition of a very famous book—*The Cell in Heredity and Development* by E. B. Wilson. In that book Wilson drew together the threads of cytological research up to that time. Microscopes had by then been improved up to the limits set by the wavelength of visible light; preparatory techniques were well established and the lesson of fixation artefacts had been taken to heart. No great advances were to be made in the next 20 years. What then did we know about the cell in 1925?

In illustration of the first chapter of Wilson's book there is a diagram of a generalised cell. This diagram is reproduced here as Fig. 5.1. First, what are we to understand by a generalised cell? If we make a survey of plant and animal cells we find that while they show a very great deal of diversity and a high degree of specialisation in relation to function, as in nerve and muscle, there are some features, such as the presence of mitochondria, which are common to them all. We may also allow our attention to rest upon the apparently unspecialised cells of the plant meristem or the regenerating tissue of a wound; these all look very much alike, notwithstanding that their progeny will later develop characteristically specialised structures. Yet even these cells can be regarded as specialised—they are specialised for growth and cell division. There is no such thing in nature as a generalised cell; it is a creation of man's mind, arrived at by surveying a wide range of cells, disregarding their specialised features and abstracting the features which they have in common. The generalised cell, then, is an abstraction, but it was an abstraction which in 1925 was found to be useful.

Let us now turn to the generalised cell and consider what was known about some of the features labelled in Fig. 5.1. First, the cell wall. This is taken to refer to what appears to be a container, such as the cellulose cell wall in plants, from which the cytoplasm can be temporarily detached without disastrous consequences. The cell wall is to be distinguished from the cell membrane which is the physiological cell surface, the seat of resistance to penetration. Probably more was known in 1925 about the properties of the cell membrane than about any other part of the cytoplasm, but nothing was known about its structure except

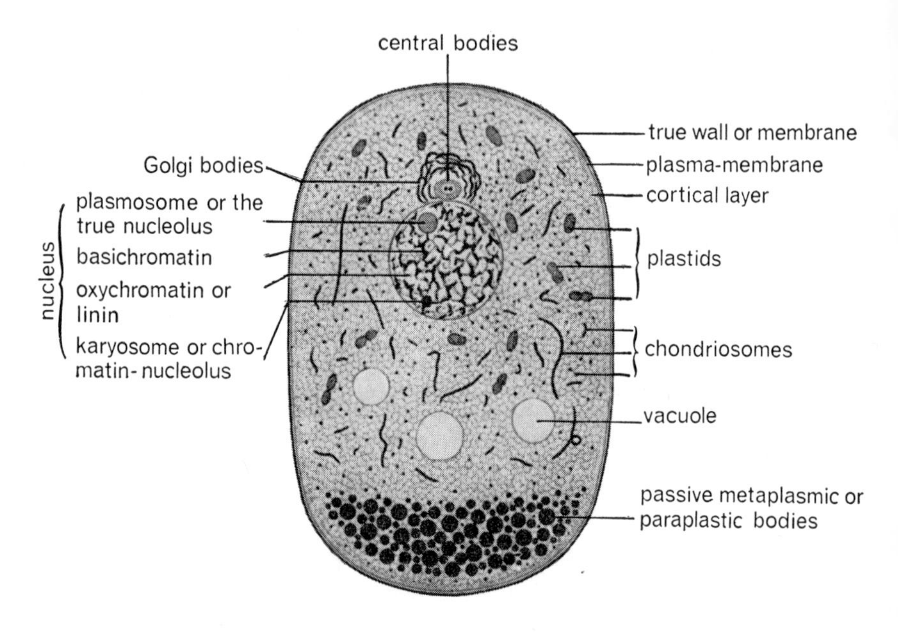

Fig. 5.1. E. B. Wilson's general diagram of a cell.

for what, by a strange inversion of the usual situation, could be deduced from its physiological properties. Within the cytoplasm there are vacuoles, at least in plant cells where they are united into one large vacuole, but vacuoles are not a characteristic feature of animal cells. Nor are plastids, which, being intended to include chloroplasts, are another concession to the plant cell. The central bodies, on the other hand, are better seen in animal cells. Their arrangement varies; typically there is a single relatively large centrosome enclosing two small centrioles. The central bodies are concerned in cell division, the two centrioles separating and forming the poles of the mitotic spindle. Mitochondria are found both in plant cells and in animal cells and are well seen in the living cell under dark-ground illumination. They appear to move about in the cytoplasm with apparent autonomy, and in the past it had been suggested that they were parasitic micro-organisms. The Golgi bodies, or Golgi apparatus, are named after their discoverer who demonstrated their presence in fixed cells stained with osmium or silver. They seemed to be a collection of branched and anastomosing canals. It would be wrong to say that cytologists had no idea about their functions, but correct to say that they had little evidence of them. Besides these larger organelles a great variety of granules were recognised and given names. The larger ones seemed almost certainly to be food reserves—fat, glycogen, starch—since their presence depended upon the nutritional state of the organism. There was very little to be said about the smaller ones. Finally, there was the ground substance of the cytoplasm, about which controversy had been silenced by the recognition of fixation artefacts.

The cell in Fig. 5.1 is in the non-reproductive condition. The nuclear membrane is intact and the nucleus contains a sap whose refractive index in life is distinctly different from that of the cytoplasm. Within the nucleus itself is the nucleolus, also to be seen in life in consequence of refractive index difference. In fixed and stained nuclei there is, under the nuclear membrane, a reticulum of material which takes up basic stains and probably represents the chromosomes.

The foregoing could be seen in the non-dividing cell with the resources available in 1925—quite a lot of structures, but very little information about what parts they played. As significant and fundamental knowledge it perhaps does not seem very much to show for a hundred years of research. But we have only dealt with the non-dividing cell in this account and have thereby left out the really exciting part of the contemporary story. The outlines of the processes of mitosis and meiosis had been discerned by the end of the nineteenth century, and details were still being filled in. T. H. Morgan had begun his work on *Drosophila* and in 1919 he had summarised in a book called *The Physical Basis of Inheritance* the results of 10 years of work which established the linear order of the genes upon the chromosomes. The decade prior to 1925 had witnessed the birth of the science of cytogenetics. After nearly a hundred years of slow growth cell biology had now come into flower.

6

THE NUCLEUS

As we have already had occasion to remark (p. 12), the elaborate ritual of mitosis, and the striking similarity of this process throughout the animal and plant kingdoms as well as in the Protista, are the strongest evidence for our belief, implicit in the cell concept, that all cells share a great many fundamental similarities in their basic organisation. The same basic similarities are also found in the process of meiosis which in most organisms immediately precedes the formation of the gametes. Both these processes had been very fully described by 1925 and some important details concerning the behaviour of the chromosomes during meiosis were added during the 1930's; but in comparison with the general advance in cytology little new knowledge of mitosis and meiosis has been added since that time. Applied to these problems the EM has given rather disappointing results, the reason being that whereas chromosomes are pre-eminently stainable structures for purposes of light microscopy they do not show up well in the EM. The principal events of mitosis and meiosis can be observed in the living cell with the phase-contrast microscope, and for the finer details we have to rely upon fixed and stained preparations.

In the non-dividing cell the nucleus is in what we call interphase. No recognisable chromosomes are visible even in fixed and stained preparations; the material of the chromosomes appears to be swollen and is often gathered into a reticulum under the nuclear membrane.

1. MITOSIS

It is convenient to divide the process of mitosis into four phases: prophase, during which the chromosomes appear; metaphase, during which they arrange themselves on the spindle; anaphase, during which the daughter chromosomes move to opposite poles; telophase, during which the changes of prophase are seen in reverse. These phases are illustrated in Figs. 6.1 and 6.2.

(i) Prophase

The first indication of the chromosomes is the appearance of long, very slender threads known as chromonemata. Each chromosome from the moment of its first appearance can be seen to be double, the two chromonemata being loosely wound one about the other. At one point the two chromonemata are united in a granule, the centromere (Fig. 6.3).

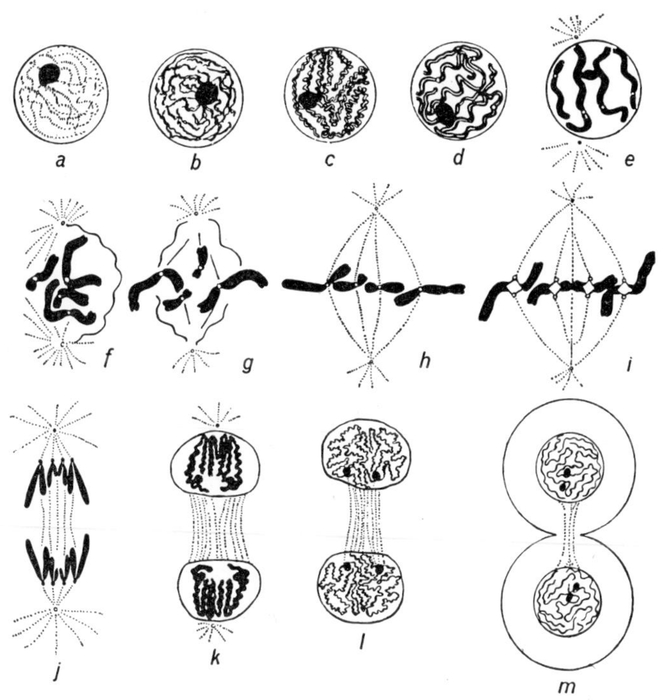

Fig. 6.1. Mitosis.
a, Interphase; *b*, *c*, *d*, *e*, prophase; *f*, *g*, *h*, *i*, metaphase; *j*, *k*, anaphase; *l*, *m*, telophase.

Shortly after their first appearance the chromonemata begin to shorten. This shortening is brought about by their being coiled into tight helices of primary, secondary and even higher orders, like the coiling of the tungsten filament of an electric light bulb (Fig. 6.4). At the same time they begin to stain more deeply with basic stains. By the end of prophase each chromosome is relatively short and compact. It consists of two chromatids, derived from the original chromonemata which are still held together by the centromere (Fig. 6.5). There is a suggestion, too, that the coils are embedded in some sort of matrix.

In order to avoid confusion later when we go on to consider meiosis, it may be helpful to recall certain basic facts. The ordinary somatic cells

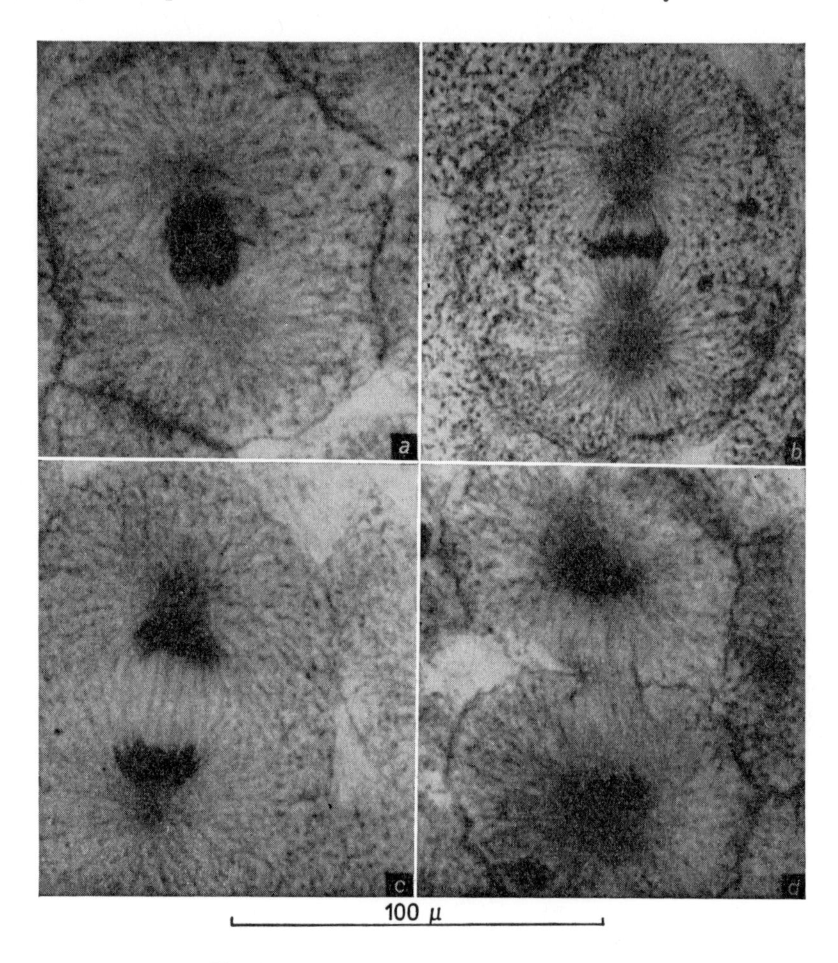

Fig. 6.2. Mitosis in the whitefish, *Coregonus*.

of the higher animals and plants are diploid. They contain an even number of chromosomes, the chromosomes being present in pairs; one member of each pair comes from the male parent, the other from the female parent. We have just traced the process by which a single chromosome appears, shortens and is seen to be made up of two chromatids.

Somewhere else in the diploid nucleus there will be an exactly similar chromosome, likewise made up of two chromatids. In mitosis there is no association between the homologous chromosomes, i.e. between paternal and maternal chromosomes of a pair.

During prophase the nucleolus grows smaller and eventually disappears. Within the centrosome the centriole divides—if there are not

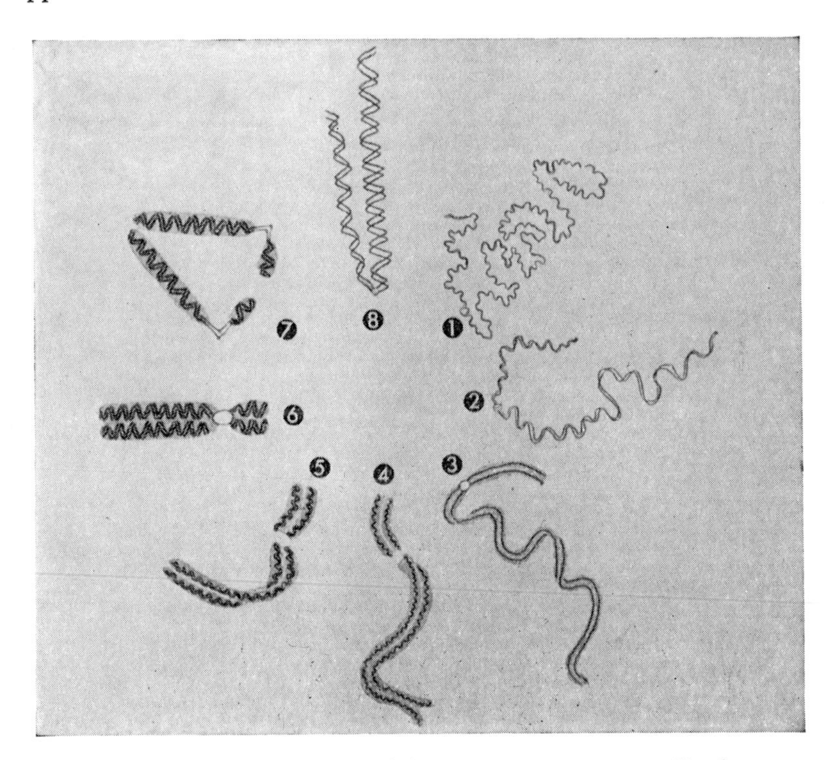

Fig. 6.3. Coiling and shortening of chromonemata. 1, 2, 3, 4, 5, Prophase; 6, metaphase and anaphase; 7, 8, telophase.

already two centrioles—and division of the centrosome follows. The centrosomes migrate to opposite sides of the nucleus, and radial striations, known as astral rays, begin to appear around them. These rays are the expression of a local fibrillar structure which develops in the cytoplasm, radially arranged about the centrosomes.

(ii) Metaphase

The nuclear membrane usually disappears at this stage and the fibrils of the astral rays extend across the nuclear region, forming the mitotic

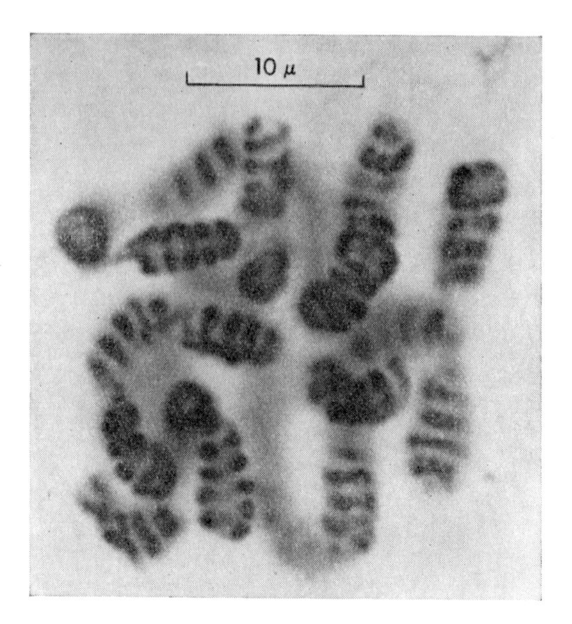

Fig. 6.4. Photomicrograph of chromosomes of *Tradescantia* showing helical coiling.

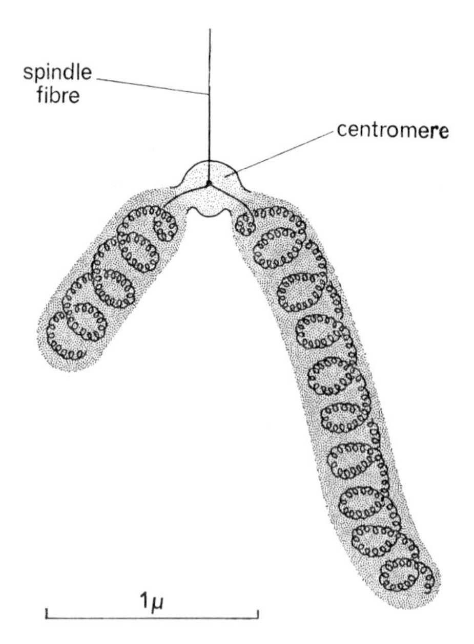

Fig. 6.5. Anaphase chromosome.

Schematic representation of morphology. The spirally coiled chromonema is disclosed by reagents (e.g. KCN) which cause the coils to relax (see Fig. 6.4). The existence of a matrix in which the coils lie is problematical.

spindle. The chromosomes take up positions on the equator of the spindle and seem to be attached to the spindle fibrils by their centromeres, their arms being free. They become accurately lined up on the equator and are spaced apart as though in mutual repulsion.

(iii) Anaphase

On each chromosome the centromere divides. Instead of having one chromosome made up of two chromatids held together by a single centromere we have now two daughter chromosomes each being derived from one chromatid and each having its own centromere. These daughter chromosomes now move rapidly to the poles, a movement which is accurately synchronised throughout the whole chromosome complement.

(iv) Telophase

As the chromosomes approach the poles their rate of movement slows down and even before they reach their final positions the changes observed in prophase are seen to undergo reversal. The coils of the chromonema relax and staining weakens; the nucleolus reappears; the nuclear membrane is reconstituted.

Starting during telophase and continuing after its end, the division of the cytoplasm takes place. In plant cells this is seen to be the result of the formation *in situ* of a membrane which ultimately becomes a cell wall separating the two daughter cells. In animal cells, which are not enclosed in a rigid cell wall, the process of cytoplasmic division is more variable. Sometimes, as in amoeboid cells, the daughter cells seem just to walk away from each other. But in other cases it appears that the daughter cells are separated by the ingrowth of the cell membrane and it is therefore likely that this is the basic method of cytoplasmic division.

The duration of mitosis varies greatly from one organism to another, taking anything from 5 to 200 minutes. Proportionately, prophase is the longest and metaphase the shortest. The following figures will serve as a rough guide:

	% of total time
Prophase	55
Metaphase	10
Anaphase	15
Telophase	20
	100

When one watches this process under the microscope, even if it is all over in 5 minutes, one is scarcely aware that anything is going on, so slowly does it seem to happen. A very much more lively impression is gained by the method of time-lapse cinematography. The cell is photographed through the microscope with a special ciné camera which takes single frames at regular intervals, controlled by a timing device. Suppose that mitosis takes 20 minutes. We set the timing device to take a picture every second. We then run the film through a projector at normal speed and see the whole thing happen in less than a minute. This brings it all very much to life, and gives such a strong impression of orderly purposefulness that one cannot resist the temptation to ask, teleologically, what it is all about.

A cell divides into two daughter cells and each daughter cell contains a replica of the genetic material—DNA—of the parent cell. DNA exists in the form of molecular helices of indefinite length, and the DNA content of a single nucleus, if it were all in a single strand, would have a length of millimetres, if not centimetres—very much longer than the diameter of the nucleus. Granted that the parent DNA helix has split into two daughter helices—we shall consider this elsewhere (Chapter 23)—the problem is to gather up the long daughter helices into compact bundles for dispatch to their appropriate destinations, otherwise parts of them might get cut off by the formation of the new cell wall. One cannot say this for certain, but it does look as if the significance of the successive orders of coiling seen during prophase is that thereby the previously extended DNA is contracted down into a few bodies which are much shorter than the length of the cell. Separation then follows with swift precision.

It may be noted, before we leave mitosis, that since each chromosome when it first appears has two chromonemata, replication of the chromosomes must take place before prophase.

2. MEIOSIS

The word 'meiosis' means 'a making less', and refers to the reduction in the number of chromosomes from the diploid condition, in which there are two of each kind, to the haploid condition in which there is only one of each kind. The whole process of meiosis (Fig. 6.6) involves two successive divisions which employ the same mechanism—centrosomes, spindle, etc.—as does mitosis; but meiosis differs from mitosis in certain very important features which we shall pause to underline. The principal differences are to be seen in prophase of the first division; this is a much longer process than in mitosis and has four stages.

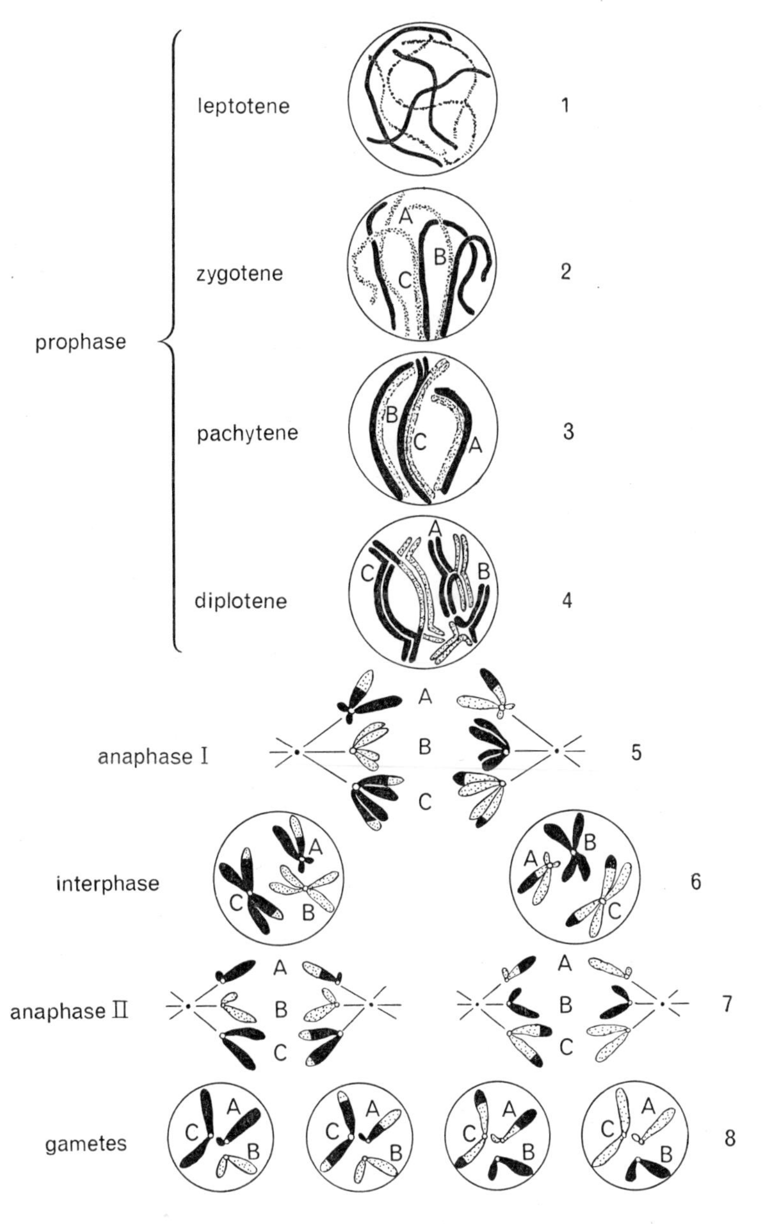

Fig. 6.6. Meiosis.

(i) Prophase I

(*a*) *Leptotene.* Long fine chromonemata appear in the nucleus, as in mitosis, but these differ from the chromonemata of mitosis in two features: first, they are single; secondly, they bear small densely staining granules, chromomeres, along their lengths, resembling beads on a string.

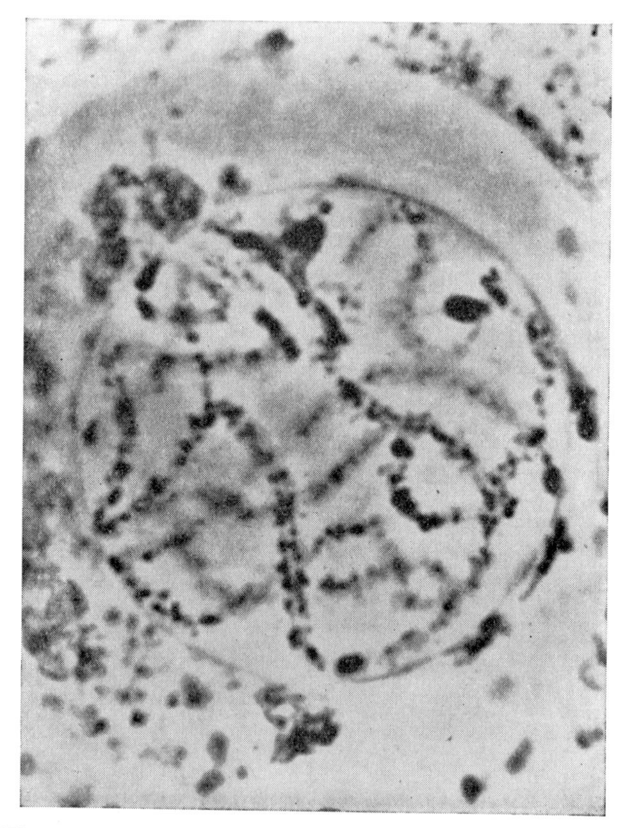

Fig. 6.7. Photomicrograph of chromosomes in synapsis, showing accurate pairing of chromomeres.

(*b*) *Zygotene.* The members of each pair of homologous chromosomes approach and become apposed. This process, known as synapsis, appears to be very precise, taking place chromomere by chromomere along the whole length (Fig. 6.7). Where formerly we had two independent units, the homologous chromosomes, we now have a single unit, known as a bivalent, in which the two homologous chromosomes are very closely associated together. There is at this stage a superficial similarity with the early prophase of mitosis but the important differences are: the units are bivalents, not chromosomes; each unit has two centromeres, not one; the number of units is halved.

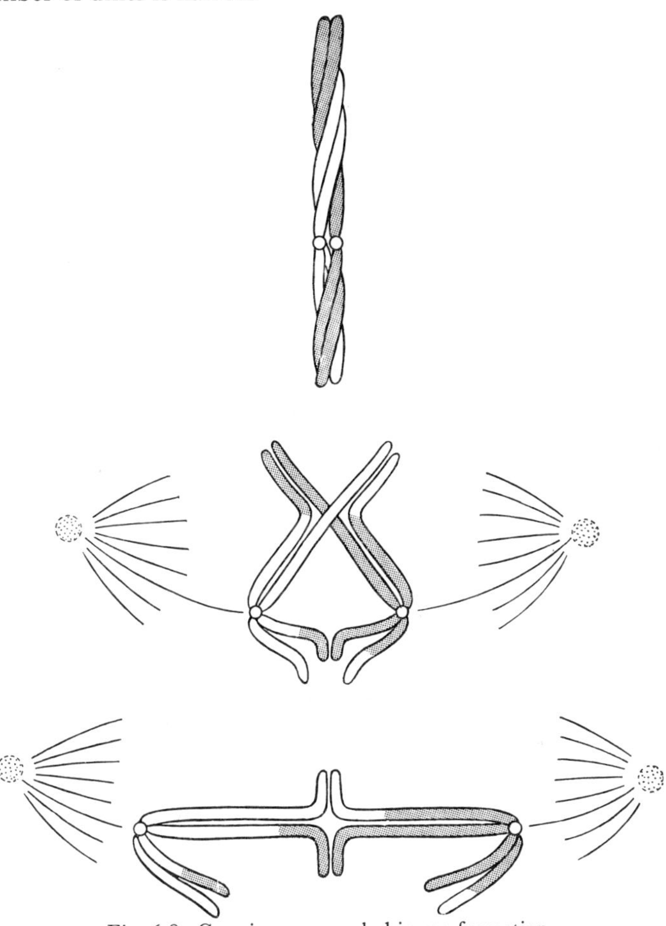

Fig. 6.8. Crossing-over and chiasma formation.

Sister chromatids are shown either stippled or white. Crossing-over takes place only between non-sister chromatids. As the chromosomes separate the characteristic appearance of chiasma is seen.

(*c*) ***Pachytene.*** Synapsis having been completed, and the paired chromosomes being closely twisted one about the other as in a piece of flex, shortening begins. At this stage we observe that each chromosome splits into two (sister) chromatids. Each unit (bivalent) now has four chromatids.

(*d*) ***Diplotene.*** At this stage the forces operating during synapsis, which bring the homologous chromosomes together and hold them so, seem to weaken. The chromosomes begin to come apart and as they do so we observe the appearance known as chiasma, which is illustrated in Fig. 6.8. The essential feature of a chiasma is that two non-sister chromatids lie one over the other, making a sort of cross. As we shall see later (Chapter 21), the evidence of breeding experiments suggests that at some stage lengths of chromosome are exchanged between members of a pair during the formation of the gametes. It was proposed by Janssens in 1909 that in chiasma an exchange of lengths of chromatid took place by breakage of the chromatids at the point of contact and reunion of each with its opposite number, as indicated in Fig. 6.8. It is important to realise that this process of crossing-over by breakage and reunion has never been observed directly. Ordinarily, the two chromatids are indistinguishable and if they exchanged parts we would not be able to see that this had happened. Cytological evidence of crossing-over was first obtained in 1937 from the study of strains of *Drosophila* having certain chromosomal abnormalities, visible in fixed preparations, which could be observed to have been interchanged.

As the chromosomes drift apart they undergo further shortening, and these two processes together have the effect of shifting the chiasmata towards the tips of the chromatids. As in mitosis, the nucleolus disappears at this stage.

(ii) Metaphase I

This is much the same as in mitosis, but the alignment at the equator is less precise. In meiosis each unit is a bivalent with two centromeres already widely separated, whereas in mitosis each unit is a single chromosome with a single centromere as yet undivided.

(iii) Anaphase I

This is not significantly different from mitotic anaphase.

This effectively completes the first meiotic division. In some organisms there is a typical telophase followed by a short interphase before prophase of the second division starts. Others go straight from anaphase I to metaphase II.

The second meiotic division resembles mitosis in that the units are chromosomes, each consisting of two chromatids held together by a single centromere; it differs from a normal mitosis of the same organism in that only half the number of chromosomes is present, one representative of each pair of homologous chromosomes.

The genetically important features of meiosis are to be seen in prophase of the first division. These features are synapsis and crossing-over. Synapsis is the essential act in the reduction of the chromosome complement from the diploid to the haploid condition. Crossing-over makes possible the recombination of heritable characters and is the basis of much of the variation which sexual reproduction provides. These are matters to which we shall return in Chapter 21.

3. CHROMOSOMES

As mentioned, the EM is of little help in the study of chromosomes, and the examination of ordinary chromosomes with the light microscope would not have taken us very far. But fortunately in some animals there are chromosomes of relatively gigantic size.

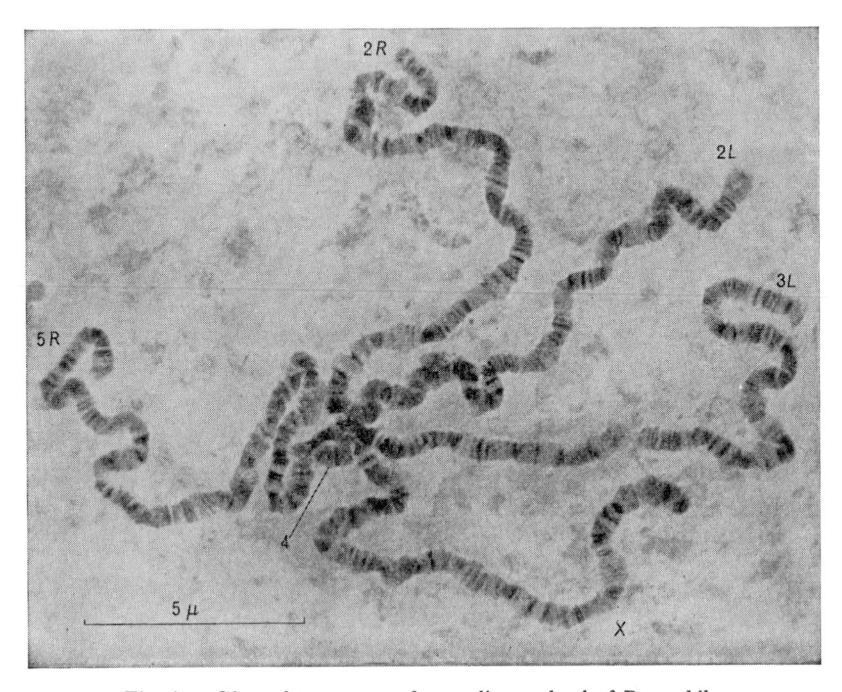

Fig. 6.9. Giant chromosomes from salivary gland of *Drosophila*.
Photomicrograph of a Feulgen-stained preparation. There are 4 pairs of chromosomes in *Drosophila*. In the salivary gland the members of each pair are associated together as in synapsis and all are united by their centromeres. The centromeres of chromosomes (1)(=X) and (4) are near their ends; those of chromosomes (2) and (3) are near the middle. Consequently there appear to be 6 chromosomes.

The salivary glands (and other tissues also) of flies belonging to the order Diptera, to which *Drosophila* belongs, contain giant chromosomes which are 100–200 times longer than the ordinary metaphase chromosomes. They were first described by Balbiani in 1881, but his account of them was overlooked by geneticists and it was not until 1930—20 years after Morgan had started work on *Drosophila*—that Kostoff drew attention to their possibilities as cytogenetic material.

Stained by Feulgen's method, which is a cytochemical technique specific for DNA, they show a banded appearance as in Figs. 6.9 and 6.10. By various treatments the ends can be made to fray out, and this discloses that the giant chromosome is made up of many fibrils. It has

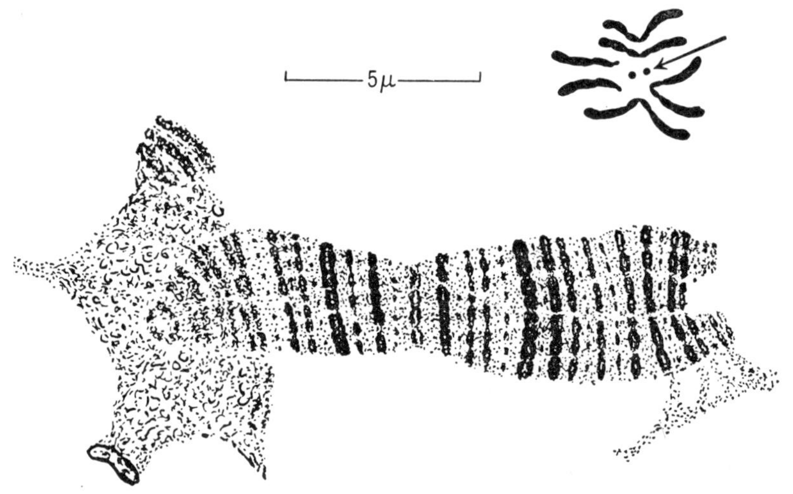

Fig. 6.10. The relative sizes of salivary and ordinary metaphase chromosomes. The drawing is of the fourth chromosome, indicated by an arrow on the metaphase group.

been suggested that the giant chromosomes are derived from the interphase chromosomes by replication of the chromonemata without subsequent mitosis, a condition which is described as polytene. The thickness of the fibrils into which they can ultimately be resolved is not agreed upon; the fibrils are certainly not more than 200 Å in thickness, and according to one authority, not more than 50 Å. The banded appearance of the giant chromosome is a consequence of the accurate alignment of granules on the individual fibrils. This is reminiscent of synapsis, but we cannot identify the bands of the giant chromosome with the chromomeres of meiotic prophase because there are more bands than chromo-

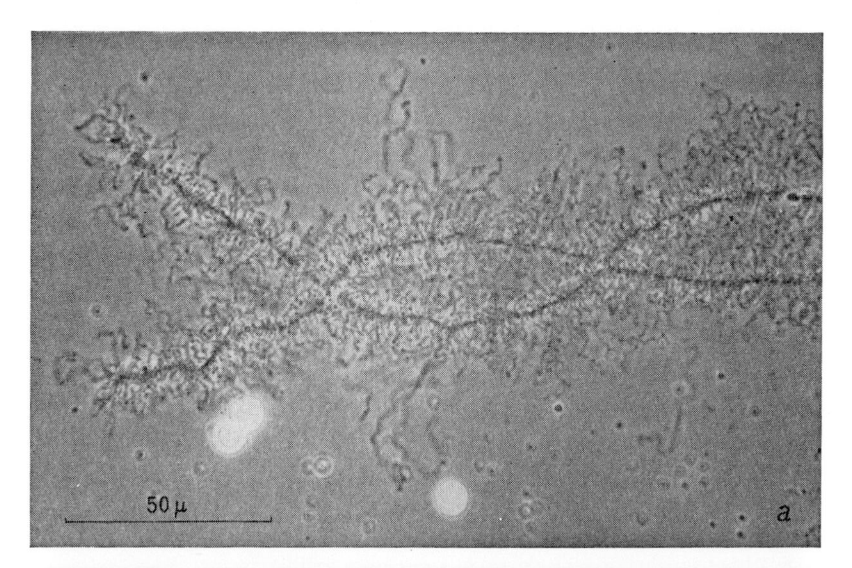

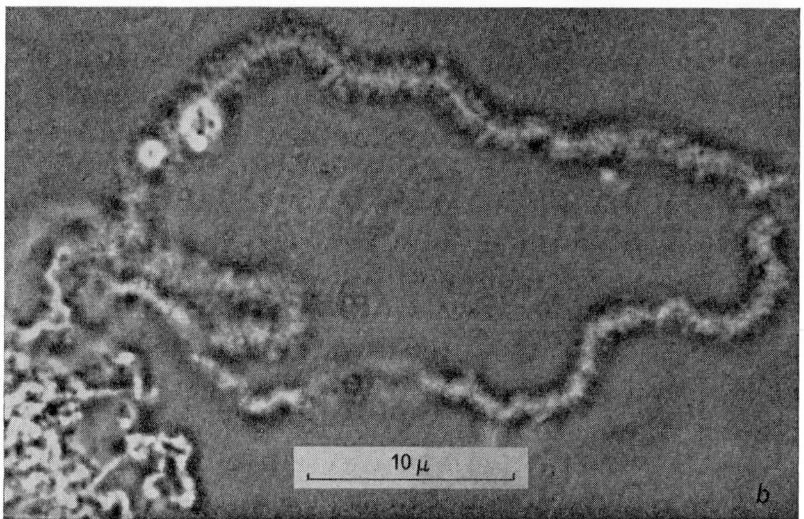

Fig. 6.11. Lampbrush chromosomes.

(*a*) A pair of lampbrush chromosomes in diplotene, showing loops. (*b*) A single loop at higher magnification, showing the 'fluff' (RNA). (*c*) An interpretation of the lampbrush chromosome, according to which there are two chromonemata continuous from end to end, coiled in the chromomeres and extended in the loops.

meres. By the use of a micromanipulator it has been found possible to stretch the giant chromosome between two glass microdissection needles, and it turns out to be a highly extensible structure, being capable of extension up to 10 times its resting length with complete elastic recovery.

Another type of chromosome of relatively vast size, about 200 μ in length, can be found in the developing eggs of amphibia. These are known as 'lampbrush chromosomes' from their resemblance to the brushes formerly used for cleaning the glass chimneys of paraffin lamps. Lampbrush chromosomes are in fact bivalents, but the two chromosomes are not closely synapsed (Fig. 6.11*a*). Each chromosome is formed of two chromonemata which bear chromomeres at intervals. From each chromomere there extends a pair of loops, members of each pair being of similar appearance, but size and appearance differing from one chromomere to another (Fig. 6.11*b*). The loops appear to be garnished with some material of 'fluffy' appearance, and Callan has made the interesting observation that the 'fluff' disappears after treatment with ribonuclease—presumably it is mainly RNA. Treatment with deoxyribonuclease has no effect on the 'fluff' but causes the loop itself to disintegrate; this suggests that DNA is an important structural component of the loop. Like the salivary gland chromosomes of Diptera these lampbrush chromosomes are highly extensible and some authorities believe that the chromomeres are regions where the chromonemata are coiled or folded, whereby the extensibility might be explained (Fig. 6.11*c*).

All the evidence, from the helical coiling of ordinary chromosomes as they shorten during prophase, from the extensibility of the salivary gland chromosomes and from the form and extensibility of the lampbrush chromosomes, suggests that the basis of chromosome structure is a very long fine fibril—or a bundle of fibrils—probably of nucleoprotein. This view is certainly not in conflict with current conceptions of the nature of the genetic material. But it will be better to postpone discussion of this topic to a later chapter (Chapter 26).

7

THE STRUCTURE OF SPECIALISED CELLS

At a time when intracellular organelles were only just within the limits of resolution, when we had few means of discovering their functions and when many of the structures with which we are now familiar were as yet undiscovered, the conception of the generalised cell was a convenient means of unifying our rather meagre knowledge. We now know so much more about cells and the various ways in which their internal arrangements may be specialised that in attempting to abstract what is common to all we have perforce to leave so much behind that the abstraction becomes rather unreal. For this reason we will henceforward abandon the conception of the generalised cell and instead we will take three examples of cells which are admittedly specialised for particular functions in the organisms to which they belong. These are: the exocrine cell of the pancreas, the cell from the proximal tubule of the mammalian kidney and the palisade cell of the leaf. Each of these cells is specialised in its own way, but none is to be ranked as a very highly specialised cell. Nerve cells and muscle fibres would qualify as being very highly specialised and with these we will not be concerned.

Although we are supposedly restricted to only three types of cell we would have much to lose by interpreting this restriction too narrowly. Many intracellular organelles, notably the mitochondria, are remarkably uniform throughout a wide range of cell types. It so happens that the most revealing electronmicrographs of cell membranes have been obtained from cells of tissues other than pancreas, kidney and leaf, and it would be a pity not to avail ourselves of them since we have as yet no reason to believe that the unit cell membrane has a different structure in those cells which we have chosen primarily to study. We shall therefore allow ourselves this and comparable liberties in the descriptions which are to follow.

1. PANCREAS

The mammalian pancreas contains cells of two types. First, there are the exocrine cells which secrete digestive enzymes; these are grouped into little clusters surrounding a space which communicates with the pancreatic duct. Secondly, there are the endocrine cells which are found in the islets of Langerhans and which secrete the hormone insulin into the blood. We shall deal only with the exocrine cells.

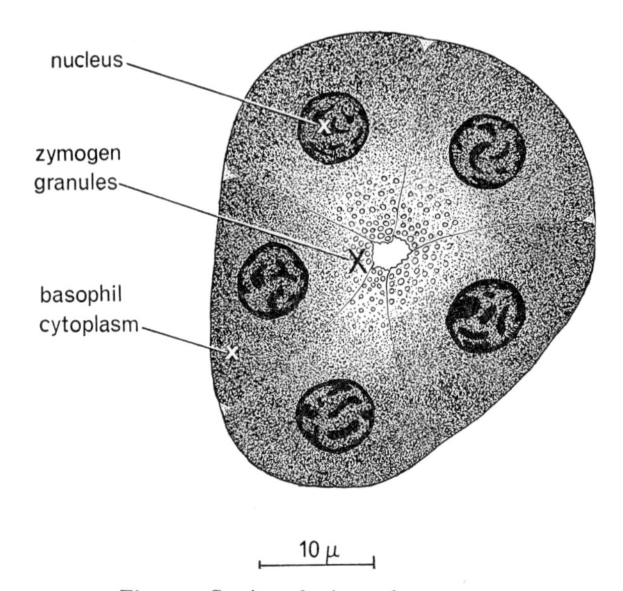

Fig. 7.1. Section of acinus of pancreas.

To show the appearance under the light microscope after treatment with a basic stain. At the base of the cell the cytoplasm is deeply stained by reason of its content of RNA. At the apex (next to the lumen) the cell contains granules of zymogen (enzyme precursor).

Fig. 7.1 represents a transverse section of a group of exocrine cells as they might appear if stained with haematoxylin. As compared with other cells of the body we would first notice that the cytoplasm, particularly at the basal end of the cell, has a very strong affinity for the basic stain which as a rule is taken up only by the nucleus. This staining reaction is abolished by pre-treatment with the enzyme ribonuclease and is therefore presumably due to the presence of RNA. Fig. 7.2 is an electronmicrograph and Fig. 7.3 is a diagram to indicate the general arrangement of the features to be seen under the EM, and is to be considered in conjunction with the electronmicrographs of the various regions reproduced in Figs. 7.4, 7.7, 7.8 and 7.10.

Basement membrane

There does not seem to be any cell wall completely enclosing the cell, but applied to the basal parts of all the cells and enclosing the whole

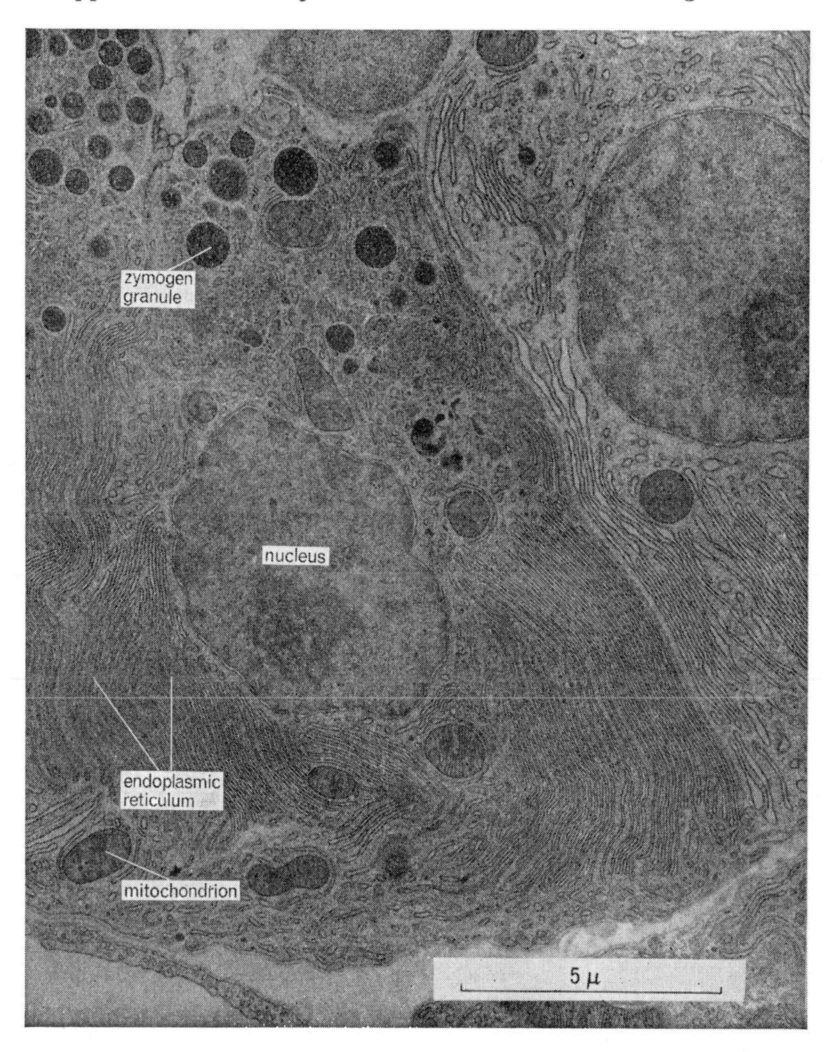

Fig. 7.2. Low-power electronmicrograph of pancreas cell.

group there is an extracellular sheath known as the basement membrane, which we may tentatively compare with the cell wall in other cells. In the pancreas the basement membrane has no intrinsic structure but in other tissues and in other animals it is often seen to be much thicker, containing fibrils and probably contributing to the maintenance of shape.

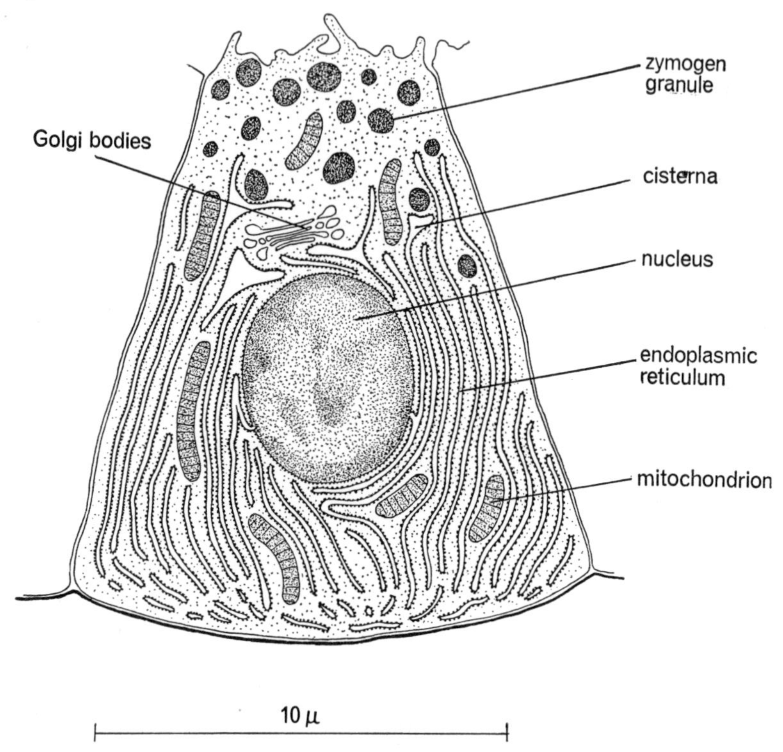

Fig. 7.3. Interpretation of electronmicrograph of pancreas cell.

Cell membrane

Within the basement membrane is what we believe to be the cell membrane, and the separate cell membranes of adjacent cells turn inwards from the basement membrane. In good preparations each cell membrane appears as two electron-opaque lines about 75 Å total width (Fig. 7.4). This double-line structure has aroused much interest. We have already had occasion to mention (p. 50) the properties of the cell membrane and the possibility of its lipid nature, and we considered the properties of polar lipid molecules spread as monolayers over surfaces. A monolayer such as that represented in Fig. 3.14*b* (p. 51) is stable at an air-water interface, but this arrangement would not be stable at the cell surface, which has water on both sides. An indication of how the lipid component of the cell membrane might be arranged is gained from the following experiment, illustrated in Fig. 7.5.

A thoroughly cleaned microscope slide is placed in a vessel of pure water. A monolayer of lipid is spread over the surface. The slide is then picked up with forceps and lifted out of the vessel. The glass, being clean, is 'wetted' by the water and carries the monolayer with it as it is lifted out. We now allow the slide to dry, whereupon the polar groups of the lipid molecules attach themselves to the glass surface. The slide

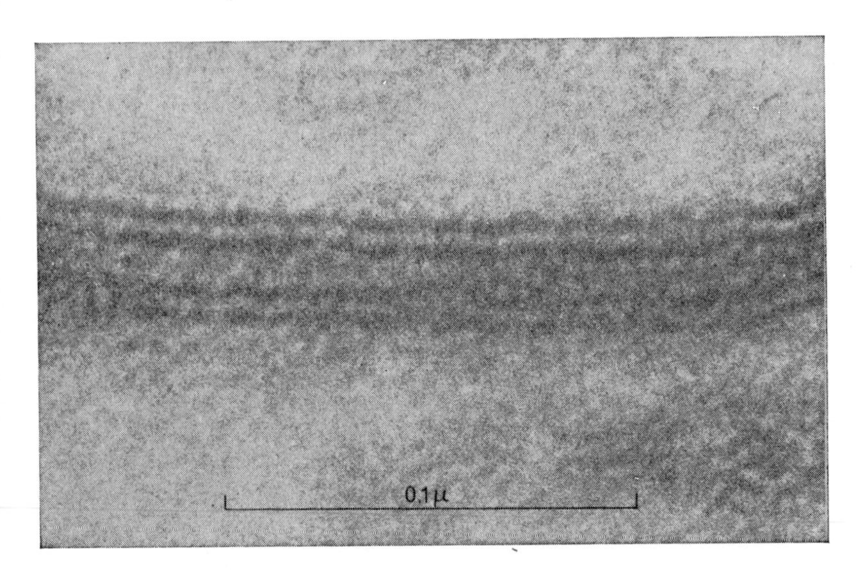

Fig. 7.4. The cell membranes of two contiguous muscle cells, each membrane showing two electron-dense lines.

now presents a surface composed of the non-polar ends of the lipid molecules and so it is not 'wetted' by water; if we insert it through a clean surface of the water a convex meniscus is formed. If now we re-insert it through the surface which is covered by the monolayer the non-polar ends of the lipid molecules come together and a bimolecular layer of lipid now covers the surface of the glass. Provided that it has some mechanical support, for example, from a protein framework, a bimolecular leaflet of this type is stable with water on both sides and so may be considered to be a possible basis for the lipid component of the cell membrane, as was suggested nearly 30 years ago by Danielli, and a possible basis for the double line which is seen in the electronmicrograph; the separation of the lines is of the right order of distance for a bimolecular leaflet of phospholipid.

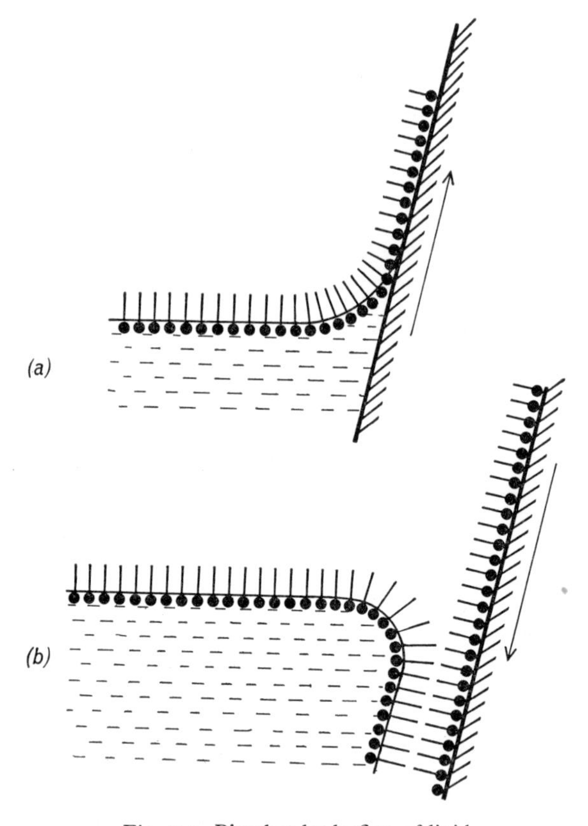

Fig. 7.5. Bimolecular leaflets of lipid.

(*a*) A clean glass surface is withdrawn through an air-water interface covered with a monolayer of polar lipid.

(*b*) After having been dried the glass surface is reinserted through the monolayer and a stable bimolecular leaflet is formed.

Crude extracts of phospholipid, contaminated with protein, show a very characteristic reaction when placed in contact with water. They form layers which are called myelin figures and when these are examined by a variety of physical techniques it is found that the basis of the layered structure is a bimolecular leaflet of the type which we have been considering, stabilised by being covered on both sides with layers of protein. When stained with osmium tetroxide and examined in the EM a bimolecular leaflet of this type shows two electron-opaque lines, each

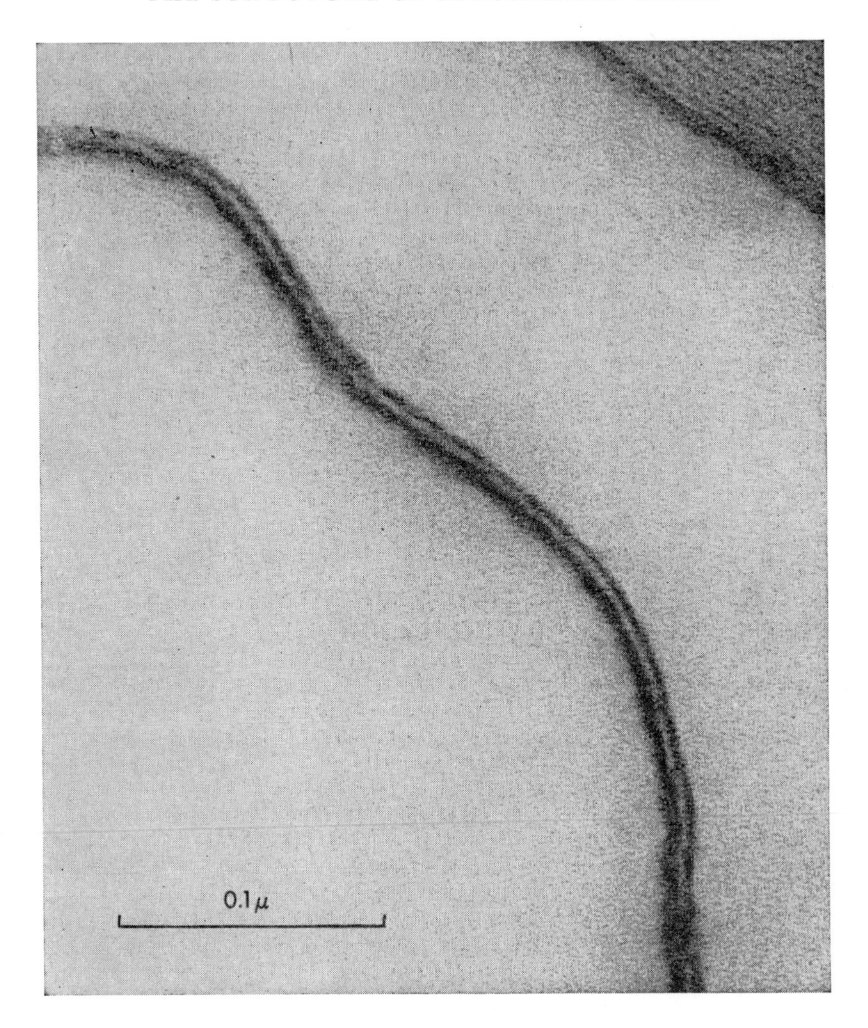

Fig. 7.6. Spontaneously appearing myelin figure in a lipid-protein-water system, showing two electron-dense lines.

25–50 Å wide, separated by a less opaque space 20–25 Å wide (Fig. 7.6). The average overall width of the double line is thus 95 Å, which is reasonably close to the value of 75 Å measured on the cell membrane.

Taking all this evidence together—permeability to lipid-soluble substances, effect of lipid solvents, surface behaviour of lipid-protein-water systems—it might seem that biophysicists have come to very close grips with this problem. But this is not the whole story. The cell membrane has other properties. It is no mere passive barrier to diffusion but is also the seat of active processes whereby differences in the concentrations of ions on the two sides of the membrane are produced and maintained. The bimolecular leaflet of lipid, although it appears to account for the 'lipid' properties of the cell membrane and can be reconciled with its appearance in the EM, provides no basis for active transport across it. No molecular basis for active transport so far suggested has found general acceptance.

Golgi bodies

The Golgi bodies do not appear in Fig. 7.1 because this preparation is supposedly stained with haematoxylin whereas Golgi bodies stain by metallic impregnation. They are present, none the less, on the apical side of the nucleus, and in this position the EM shows the structure

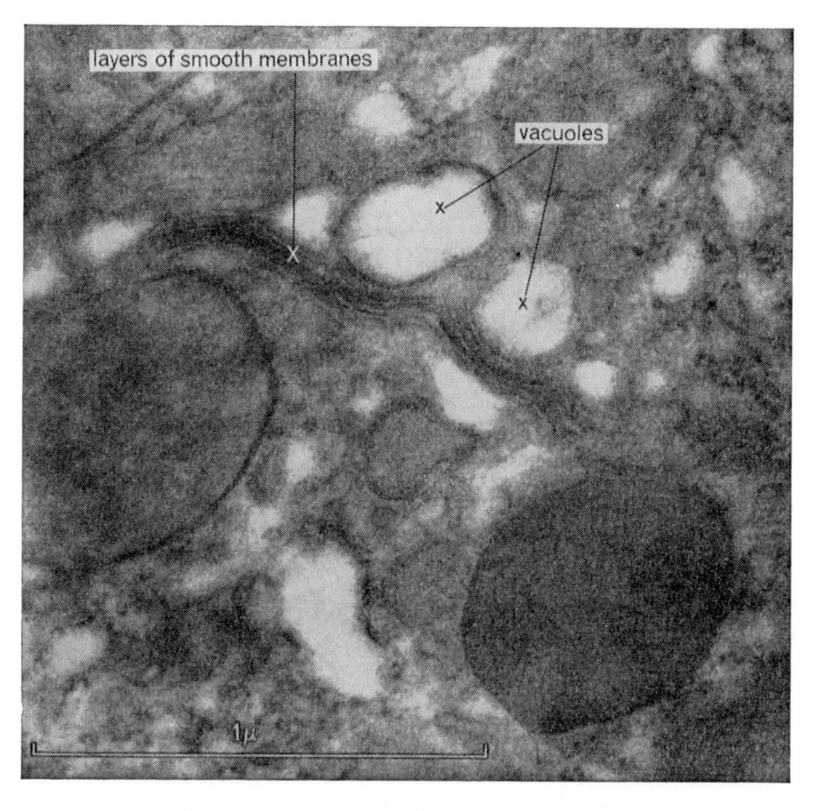

Fig. 7.7. The Golgi bodies in a pancreas cell.

reproduced in Fig. 7.7: a limited number of vacuoles and alongside them a few layers of smooth membranes which seem to have the form of flattened bags. We have little information as to the function of this organelle except that it is particularly well developed in secretory cells.

Mitochondria

By contrast with the Golgi bodies there is no cytoplasmic organelle which we know more about than the mitochondrion. Mitochondria are somewhat variable in size, but generally they are sausage-shaped, about 3 μ long and 0·5 μ in diameter. They have a very characteristic appearance under the EM, having an outer membrane which encloses the whole organelle and an inner membrane which is folded inwards to produce internal plates or partitions known as cristae (Fig. 7.8).

Mitochondria were among the first of the cytoplasmic organelles to which a definite functional role could be ascribed, and this is a convenient point at which to look more closely at some of the many investigations in which the method of bulk separation has been used to complement the observations made with the EM. The method of bulk separation by differential centrifugation (p. 73) was developed by Claude during the Second World War. His first separations were carried out on mammalian liver. The liver was minced and was then suspended and homogenised in 0·85% NaCl. The homogenate was centrifuged at 1500 **g** for 3 minutes; this brought down a fraction which contained various relatively large cell fragments and also a large number of apparently intact nuclei. The supernatant was then taken and centrifuged at 2000 **g** for 25 minutes, which brought down a fraction consisting of large granules, ranging in size from 0·5 to 2·0 μ, including mitochondria. The supernatant from this second centrifugation was then centrifuged at 18,000 **g** for 90 minutes and yielded a fraction consisting of small granules ranging in size from 0·06 to 0·15 μ whose nature could not be revealed by the light microscope; these small granules came later to be known as microsomes. There remained the final supernatant, believed to contain particles of even smaller size.

The biochemical study of these fractions was taken up by Hogeboom and others, who began by exploring the distribution of various types of enzyme activity among the fractions. One of the earliest of these activities to be investigated was that of cytochrome oxidase (p. 194) and this was found to be associated almost exclusively with the large granule

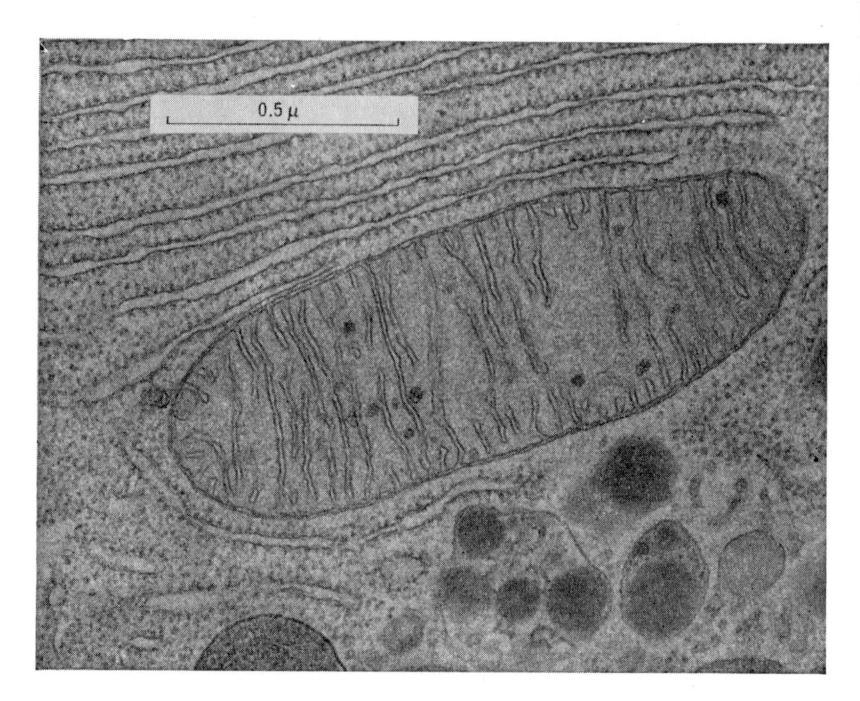

Fig. 7.8. The mitochondrion is surrounded by a membrane made up of two electron-dense lines. The inner line is extended inwards to form the cristae. The endoplasmic reticulum bearing ribosomes is well shown.

Fig. 7.9. Model showing mitochondrial structure.

fraction, as is shown by the figures in Table 7.1. As these investigations proceeded, further activities of respiratory enzyme systems were traced to the large granule fraction, and with improved methods of preparation, involving the use of 0·88 M sucrose instead of 0·85 % NaCl as suspending medium, it was possible to show that the large granule fraction could accomplish the complete oxidation of fatty acids to water and carbon dioxide. Some doubt still remained, however, about the identification of the large granules as mitochondria until the EM could be brought to bear upon the problem. Pellets of the various fractions were sectioned and examined under the EM and the characteristic appearance of the double membranes and cristae confirmed that nearly all the mitochondria were present in the large granule fraction.

Table 7.1

	Cytochrome oxidase activity (%)
Entire homogenate	100
Nuclear fraction	1·0
Large granule fraction	94
Small granule fraction plus supernatant	1·5

Mitochondria have been described as the powerhouses of the cell because they are responsible for all the oxidative metabolism and for nearly all the energy production. What goes on within them will be described in more detail in Chapters 17 and 18.

Endoplasmic reticulum

We now come to those features of the exocrine cell which are most directly concerned with its principal function, that of secreting enzymes. All enzymes being proteins, the intracellular arrangements now to be described may be taken as characteristic of protein-secreting cells. First, we may note the large granules, 0·5–1·0 μ in diameter, seen near the apical end of the cell. These are the products of the cell's activities, the zymogen granules containing the precursors of the digestive enzymes which will become active in the alimentary canal. The zymogen granule is surrounded by a membrane but shows no internal structure in the EM. Secondly, there is the endoplasmic reticulum,† which is a system of membranes densely packed into the basal part of the cell.

† Also known as the ergastoplasm and the α-cytomembrane system.

The endoplasmic reticulum is further illustrated in Fig. 7.10. The nuclear membrane is double and its outer layer is seen to be continuous with the membranes of the endoplasmic reticulum. The characteristic feature of these membranes is that they bear granules upon one side, for which reason they have been called the rough membranes, to distinguish them from others. The granules are about 150 Å across; they

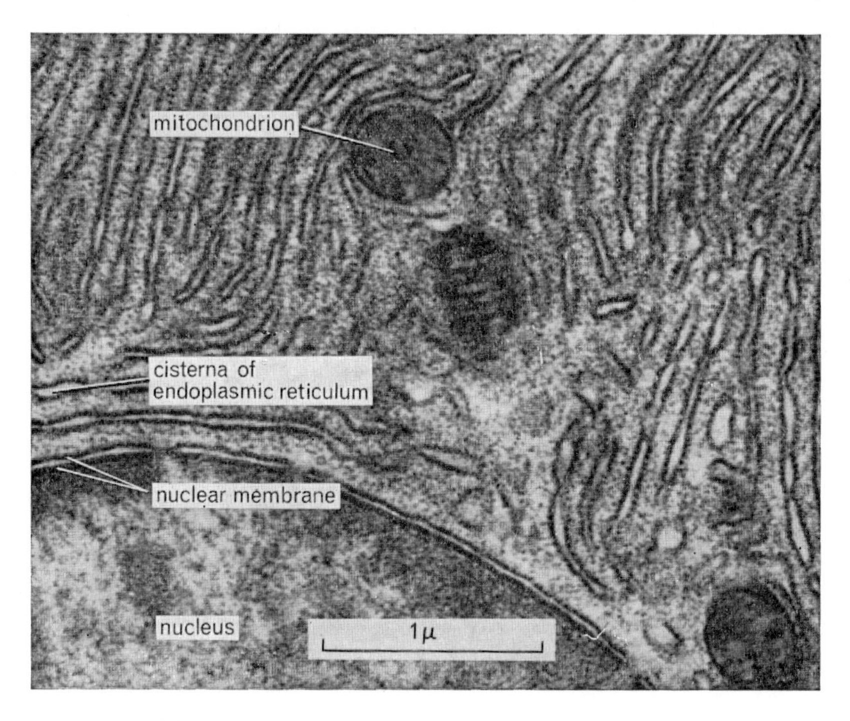

Fig. 7.10. The endoplasmic reticulum of a pancreas cell.
The nucleus is surrounded with a double membrane, pierced by pores. The outer part of the membrane resembles the membranes of the endoplasmic reticulum in bearing ribosomes.

are not seen in sections pre-treated with ribonuclease and are therefore believed to be mainly composed of RNA. For this reason they are called ribosomes. They are responsible for the high affinity of the cytoplasm for basic stains.

The spaces, known as cisternae, enclosed between the smooth sides of the membranes are nowhere in continuity with the spaces towards which the ribosomes project. Some authorities have claimed that the cisternae communicate with the exterior, the membranes of the endoplasmic reticulum becoming continuous with the cell membrane; but this is denied by others.

Our knowledge of the endoplasmic reticulum and its role in the production of enzymes is mainly due to Palade and Siekevitz, and provides one of the best examples of the strikingly successful collaboration of an electron microscopist and a biochemist in a fully integrated attack upon the problem of relating function to structure at the subcellular level. In this work pancreatic cells from guinea pigs were homogenised in 0·88 M sucrose and were separated by differential centrifugation into five fractions: nuclei, zymogen granules, large granules (mitochondria), small granules (microsomes) and supernatant. The identity of the material in these fractions was first established by examining sections of the pellets under the EM. As has been stated elsewhere (p. 74) the fractions obtained by differential centrifugation are by no means pure preparations of the components after which they are named. Besides nuclei the nuclear fraction contains zymogen granules and fragments of cells and tissue; zymogen granules are also found in the mitochondrial fraction, and so on.

The fractions were then studied by biochemical methods. The figures in Table 7.2 are reproduced as an example of the results which such investigations provide. They show that proteolytic enzyme activity is found predominantly in the zymogen granule fraction but is also appreciable in the nuclear and mitochondrial fractions which, as we have seen, are contaminated with zymogen granules.

Table 7.2

Fraction	Proteolytic enzyme activity (%)
Nuclear	26
Zymogen	36
Mitochondrial	18
Microsomal	12
Supernatant	8

At this time the question of the nature of the microsomes was as yet unsettled. Under the EM they are seen to be small vesicles bearing granules upon their walls (Fig. 7.11). These granules are of the same size as ribosomes, but the small vesicles to which they are attached have no counterpart in the undisrupted cell. Palade concluded that microsomes are artefacts, being fragments of the endoplasmic reticulum whose broken edges have closed together. In support of this view he noted that this process of fragmentation could be observed in damaged cells (Fig. 7.12).

Further resolution of the microsome fraction was achieved by the use of a detergent, the bile salt sodium deoxycholate. This causes breakdown of the lipid-protein membrane of the microsomal vesicle, liberating

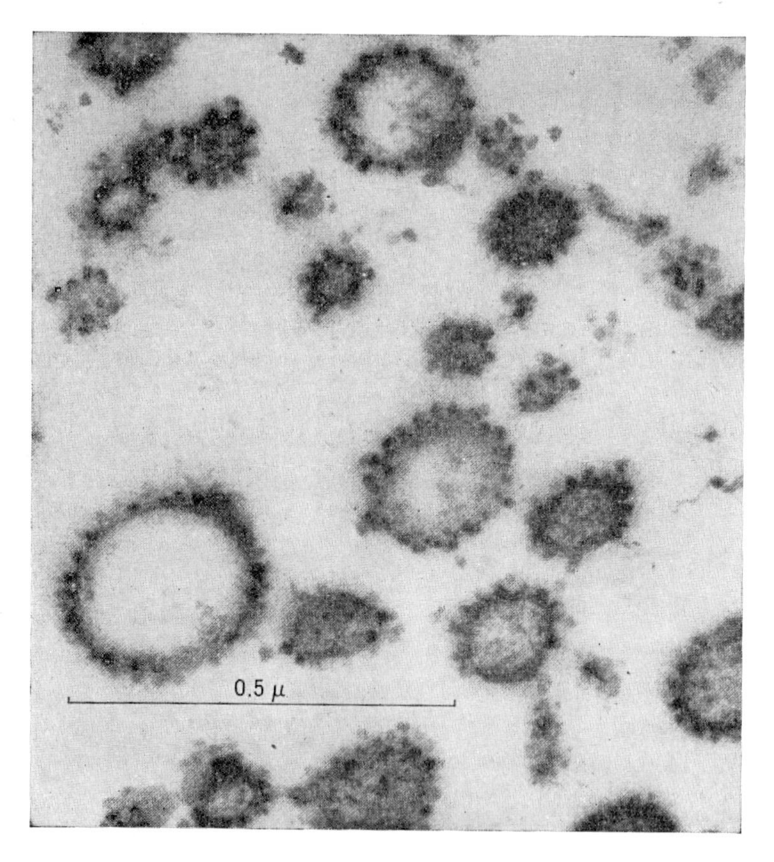

Fig. 7.11. Microsomes.

Microsomes are not organelles. They are artefacts produced by disruption of the endoplasmic reticulum. The pieces of the endoplasmic reticulum round up into little spheres with the ribosomes on the outside.

the ribosomes which can then be collected in a pellet by centrifuging at 100,000 *g* for 2 hours. The evidence which attributes to the ribosomes a central role in the synthesis of protein will be discussed in Chapter 24. For the moment we will take this on trust and will continue to follow the investigations of Palade and Siekevitz into the manner in which the endoplasmic reticulum with its attached ribosomes brings about the manufacture of the enzyme precursors which later appear in the zymogen granules. In these investigations Palade and Siekevitz obtained further insight into the operations of the endoplasmic reticulum by observing the changes which it underwent during its natural cycle of activity. If the animal is starved for 48 hours the apical ends of the cells are found to be packed with zymogen granules and there is no secretion of enzyme from the pancreatic duct. One hour after feeding

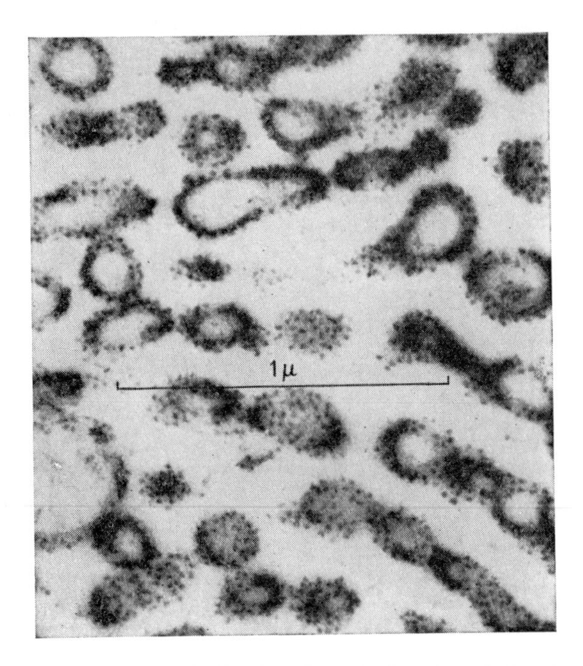

Fig. 7.12. A damaged cell, showing the endoplasmic reticulum breaking up into microsomes.

many of the zymogen granules have been discharged into the lumen and the cell appears to be actively engaged in synthesis.

The onset of synthetic activity is associated with the appearance of relatively large (0·1–0·2 μ) granules in the cisternae of the endoplasmic reticulum (Fig. 7.13). These granules appear to have the same texture as zymogen granules, which suggests that they may be the first products of synthetic activity. It proves possible to concentrate these intracisternal granules by a refinement of the differential centrifugation technique and in this way their high enzyme content can be confirmed.

The intracisternal granules differ from the zymogen granules, however, in their situation within the cisternae and in the absence of any limiting membrane such as is found around the zymogen granules. About 2 hours after feeding the intracisternal granules are no longer to be seen, but now small and irregularly shaped zymogen granules make their appearance. It is not clear how the material is transferred from the intracisternal granules to the zymogen granules. Palade and Siekevitz have

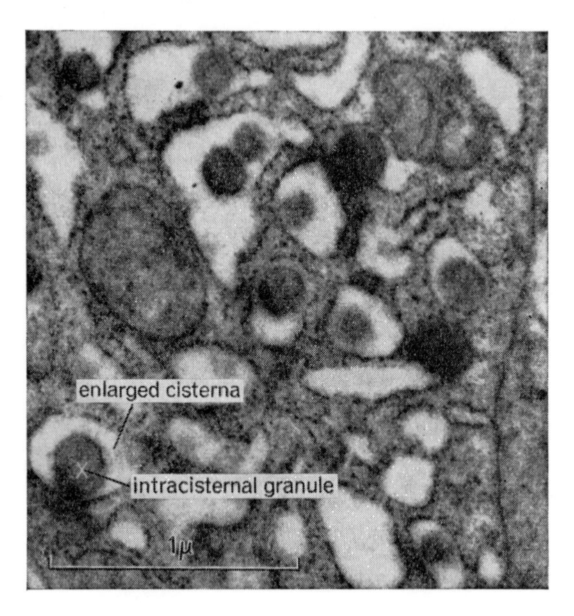

Fig. 7.13. Part of a cell fixed 1 hour after feeding, showing enlarged cisternae containing intracisternal granules.

suggested that the intracisternal spaces are continuous with the spaces of the Golgi vacuoles and that the intracisternal granules pass into the Golgi vacuoles wherein they are accumulated to form zymogen granules, the membrane surrounding the zymogen granule being the membrane of the Golgi vacuole. Decisive evidence in favour of this interpretation is yet to seek.

2. KIDNEY

In the mammalian kidney the unit structure—the nephron—has the following parts: Bowman's capsule, proximal tubule, loop of Henle, distal tubule, collecting tubule. We will here consider only the proximal tubule, shown in transverse section in Fig. 7.14.

Ordinary staining procedures do not show up any strongly basophil or acidophil regions of the tubule cell. At the apical side there is a striated region known as the brush border, and at the basal side much fainter striations can sometimes be made out. Under the EM (Fig. 7.15)

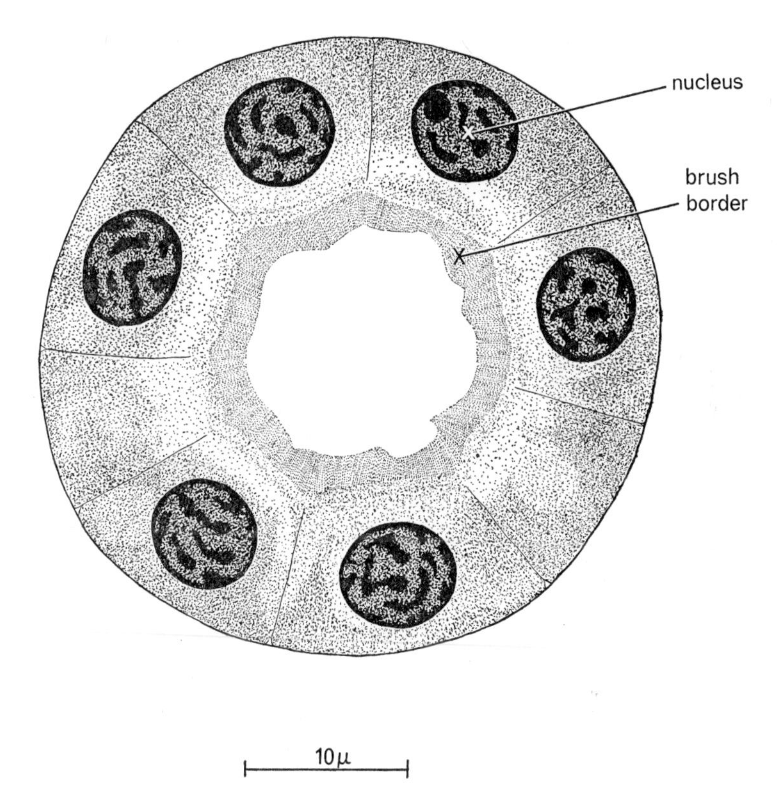

Fig. 7.14. Section of proximal tubule of kidney.

To show the appearance under the light microscope after treatment with a basic stain. The stain is not strongly taken up by the cytoplasm. There are faint indications of radial striation at the base of the cell; at the apex (next to the lumen) the brush border can be seen.

the internal structure is seen to be very different from that of the pancreas cell. There is no endoplasmic reticulum and, naturally, there are no zymogen granules, but there are a few ribosomes free in the cytoplasm and the Golgi bodies are present. The brush border, not found in the pancreas cell, is seen to be made up of an array of finger-like processes, about 1 μ in length and 0·08 μ in diameter, fairly closely packed together (Fig. 7.17). These are known as microvilli. Between them the cell membrane is deeply invaginated and small vacuoles are seen near these invaginations. They remind one of the food vacuoles formed by an amœboid cell during phagocytosis; but since the kidney cell does not 'eat' particles but may 'drink' fluids the word 'pinocytosis' is more appropriate here. At the basal side of the cell the cell membrane is deeply infolded and this region is very well supplied with mitochondria

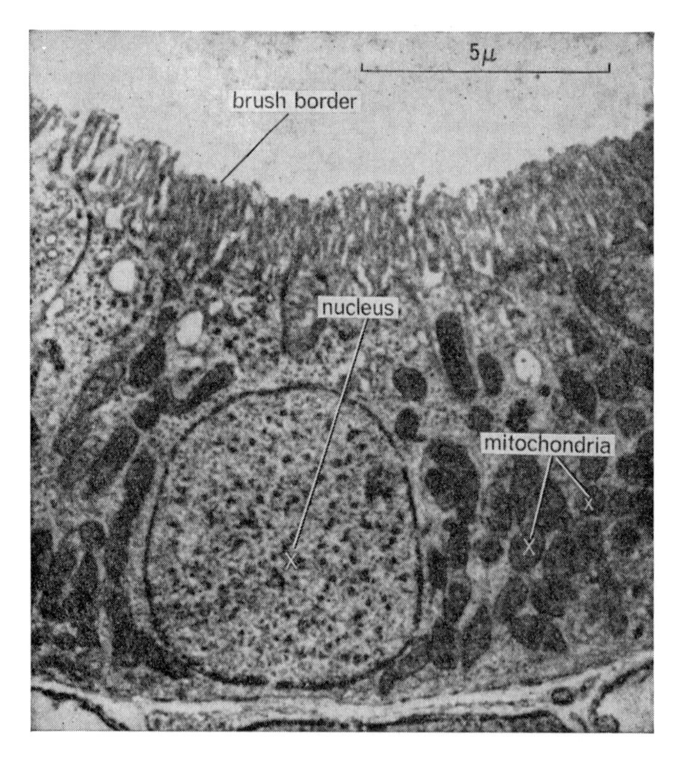

Fig. 7.15. Low-power electronmicrograph of cell from kidney tubule.

(Fig. 7.18). Careful examination reveals that some of the folds have no continuity with the rest of the cell; they belong to neighbouring cells, the bases of the cells interdigitating as shown in Fig. 7.19. The basement membrane, the cell membrane, the mitochondria and the Golgi bodies, all are very much the same as in the pancreas cell, except that the Golgi bodies are smaller.

Once again there is a quite well marked correlation between structure and function. The main function of the proximal tubule is the reabsorption of water and physiologically useful substances, such as glucose, from the glomerular filtrate. Its traffic in water and dissolved substances is considerable. In correlation with this we may point to the great extension of the area of cell membrane, as microvilli at the apical end and as infoldings at the basal end. This transport of materials requires energy and it is known that energy is provided by mitochondria. The mitochondria are concentrated in close relation to the

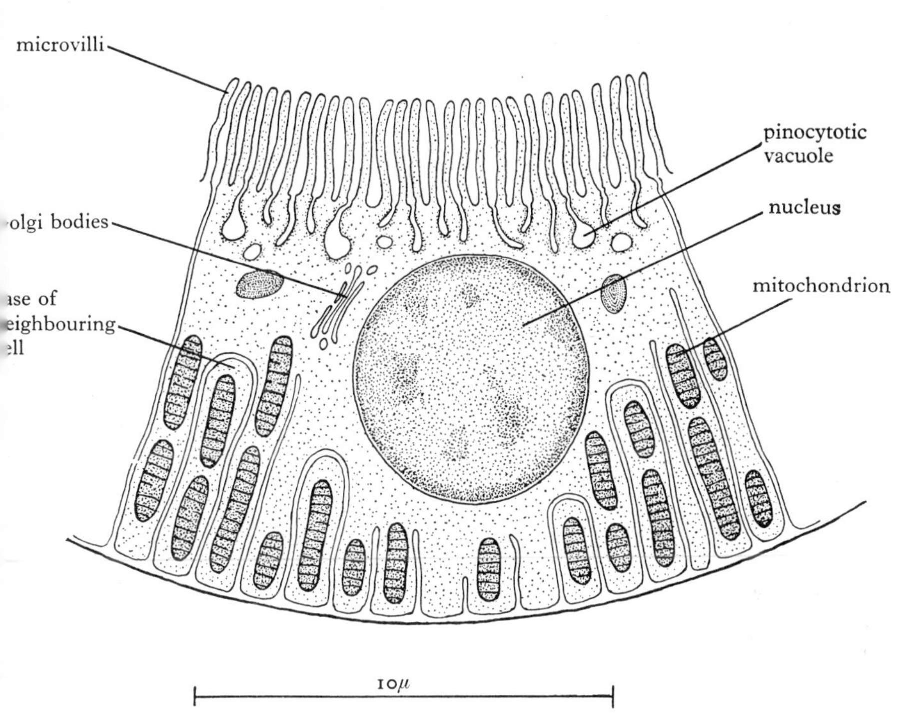

Fig. 7.16. Interpretation of electronmicrograph of kidney cell.

basal folds and it is not unreasonable to interpret this as suggesting that water and solutes enter the cell at the apical end by a process (possibly pinocytosis) which does not require much energy, and are then pumped out through the base of the cell by some process for which energy is required. Let it be remembered, however, that there is scarcely any physiological evidence to tell us about the mechanism of active transport in these cells at the subcellular level. All we know is that water and solutes move from the lumen of the tubule to the blood.

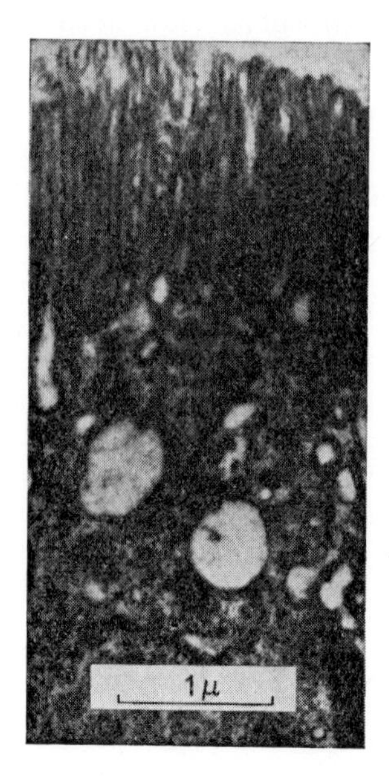

Fig. 7.17. Apical end of kidney cell showing microvilli and pinocytotic vacuoles.

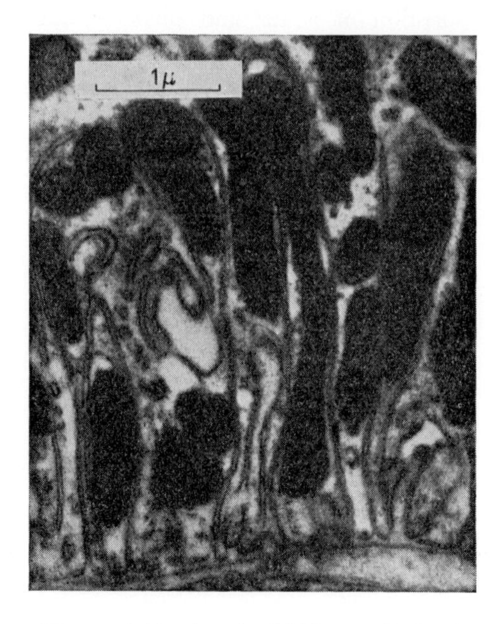

Fig. 7.18. Basal end of kidney cell showing infoldings of membrane with many mitochondria.

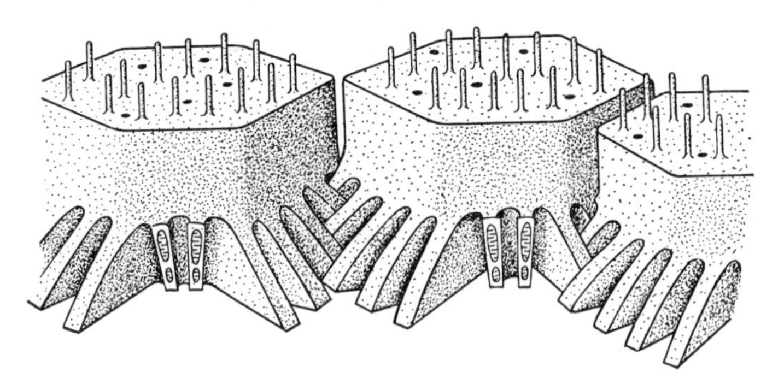

Fig. 7.19. Diagram of kidney cells to show interpenetration of basal regions.

3. LEAF

A leaf is covered on both sides with a single layer of epidermal cells, pierced by stomata which lead into the air spaces of the mesophyll. Just under the upper epidermis the mesophyll cells are elongated at right angles to the surface and these are known as the palisade cells. Of all the cells in the leaf they contain the most chlorophyll and are presumably the most active in photosynthesis.

The palisade cell (Fig. 7.20) has, of course, a cell membrane and a nucleus, and like the other cells we have considered it contains mitochondria. It resembles the kidney tubule cell in having Golgi bodies

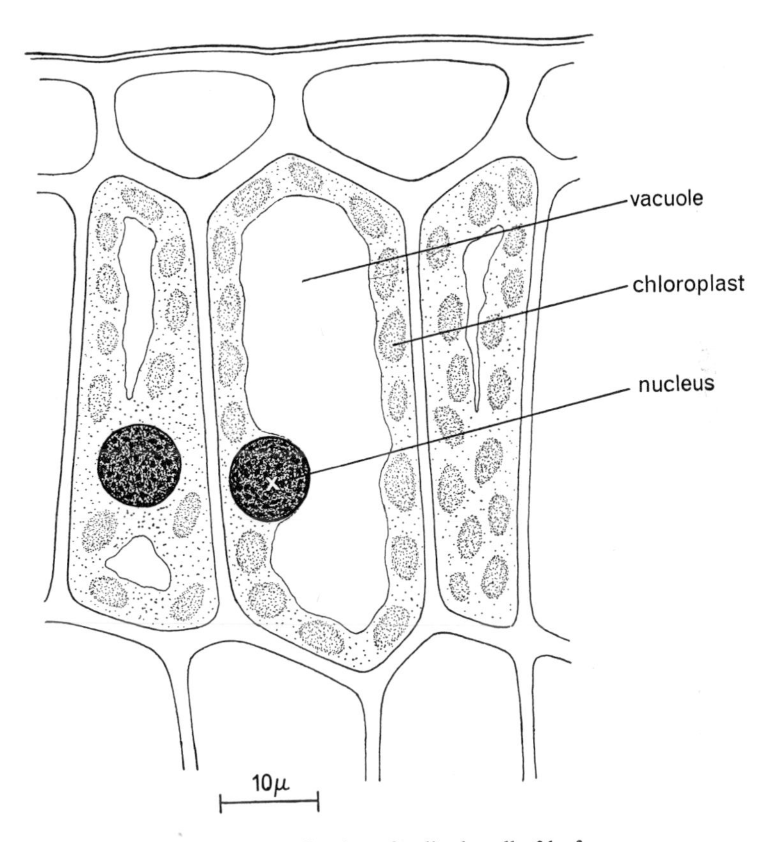

Fig. 7.20. Section of palisade cell of leaf.

To show the appearance under the light microscope after treatment with a basic stain. The stain is not strongly taken up by the cytoplasm or by the chloroplasts.

and free ribosomes, but no well-developed endoplasmic reticulum. It differs from the cells of animal origin principally in three features: the cell wall, the vacuole and the chloroplasts.

The structure of the cell wall is best understood by tracing its development from the stage when it first appears as a septum which separates the two daughter cells immediately after cell division. The first sign of the new cell wall is a thin membrane, known as the middle lamella. It does not contain cellulose, being made of another polysaccharide, pectin, which does not show the highly oriented structure of cellulose. Having laid down the middle lamella as a sort of no-man's-land between them, the daughter cells begin to lay down their primary cell walls on either side of it. Here cellulose appears for the first time. Earlier (p. 48) we saw that cellulose consists of long straight chains of glucose molecules, hydrogen-bonded together to form crystalline micelles, each of which may contain up to 200 molecular chains. Not all

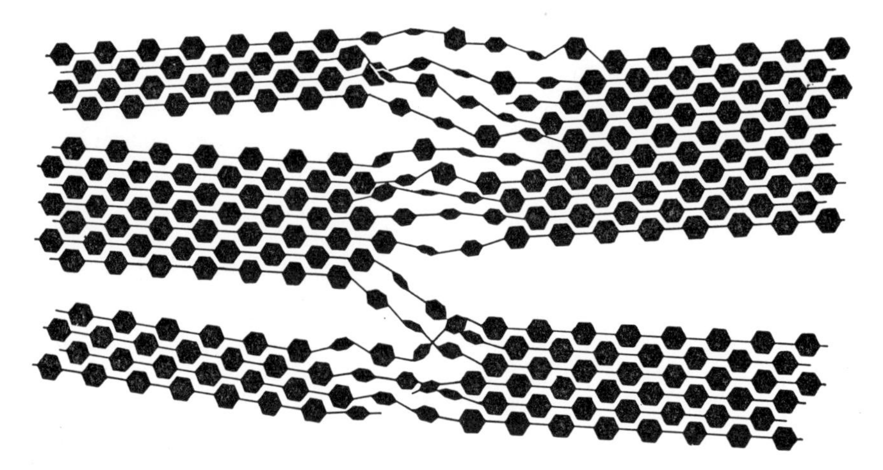

Fig. 7.21. Transitions between crystalline and amorphous cellulose.

the cellulose is crystalline, in fact in some cells the first cellulose to be laid down is amorphous; but where long molecular chains are concerned the distinction between the crystalline and the amorphous states is not hard and fast, as Fig. 7.21 indicates. In the primary cell wall the cellulose molecules and micelles form a loose network with the interstices filled with pectin (Fig. 7.22). This structure is readily extensible and as the cell grows in size more cellulose is laid down in the spaces between pre-existing micelles. When the cell has reached its full size the secondary cell wall is laid down inside the primary cell wall. This secondary cell wall (Fig. 7.23) is thicker and much more highly organised. Most of the cellulose in it is in the form of crystalline micelles and these associate together into coarser fibrils which sometimes attain to thicknesses which could be seen in the light microscope. These fibrils, often in bundles, wind spirally around the cell and are laid down in

Fig. 7.22. Electronmicrograph (shadowed) of the primary cell wall of the alga *Valonia* showing cellulose micelles forming loose network.

Fig. 7.23. Electronmicrograph (shadowed) of the secondary cell wall of the alga *Valonia* showing cellulose fibrils.

successive layers, the general direction of the fibrils changing between one layer and the next. Since the cellulose molecular chains, and therefore also the fibres, are highly inextensible, this arrangement of fibrils running in all directions confers a high degree of rigidity upon the secondary cell wall.

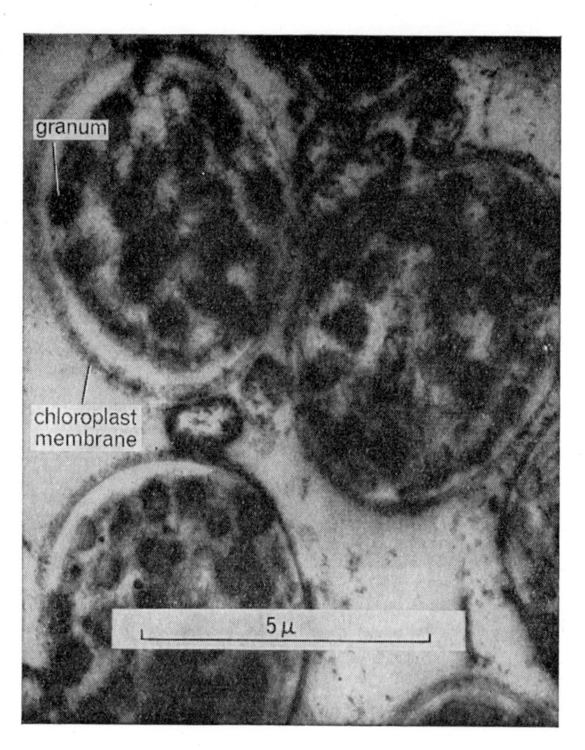

Fig. 7.24. Low-power electronmicrograph showing chloroplasts containing grana.

The central vacuole, so characteristic of plant cells, develops by the fusion of smaller vacuoles, and the growth in size of the cell up to the time when the secondary cell wall begins to be laid down is mainly due to the increase in size of the vacuole.

Chloroplasts, being of the order of 5 μ across, are readily visible in the light microscope. Seen under the EM the chloroplast has a bounding membrane and contains electron-opaque bodies, the grana, dispersed in a less opaque matrix, the stroma (Fig. 7.24). At higher magnifications both the grana and the stroma show themselves to be made up of layers, but in the grana the layers are more regular and more closely spaced (Fig. 7.25). Each granum consists of a series of flat plates, stacked like a pile of coins. The dimensions of the spacing are indicated in Fig. 7.26. We have as yet only very indirect evidence as to how chlorophyll molecules are arranged within these structures; this is the sort of problem which could be tackled by X-ray diffraction if only we were able to obtain crystalline grana.

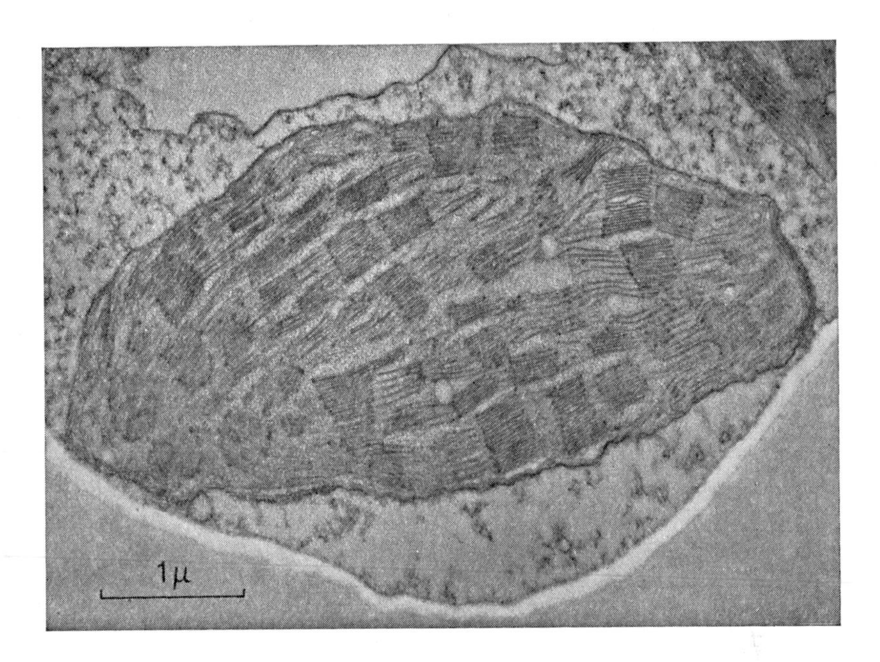

Fig. 7.25. Electronmicrograph of a section of chloroplast showing the grana (with closely spaced electron-dense membranes) connected by stroma (with less closely spaced membranes).

In bringing this chapter to a close it is perhaps necessary to stress once again its incompleteness and to remind the reader of how much has been left unsaid. We have dealt only with three types of cell. These have been chosen because, though they are specialised, they are not highly specialised and because they best illustrate how structure can be related to function at the subcellular level. We have been able to see how the secretion of protein is related to the development of the endoplasmic reticulum and how the extension of the cell membrane in the form of microvilli and basal folds reflects the kidney's activity in transporting fluids. Many points of detail have been omitted. Some organelles have

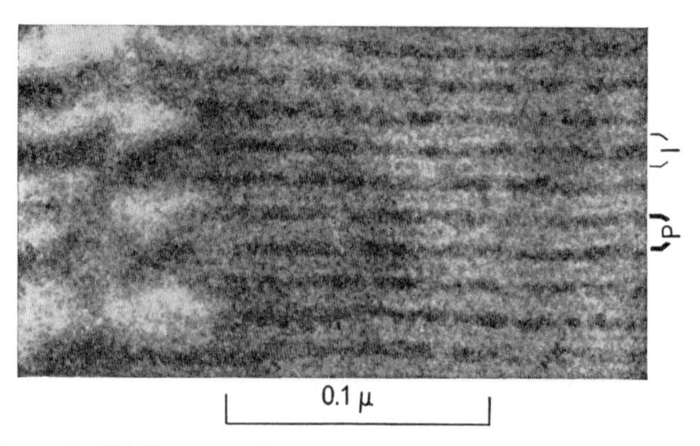

Fig. 7.26. High-power electronmicrograph of section through the transition between stroma (left) and grana (right).

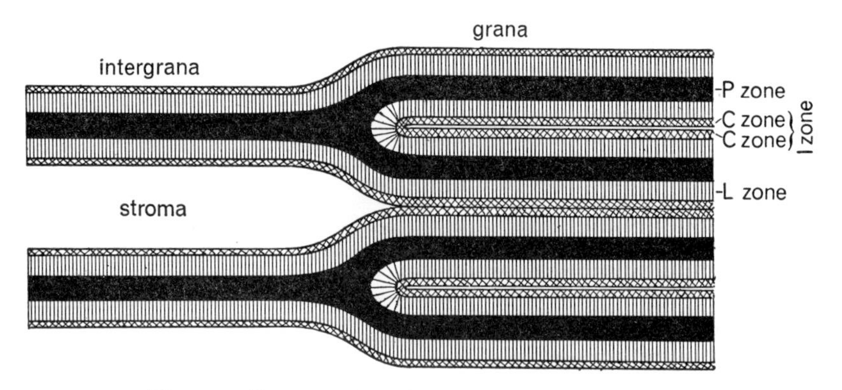

Fig. 7.27. Interpretation of appearance seen in Fig. 7.26.

been left out of consideration because their function is unknown or because their separate identity has been questioned. Other organelles, such as cilia, whose fine structure is well known, are not found in these three cell types and this has been, frankly, a welcome excuse for passing them over; a book such as this would defeat its aims if it became too compendious.

The EM has revealed a whole new world of subcellular structure some of whose features were previously only guessed at. Yet it stops tantalisingly short of disclosing the molecular arrangements upon which ultimate understanding of function must rest. We shall see in later chapters how, for example, the chemical activities of the mitochondrion do not survive the destruction of the double membrane system which finds expression in the cristae, and how all the indications point to the existence of some orderly arrangement of enzymes and respiratory carriers in the mitochondrial membranes. To explore this as yet hidden world it seems that we must await the discovery of some still more powerful method of investigation. The EM has served us well and has much service still to offer. But even if its resolution were to be improved by a factor of ten there would still remain the stubborn fact that it can be used only on fixed, dead cells; and it is as we approach the molecular level that the interpretation of artefacts becomes so very difficult.

8

THE STRUCTURE OF CERTAIN MICRO-ORGANISMS

Bacteria show great variety of form and any attempt to treat their structure comprehensively must inevitably become a catalogue of their external features, since the internal structure is known for very few. We shall therefore continue to restrict ourselves to *Escherichia coli*, the example which we discussed earlier (p. 15), and will digress only to

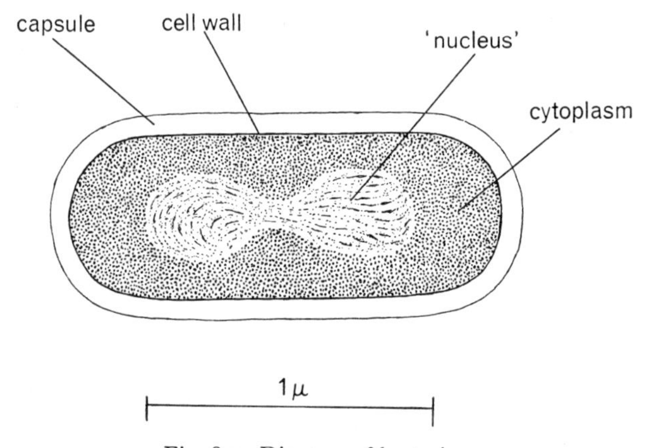

Fig. 8.1. Diagram of bacterium.

The cytoplasm contains a great many granules. The 'nucleus' is less dense than the cytoplasm and is not separated from it by any nuclear membrane.

describe some of the structures which are seen in other bacteria of the same general type. Among viruses there is again a very great deal of variation, and here we will restrict ourselves to the single example of the bacteriophage known as T_4.

1. ESCHERICHIA COLI

This organism is about 1·5 μ long and 0·5 μ in diameter. It is enclosed in a more or less rigid cell wall which gives it its characteristic shape. The chemical nature of the cell wall is not fully understood; its main component is polysaccharide which, however, is not cellulose, and it also contains protein. Functionally it resembles the plant cell wall in so far as it restrains the tendency of the protoplast to swell; this can be shown by plasmolysis with concentrated sucrose solutions. Outside the cell wall there is often a relatively thick capsule which may be of firm or of slimy consistency.

The question of the existence in bacteria of any structure corresponding to the nucleus was for long a matter of speculation. When a bacterial cell is treated with stains the stain may be taken up by the cell wall or the capsule and the inner parts of the cell are thereby obscured. The protoplast itself usually stains very deeply and uniformly with basic stains. This we now know to be due to its high content of RNA. The cytoplasm contains a very great many ribosomes which are dispersed through it and are not attached to the membranes of an endoplasmic reticulum as in the pancreas cell. The Feulgen reaction, which is specific for DNA, shows a deeply staining body in the middle of the cell; and the presence of this body is confirmed by the use of basic stains after the RNA has been removed from the ribosomes by pretreatment with ribonuclease (Fig. 8.2). In sections under the EM this body shows up as an area of low electron-opacity, without any constant structural features. No chromosomes have been identified with certainty. There are no Golgi bodies. Even more surprising perhaps is the absence of anything resembling mitochondrial structure in an organism which has all the respiratory facilities which we associate with the mitochondria in other organisms.

2. BACTERIOPHAGE T_4

Much of what is known about the fine structure of viruses has been obtained by the method of X-ray diffraction, to which viruses lend themselves by reason of their ability to form crystals. This ability is a consequence of their simple and regular shapes. Bacteriophage, on the other hand, does not crystallise and, as we are about to see, its shape is very complicated. The remarkable amount of detail which has been discovered in the structure of bacteriophage has come from the EM used in conjunction with the method of negative staining. The phage particles, suspended in a solution of phosphotungstic acid, are sprayed upon a collodion film and dried down. The phosphotungstic acid does not enter the phage but is displaced by it, so that we see the phage as a relatively transparent area on an electron-opaque background.

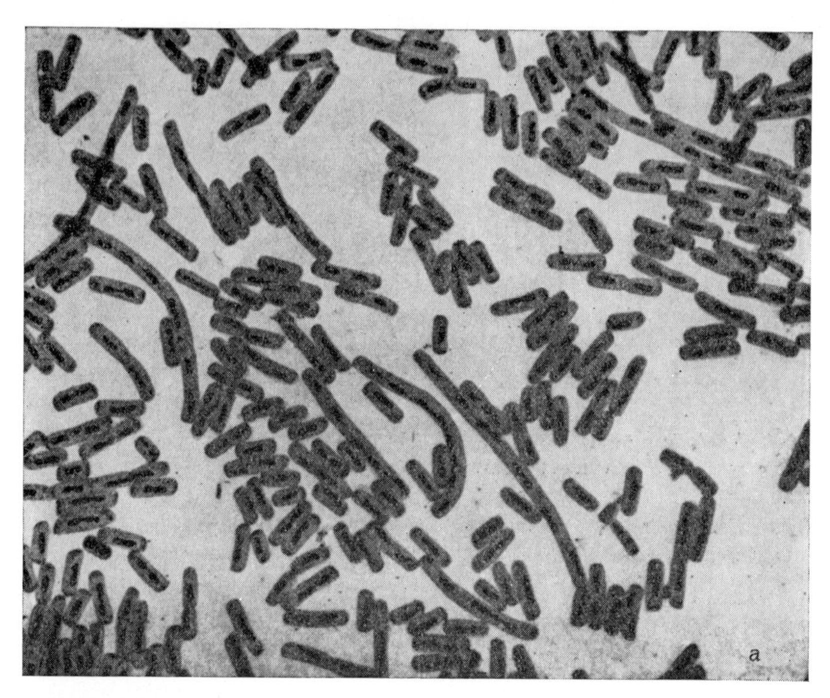

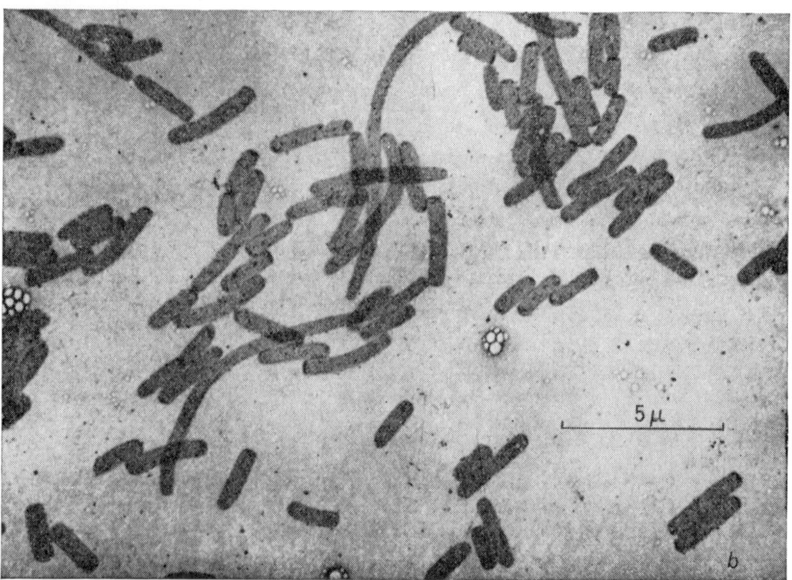

Fig. 8.2. The 'nucleus' of *Escherichia coli.*

The upper figure shows bacteria which have been treated with ribonuclease prior to staining with a basic dye: the ribonuclease removes the deeply staining RNA which would otherwise obscure the DNA in the 'nucleus'. The lower figure shows the further effect of treatment with deoxyribonuclease, after which the 'nucleus' no longer stains.

Fig. 8.3. Electronmicrograph of a section through *Escherichia coli*.

The phage particle has been shown to be built up of a number of separate component parts (Fig. 8.4). The head of the phage is a hexagonal prism. It contains the phage DNA. The DNA can be removed from the phage by subjecting it to the treatment known as osmotic shock. This treatment consists in suspending the phage particles in 3 M-NaCl and then suddenly diluting the solution by mixing it rapidly with 40 volumes of distilled water. After this treatment the phage

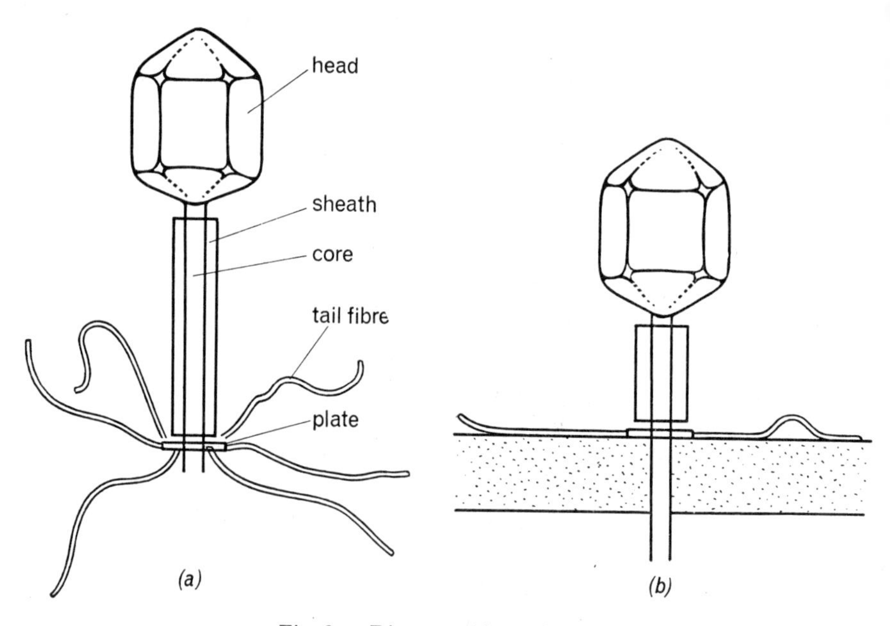

Fig. 8.4. Diagram of bacteriophage.
(*a*) Free 'untriggered'. (*b*) Attached to surface of bacterium, 'triggered', with core driven through bacterial cell wall.

particles are seen to have collapsed heads—they are called 'ghosts'—and they no longer contain DNA, which can be recovered from the medium (Fig. 8.8). Projecting from one end of the head is the core, apparently a simple tube. Around the core is the sheath, a collar-like structure with spiral ridges. At its distal end the sheath is attached to the plate, a hexagonal disc from whose corners extend the six tail fibres. These components can be dissociated by various carefully graded treatments, such as exposure to media of pH 2, and once dissociated they can be separated and purified by various standard techniques for purifying proteins.

Electronmicrographs, such as that in Fig. 8.9, show phage particles attached to the bacterial surface by their tails. This attachment is brought about by a serological reaction between the tail fibres and the bacterial surface. Once the phage particle is so attached and oriented

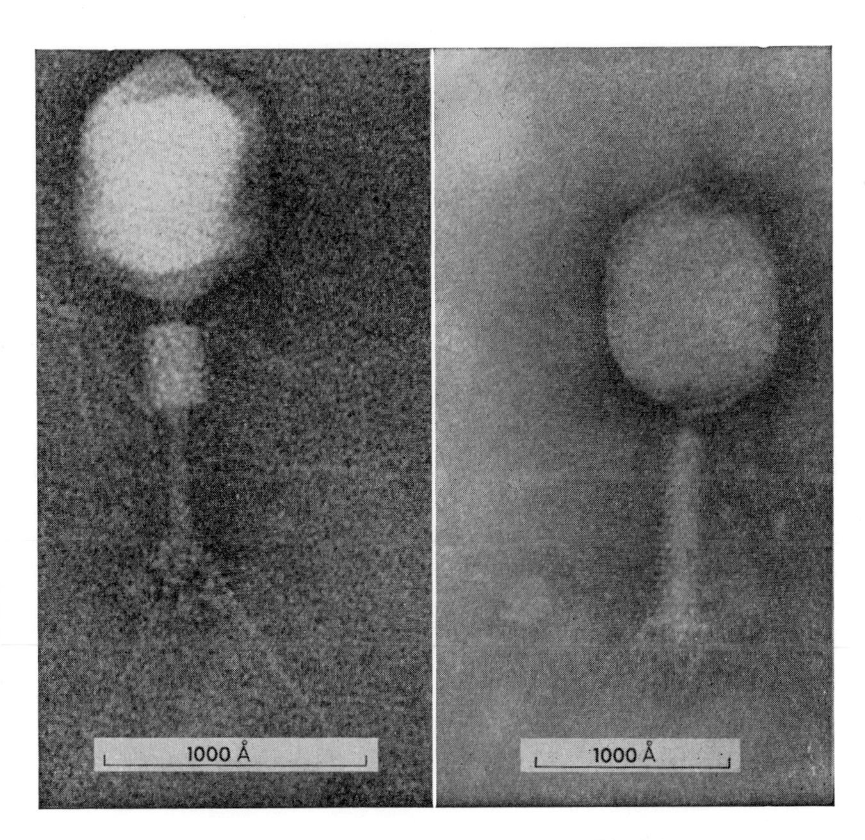

Fig. 8.5. Fig. 8.6.

Fig. 8.5. Electronmicrograph of bacteriophage negatively stained with phosphotungstic acid, in 'triggered' condition with contracted sheath.

Fig. 8.6. Electronmicrograph of bacteriophage negatively stained with phosphotungstic acid, in 'untriggered' condition.

perpendicular to the surface, the sheath contracts (Fig. 8.4*b*). This has the effect of driving the core through the bacterial cell wall. So much can be seen in the electronmicrographs. The DNA of the phage head then enters the bacterium, presumably by passing through the core, while the coat remains outside.

The evidence that only the DNA enters the bacterium comes from a well-known experiment carried out by Hershey and Chase in 1952. From the analytical point of view the possibility of tracing separately the movements of DNA and of protein depends upon the fact that DNA

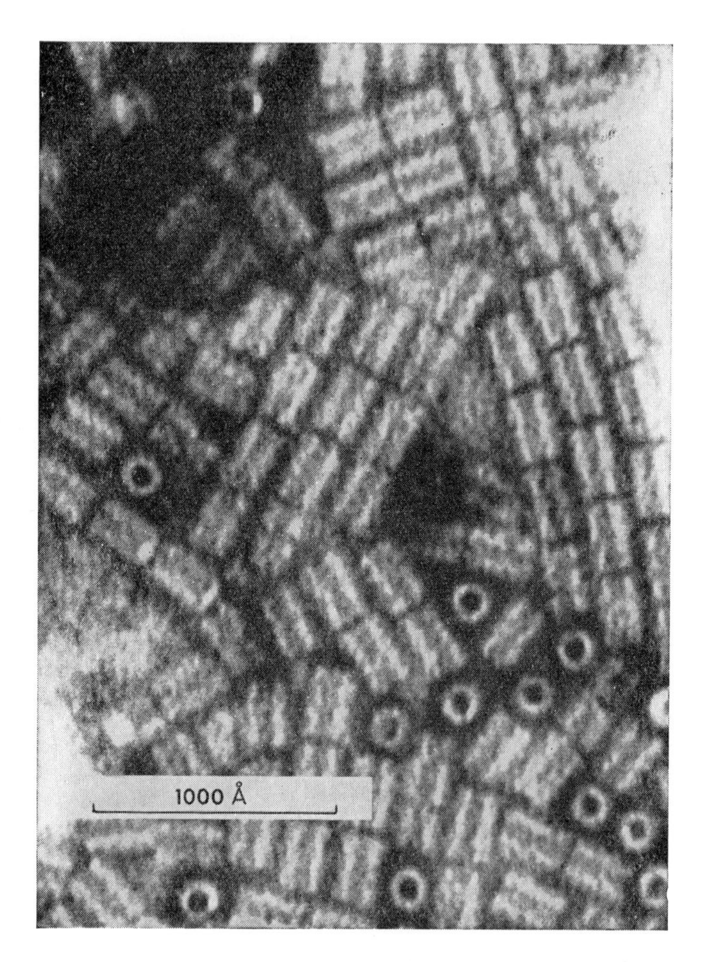

Fig. 8.7. Electronmicrograph of purified preparation of sheaths from bacteriophage, negatively stained with phosphotungstic acid.

contains phosphorus but no sulphur, whereas protein contains sulphur but no phosphorus. The phage DNA can therefore be labelled with the radioactive isotope ^{32}P and the phage protein with the radioactive isotope ^{35}S. This can be done by growing the phage on a culture of bacteria in a medium containing either $KH_2{}^{32}PO_4$ or $Mg^{35}SO_4$ as sole sources of phosphorus or sulphur. The possibility of tracing the fate of the phage DNA and of distinguishing it from the fate of the phage protein depends also upon the fact that the phage ghosts can be detached from the bacterial surface by running the suspension of bacteria in a Waring blendor.

The experiment was conducted in the following way. A suspension of unlabelled bacteria was heavily infected with labelled phage. The

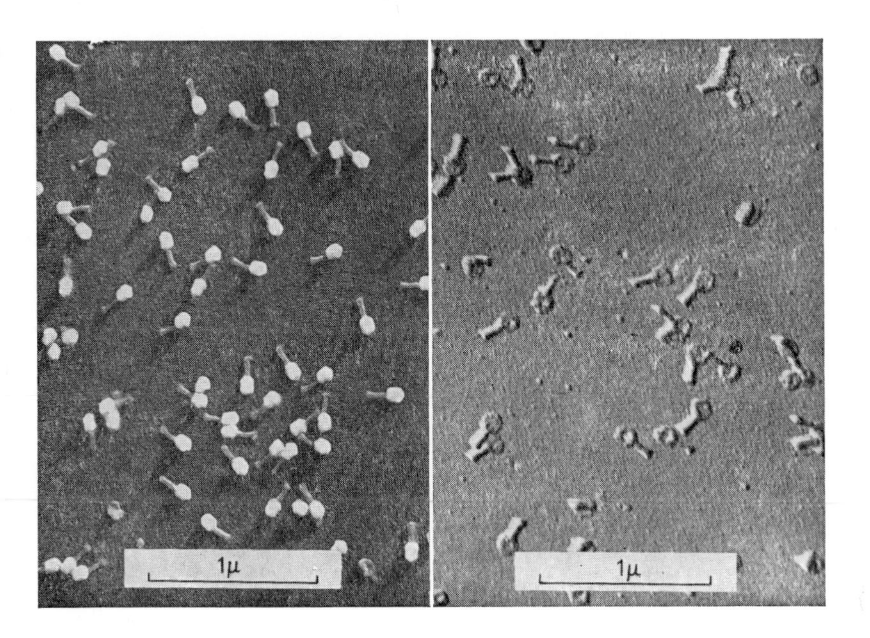

Fig. 8.8. Electronmicrographs of shadowed bacteriophage. On left, normal. On right, after osmotic shock, showing collapse of head.

suspension was centrifuged at a speed which threw down the bacteria but not any free unattached phage particles, and in this way the free phage particles were separated from the bacteria and discarded. The bacteria were re-suspended in water and run in the blendor for 8 minutes, whereby the phage particles were dislodged from their attachment to the bacterial surface. After this the suspension was centrifuged once more, the bacteria being thrown down and the now detached phage coats remaining in suspension. The amount of the label present in the supernatant fluid was assayed,† i.e. the amount of label remaining in the phage coats. It was found that these accounted for 75% of the sulphur originally present in the infecting phage but only 15% of the phosphorus. The rest of the phosphorus had gone into the bacteria; and when the phage particles arising from the infected bacteria were

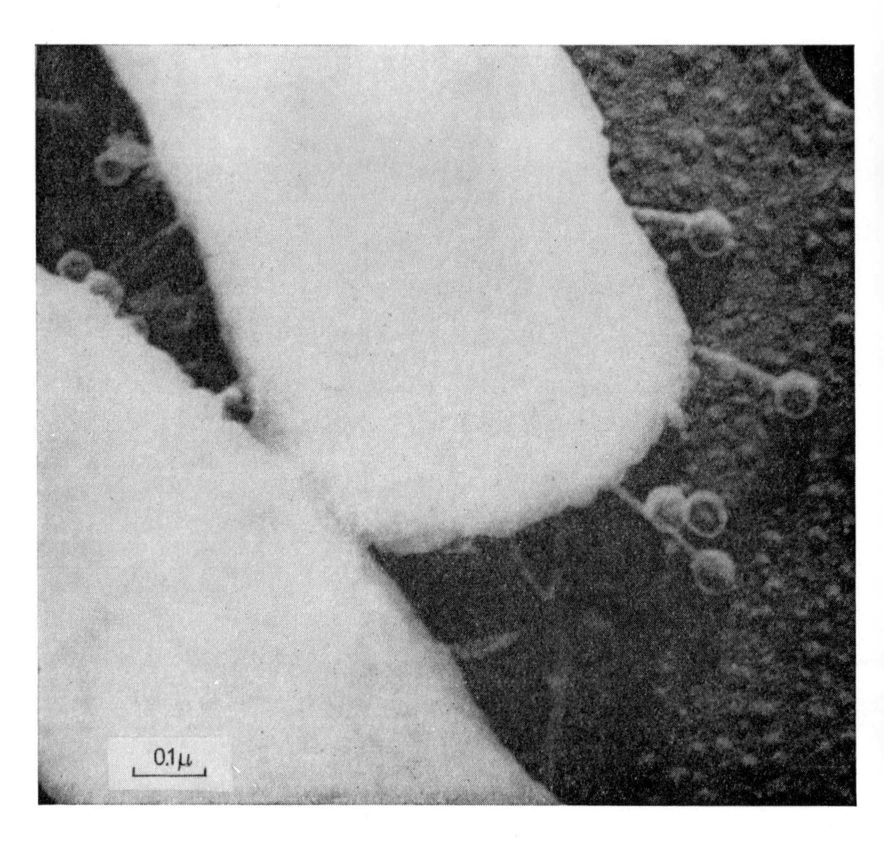

Fig. 8.9. Electronmicrograph (shadowed) showing bacteriophage particles attached to a bacterium.

assayed they were found to contain 30% of the parental ^{32}P but only 1% of the parental ^{35}S.

If we have been accustomed to think of the shapeless amoeba as a lowly organism and if perhaps we have read in old books about 'Urschleim' and 'primordial globules' we may well be a little dis-

† Methods of assaying radioactively labelled compounds are described in Chapter 14, p. 166.

concerted to find such a neat little object as this so near the bottom of the scale of beings. This phage particle makes one think not so much of an organism as of some sort of gadget. It is a little self-acting hypodermic syringe—the sort of thing one might have designed oneself and tried to make in a workshop. It seems likely on presently available evidence that all the components of the phage particle are separately manufactured within the body of the infected bacterium. How they are then assembled in the right way is still a mystery.

PART II

ENERGY

9

ENERGY IN GENERAL

The study of energy transactions is the province of thermodynamics and it is only in terms of thermodynamics that the energy transactions of organisms can be accurately described. It would be out of place (and would take far too long) to attempt a formal introduction to thermodynamics in a book such as this. On the other hand, it is hardly possible to discuss energy transactions at all without some reference to the concepts of thermodynamics and we shall therefore begin by considering some of these concepts, illustrating them by examples rather than attempting to derive them from first principles.

Thermodynamics deals with the properties of systems and with the changes in these properties when the system changes from one state into another. A gas enclosed in a cylinder by a piston is a system; and so also is a mixture of chemical compounds in a beaker. By properties we mean such attributes as volume, pressure, temperature, density, energy, refractive index, etc. By a change of state we mean that we start with the beaker containing the reactants A, B, C, ..., which is the initial state, and that a chemical reaction occurs which yields the products D, E, F, ..., representing the final state. Or we may start with a given mass of gas in the cylinder, at volume V_1, pressure P_1, temperature T_1, being the initial state, and then by moving the piston we may bring it to V_2, P_2, T_1, being the final state.

Now, as Boyle first observed, if a given mass of gas is maintained at constant temperature there is a constant relationship between P and V. If we increase the volume from V_1 to V_2 the pressure falls from P_1 to P_2, and this is true no matter how we bring about the change of volume from V_1 to V_2. We can move the piston fast or slowly, we can overshoot V_2 and then come back to it, but provided we allow sufficient time for equilibrium to be established—in the present case, provided we allow sufficient time for the gas to regain the temperature of its surroundings after a sudden change in volume—we invariably find that the pressure takes the same value P_2. If we measured the density or the refractive index instead of the pressure we would find the same thing, namely that their values depended solely upon the volumes

V_1 and V_2 and were independent of the manner in which the volume was changed.

It is a matter of practical experience, verifiable in the laboratory, that if we specify the mass of the gas and, in addition, its volume and temperature, we thereby also determine the values of its pressure, density, refractive index, etc. Or, if we specify its volume, density and temperature we thereby determine its pressure, mass, refractive index, etc. These observations are generalised in the concepts of state and properties (also known as functions of state). To specify the state of a gas we have to specify any three independently variable properties, and thereby all its other properties are determined; and the properties of the gas—or of any system—are dependent upon its state only and are *independent of the path* by which that state is reached.

This is one of the most important principles of thermodynamics and its implications are far reaching. It means that thermodynamics is concerned with ends and not with means. When applied to a chemical reaction it means that thermodynamics takes cognisance of the initial state of the reactants before the reaction starts and of the final state of the products when the reaction is over; it has no interest in, and nothing to tell us about, the mechanism of the reaction. Whether the reaction is a simple one, or whether it involves a series of component reactions, is of no significance; only the initial and final states are taken into account. By the limitation of its scope in this direction thermodynamics greatly increases the power of its methods in others. The precise mechanisms of many chemical reactions are still problematical, and by having nothing to do with these uncertainties thermodynamics can preserve the rigour of its arguments. Starting from the first and second laws and proceeding purely by deduction it provides us with a great range of mathematical relationships which connect the properties of systems and enable us to calculate the values of properties which may be very difficult to measure in certain circumstances.

The work of Joule, establishing the mechanical equivalent of heat, was generalised in the principle of the conservation of energy, according to which energy can neither be created nor destroyed but can be changed from one form into another. The first law of thermodynamics is very nearly the same thing as a statement of the principle of the conservation of energy.

Most forms of energy are in theory interconvertible without restriction. We can take a certain quantity of electrical energy, apply it to an electric motor and make it raise a weight; then we can allow the weight to fall, driving the electric motor as a dynamo, and recover, at least in theory, all the electrical energy we started with. This two-way interconvertibility does not extend to heat as a form of energy. If we convert mechanical energy (i.e. work) into heat, as in Joule's churn, we cannot, even in theory, recover all the mechanical energy we put in. Nor can we raise the weights by lighting a bunsen under the churn. There are important restrictions upon the conversion of heat into other forms of energy, and it is with these restrictions that the second law of thermodynamics is concerned. The second law has been stated in various forms; in one of the simpler forms it asserts that it is impossible *continuously* to convert heat into work within a system which is at uniform temperature. In order to convert heat continuously into work we have to use some sort of heat engine, such as a steam engine with a boiler and a condenser. A heat engine operates by taking heat from a source (the boiler), converting some of it into work and rejecting the rest to a sink (the condenser); and this is only possible if the sink is at a lower temperature than the source. From the first and second laws the following very important deduction can be made concerning the efficiency of heat engines. If T_1 is the absolute temperature of the source and T_2 is the absolute temperature of the sink and if we define the efficiency, in percent, as

$$\frac{\text{amount of heat converted into work} \times 100}{\text{amount of heat taken from source}},$$

then

$$\text{maximum possible efficiency} = \frac{100\,(T_1 - T_2)}{T_1}.$$

This simple relationship tells us at once (1) that we cannot continuously convert heat into work in any system which is at uniform temperature, i.e. $T_1 = T_2$, and (2) that we can never convert all the heat into work unless T_2 is maintained at absolute zero; in any real situation 100% efficiency is unattainable.

All forms of energy other than heat are interconvertible with efficiency which can be theoretically 100%; but heat is unique in respect of this restriction upon its conversion into other forms of energy. For this reason heat is less useful than other forms of energy and we often speak of other forms of energy being 'degraded' into heat.

Why is it that Nature places these restrictions upon the conversion of heat into other forms of energy? To understand this we have to leave the macroscopic world of steam engines for the microscopic world of atoms and molecules, and this means leaving for a moment the domain of classical thermodynamics. The quality of hotness is an expression of the random movements of atoms and molecules, known as thermal agitation. Let us consider, at the microscopic level, what happens when mechanical energy is converted into heat, as when a rifle bullet is brought to rest in a target. First, we place the bullet in the rifle. Not being at absolute zero of temperature the atoms of the bullet are in thermal agitation, which consists in vibrations about fixed positions; the directions in which these vibrations occur are completely at random. Now we press the trigger and some of the chemical energy of the cordite is used to give kinetic energy to the bullet and sends it flying towards the target. If we examine the bullet in flight we find that its atoms are still in thermal agitation, but now in addition each atom has a net movement in the direction of flight. In so far as all the atoms have the same net movement in the same direction we may recognise that the affairs of the bullet show a certain orderliness which they did not show before. Then the bullet strikes the target and is brought to rest, its kinetic energy being converted by friction into heat. The atoms of the bullet are now in more violent thermal agitation than they were before, but their movements are completely at random; there is no orderliness, no net movement in any particular direction. Now it is not impossible that by pure chance all the atoms of the bullet would happen to move in the same direction at the same moment, in which case the bullet would take off again, some of its heat energy having spontaneously reverted to kinetic energy. It is not impossible, but is so wildly improbable that we may safely assume that it will not happen. Classical thermodynamics says simply that it just will not happen, for classical thermodynamics deals only with systems containing very large numbers of atoms and molecules.

Heat has been described as disordered energy. In its most general form the second law of thermodynamics asserts the improbability of order arising spontaneously out of chaos.

10

FREE ENERGY

In a general sort of way one thinks of the energy of complex organic compounds, which so many organisms make use of, as 'chemical energy', as energy residing in the chemical bonds which hold the atoms together in the molecule. But if we try to calculate how much energy would become available as a result of some particular change in a large molecule we come up against the difficulty that we do not know enough about chemical bonds to make any useful calculations upon this basis, except in certain very simple cases. And when one opens a textbook of biochemistry one does not find much reference to 'chemical energy'; instead, all discussion of energy transactions is likely to be in terms of 'free energy'. What then are we to understand by 'free energy'?

Let us consider first a simple chemical reaction: dilute hydrochloric acid is allowed to act upon excess of calcium carbonate in a vessel which is open to the atmosphere (constant pressure) and immersed in a water bath (constant temperature). The reaction proceeds with the evolution of carbon dioxide until the acid has been neutralised and a state of equilibrium has been reached. If we test the solution at equilibrium we will find that it is slightly alkaline. Since, in the absence of restraint, the reaction is associated with an increase in volume, owing to the evolution of gas, it follows from the principle of le Chatelier that an increase of pressure will oppose the progress of the reaction. If we allow the reaction to proceed in a closed vessel (constant volume) pressure will build up inside it and if, when equilibrium has been reached, we test the solution we find that it is less alkaline than when the reaction vessel was open. Similarly, if we allow the reaction to proceed in a Dewar flask any heat produced will be retained, the temperature will rise and again the state of equilibrium will be displaced. All this is just to remind ourselves that the state of the reactants and products at equilibrium depends upon the conditions which we impose during the course of the reaction.

When a chemist or a biochemist speaks of the change in free energy associated with a chemical reaction it is to be understood that the reaction takes place at *constant temperature and pressure*. These are the conditions which prevail when a chemical reaction proceeds in a vessel which is open to the atmosphere and which is kept at constant temperature by being immersed in a water bath. These are also the conditions, by and large, which prevail within the bodies of organisms. In an energy-yielding reaction the energy may be manifested in various forms—as mechanical energy resulting upon increase in volume, as heat, as the electrical energy of electric cells, etc. But if the energy-yielding reaction takes place at constant temperature and pressure two of these forms of energy will escape from the system into its surroundings. Any energy produced in the form of heat, which might otherwise raise the temperature of the system, will simply be conducted into the water bath; any gas evolved, which in a closed vessel would raise the pressure, will simply push back the atmosphere, doing work upon it. To repeat, any energy produced in the form of heat or work (change in volume) must escape from the system and is therefore not 'free' in the sense that it is not freely available to be applied to other purposes within the system. Energy arising in any form other than heat and work (change in volume) is free energy and is available for other purposes. The free energy yield of a reaction, then, is the amount of energy which becomes available within the system when the reaction proceeds at constant temperature and pressure.

The free energy of a system, denoted† by F, is a property of the system, which means that it depends only upon the state of the system and not upon the path by which that state was reached. When the system undergoes a change of state, as it does when a chemical reaction takes place, there may be, and usually is, a change in free energy. It is changes in free energy, rather than the absolute free energies of reactants and products, that we are primarily concerned with. A change in free energy is denoted by ΔF.

Energy can be expressed in a variety of different units—calories, ergs, kilowatt-hours, etc. By convention, free energy is expressed in calories. If the reaction makes free energy available the sign of ΔF is, by convention, negative—the reactants lose that amount of free energy in becoming the products. A reaction for which ΔF is negative is known as an exergonic reaction, and exergonic reactions tend to proceed spontaneously towards a state of equilibrium. A reaction for which ΔF is

† In most textbooks of biochemistry free energy is denoted by F. In most textbooks of physical chemistry free energy is denoted by G, short for 'Gibbs' free energy' after Willard Gibbs, the founder of chemical thermodynamics.

positive is known as an endergonic reaction and will not proceed unless energy is supplied to it.

The progress of a chemical reaction, and the free energy change associated with it, depend upon the concentrations of the reactants and products as well as upon the particular conditions of temperature and pressure which we choose to impose. In order to compare one reaction with another it is therefore necessary to specify all the relevant conditions. The choice of a standard set of conditions for reference purposes is of course a purely arbitrary matter, but it makes things easier if we choose to specify our standard conditions so as not to be too dissimilar from the conditions under which we are accustomed to work. In biochemistry the standard conditions are, by convention: a temperature of 25° C, a pressure of 1 atmosphere, pH 7, and reactants and products all at unit concentration, i.e. one mole/litre. Having laid down these conditions we then define the standard free energy change of the reaction, $\Delta F'$, as the amount of free energy involved in the conversion of one mole of some specified reactant into product.

At this stage an example may be helpful. The equation for the reaction of the Daniell cell is

$$Zn + CuSO_4 \rightarrow Cu + ZnSO_4.$$

If we look at this reaction from the point of view of zinc, the significant feature is the conversion of metallic zinc into zinc ion. Let us suppose that we wish to find the standard free energy change when one mole of metallic zinc is converted into one mole of zinc ion in a Daniell cell. From the work of Faraday we know that the electrolysis or electrodeposition of one equivalent of monovalent element requires 96,500 coulombs. Since zinc is bivalent, the conversion of one mole of metallic zinc into one mole of zinc ion requires $2 \times 96{,}500$ coulombs. With a potentiometer we now measure the voltage across the cell; suppose we find that it is 1·1 volts. Then the electrical energy which is produced when one mole of zinc is converted into one mole of zinc ion is $2 \times 96{,}500 \times 1{\cdot}1 = 213{,}000$ volt-coulombs, which is equivalent to 51,000 calories. Assuming that we measured the voltage across the cell under the agreed standard conditions, i.e. 25° C, 1 atm. pressure, etc., etc., then for the reaction in the Daniell cell

$$\Delta F' = -51{,}000 \text{ cal./mole of zinc.}$$

In the case of an electric cell we are able to measure free energy changes directly, since the free energy can be realised as electrical energy. But such cases are few and far between. For the most part we have to obtain values of $\Delta F'$ indirectly, making use of the relationships between free energy and other properties, which thermodynamics provides us with. One of the most useful of these relationships expresses $\Delta F'$ in terms of K, the equilibrium constant of the reaction. More often than not even more indirect methods have to be resorted to, with increasing liability to cumulative error. Some of the listed values of $\Delta F'$ are not claimed to be more accurate than $\pm 10\%$.

Further to the relationship between $\Delta F'$ and K, just mentioned, the figures in Table 10.1 will serve as a guide to the interpretation of the values which $\Delta F'$ can assume.

Table 10.1

Values of $\Delta F'$ (cal./mole of some specified reactant)	Conditions at equilibrium
0	Concentrations of products equal to concentrations of reactants
−1000	Products exceed reactants by ×5
+1000	Reactants exceed products by ×5
−4000	Products exceed reactants by ×1000
+4000	Reactants exceed products by ×1000
−8000	Products exceed reactants by ×1,000,000
+8000	Reactants exceed products by ×1,000,000

An equation which we shall encounter many times is the equation for the oxidation of glucose

$$C_6H_{12}O_6 + 6O_2 \rightleftharpoons 6CO_2 + 6H_2O.$$

For this reaction $\Delta F' = -685{,}600$ cal./mole of glucose. Alternatively, if oxygen is to be the specified reactant, $\Delta F' = -685{,}600 \div 6 = -114{,}300$ cal./mole of oxygen. With such a high negative value of $\Delta F'$ the concentrations of reactants at equilibrium must be infinitesimal and it seems scarcely meaningful to contemplate a state of equilibrium or to write such a reaction as reversible. Yet in theory an equilibrium does exist, and from a thermodynamic point of view the reverse of this reaction is what takes place in photosynthesis.

The importance and usefulness of a knowledge of free energy changes become apparent when we come to deal with coupled reactions. It is easy to see how two reactions can be coupled if the products of one reaction become the reactants of the other:

$$A+B \rightleftharpoons C+D$$
$$D+E \rightleftharpoons F+G.$$

But even when the formal chemical equations (representing the existing state of knowledge) for the two reactions do not indicate that they have any reactants or products in common, the reactions can often be observed to be stoichiometrically related when they are allowed to take place in the same vessel in the presence of appropriate catalysts. Probably this means that they do have reactants and products in common; but exactly how the coupling comes about at the molecular level is no concern of thermodynamics—it is enough for thermodynamics that coupling exists. What thermodynamics can do is confidently to predict, on the basis of free energy changes, what the equilibrium state of the coupled reactions will be, and therefore in which direction the sequence will tend to proceed; and this is why a knowledge of free energy changes is of use in biochemistry.

Chemical reactions in living organisms are characteristically organised in sequences which we call metabolic pathways. A metabolic pathway generally consists of several reactions coupled in series. Such a situation may be represented as below, the $\Delta F'$ values all being calculated with respect to one and the same reactant (or product), say, cal./mole of A entering the sequence.

$$A+B \overset{+2000}{\rightleftharpoons} C+D \quad (1)$$
$$D+E \overset{-6000}{\rightleftharpoons} F+G \quad (2)$$
$$G+H \overset{-4000}{\rightleftharpoons} I+J \quad (3)$$
$$J+K \overset{+3000}{\rightleftharpoons} L+M. \quad (4)$$

Suppose that we know the value of $\Delta F'$ for each individual reaction. How can we find out what will happen when the reactions are coupled together in sequence? All we have to do is to take the algebraic sum of the individual values of $\Delta F'$. It works out at -5000 cal./mole of A in the case considered above. The whole sequence will therefore tend to move spontaneously from left to right and at equilibrium the concentrations of the products L and M will greatly exceed the concentrations of the reactants A and B.

A biochemist is interested in the free energy changes of reactions because they can provide him with a rough check upon the correctness of his ideas about metabolic pathways. No system of coupled reactions will proceed spontaneously unless $\Delta F'$ is negative for the system as a whole. If he is contemplating some possible scheme for a sequence of reactions and finds that it has a large positive value of $\Delta F'$, the chances are that the scheme is wrong. But the check is only a rough one by reason of the uncertainties attaching to values of $\Delta F'$ and to the concentrations of reactants and products within the cell.

As has been stated, a system of coupled reactions will proceed spontaneously only if $\Delta F'$ is negative for the sequence as a whole. In the hypothetical example considered above part of the free energy made available by the exergonic reactions (2) and (3) is used to provide energy for the endergonic reactions (1) and (4). What happens to the free energy, 5000 cal./mole of A, which is left over? The answer is that it is degraded into heat and is lost from the system to its surroundings.

It will perhaps be helpful if at this stage we recapitulate the main features of this concept of free energy. Free energy is energy which becomes available within the system when a reaction proceeds under conditions of constant temperature and pressure. Free energy is a property, depending only upon the state of the system and being independent of the path by which that state is reached. Starting with a fully charged electric cell we may discharge it completely either by connecting it to an electric motor or simply by short-circuiting it. The state of the cell when it is completely discharged is the same irrespective of the way in which we discharge it, and therefore the free energy change, ΔF, is the same. In the first case we obtain some of the free energy in the form of work done by the electric motor; in the second case all the free energy is degraded into heat. When we assign a (negative) value to the free energy change of a reaction what we mean is that this is the maximum amount of energy which we can obtain from the reaction taking place within a system maintained at constant temperature and pressure. This is not a guarantee that we are going to get the full amount in a useful form. Any that we fail to get will be degraded into heat and will escape from the system. The (negative) value assigned to ΔF therefore tells us not how much useful energy we will actually get, but *only how much useful energy will become available* as a result of the reaction.

In practice not only chemical reactions but all conversions of energy are accompanied by the degradation of useful energy into heat. Looking back to p. 135 we see that the theoretical maximum efficiency of a heat engine is determined by the temperatures of the source and of the sink. A modern power station uses superheated steam at, say, 550° C and discharges it to a condenser at, say, 30° C. Its theoretical maximum efficiency is

$$\frac{100\,(823-303)}{823} = 63\,\%.$$

The efficiency achieved in practice is more like 35 %. This difference is due to various losses which good design can reduce but cannot eliminate altogether. Some heat is lost even from lagged pipes and work is all the time being degraded into heat by friction in the bearings, by turbulence in the flow of steam, and so on. The theoretical maximum efficiency can only be approached by imposing conditions which would not only be impossible to achieve in practice but which would also be incompatible with the purpose for which a power station is built. What we expect from a power station is not just energy—we want power, that is, energy supplied at a finite (and in this case very considerable) rate. Thermodynamics tells us that we can only attain the theoretical maximum efficiency if everything happens infinitely slowly, and this of course is unacceptable. A power station is designed not for the most efficient conversion of heat into electrical energy, but, ultimately, for the most economical conversion of coal into electric power, even though this will mean accepting some loss of energy. Similar considerations apply in other energy conversions. In an electric motor some of the electrical energy is degraded into heat in the resistance of the windings and some of the work is degraded into heat by friction in the bearings. Failure to attain the theoretical maximum efficiency is no reflection upon the designer; it is the inevitable consequence of a requirement for power rather than for energy.

Equally, if we ask how we can capture all the free energy made available by a chemical reaction the answer is the same—only by allowing the reaction to proceed with infinite slowness. The degradation of some of the free energy into heat is the price we have to pay if we want to keep things moving. The free energy which is degraded into heat is used, as some writers put it, to 'drive' the reaction.

11

ENERGY-RICH BONDS

An organism's stores of energy are usually laid down in the form of insoluble substances like starch or glycogen. These stores have to undergo rather elaborate processing before the energy is immediately available for use. In the higher animals and plants the stores are first mobilised as soluble sugars and in this form they can be distributed by blood vessels or sieve tubes. Within the cells the sugars are oxidised to water and carbon dioxide liberating the stored energy. Just as soluble sugars are the form in which the stored energy can be passed from one organ of the body to another, so likewise within the cell we find special molecules whose function seems to be to act as energy carriers between the sites of intracellular processes. Of these energy carriers by far the most important is adenosine triphosphate, ATP. The full structural formula of ATP is given in the Appendix (p. 321). For the present we shall represent it as

```
                     O      O      O
                     ‖      ‖      ‖
adenosine—O—P—O—P—O—P—OH
                     |      |      |
                     OH     OH     OH
```

ATP is rather unstable and in the presence of dilute mineral acids (or a suitable enzyme) it undergoes hydrolysis, the third (or terminal) phosphate group being removed, leaving adenosine diphosphate, ADP.

$$\text{ATP} + H_2O \rightleftharpoons \text{ADP} + H_3PO_4.$$

For this reaction $\Delta F'$ is -8900 cal./mole. A similar value is found for the hydrolysis of ADP whereby the second phosphate is removed leaving adenosine monophosphate, AMP. But the remaining phosphate group is less easily removed by hydrolysis and in this case $\Delta F'$ is about -3000 cal./mole.

The terms 'energy-rich bond' and 'energy-poor bond' were introduced in 1941 by Lipmann to underline these differences, and the symbol ~ was used to indicate an energy-rich bond. Using this symbol we can then write ATP as

$$\text{adenosine}—Ⓟ \sim Ⓟ \sim Ⓟ,$$

where Ⓟ symbolises a phosphate group. Other types of energy-rich bond are known, but these we shall not have to consider.

The formula for ATP, as written just above, is in a form of shorthand which has been of great help in describing metabolic pathways. But it has also given rise to some misunderstanding. Unfortunately bond energy can mean one thing to a physical chemist and another thing to a biochemist. When the physical chemist talks about the bond energy of the hydrogen molecule he means the energy which would be required to pull the hydrogen atoms apart. When the biochemist speaks of the energy-rich bonds of ATP he does not contemplate that the terminal phosphate group is to be seized and torn off from the rest of the molecule—that and nothing more. He contemplates that the valencies exposed by the rupture of the bond are to be satisfied by the elements of water. It is important not to be misled by the use of the symbol ~ into believing that the free energy to which it refers actually resides in this one bond. When the biochemist ascribes an energy value to a bond he is in fact quoting the standard free energy change of a reaction, hydrolysis, as a result of which not only is this particular bond broken, but also various other intramolecular changes may be involved, all having influences upon the free energy change. Let us remember too, that free energy is a concept of thermodynamics which is not tied up with any interpretation of molecular structure.

At this point an example may perhaps help. In the metabolic pathway of glycolysis (p. 175) there occurs a reaction whereby glyceric acid 2-phosphate is converted into phospho-*enol*-pyruvic acid.

$$\begin{array}{c} CH_2OH \\ | \\ HC—O—Ⓟ \\ | \\ COOH \end{array} \rightleftharpoons \begin{array}{l} CH_2 \\ \| \\ C—O \sim Ⓟ + H_2O \\ | \\ COOH \end{array}$$

glyceric acid 2-phosphate   phospho-*enol*-pyruvic acid

$\Delta F' = -640$ cal./mole.

The phosphate bond of glyceric acid 2-phosphate is energy-poor, whereas the phosphate bond of phospho-*enol*-pyruvic acid is energy-rich. In this example changes in the molecule as a whole, changes which as represented in conventional structural formulae take place at some distance from the phosphate bond, can influence the latter's energy status. It is therefore better to think of energy as an attribute of the molecule as a whole rather than of any particular inter-atomic bond.

The symbol ~ indicates that the phosphate bond of phospho-*enol*-pyruvic acid is energy-rich, which is another way of saying that it is notably susceptible to hydrolysis. The terminal phosphate bond of ATP is also notably susceptible to hydrolysis and is therefore represented by the same symbol. When the phosphate group of phospho-*enol*-pyruvic acid is transferred to ADP our shorthand might seem to suggest that the symbol ~ goes with it, as though it were some kind of hook. But this is not the way to look at it. The energy-richness of a phosphate bond is not of the nature of private property, to be carried around; it is more of the nature of status which is conferred upon the bond by the society of atoms in which it happens to find itself.

12

OXIDATION

Most, though by no means all, of the energy-yielding reactions within organisms are of the nature of oxidations, and it will be convenient to make this an occasion to develop the theme of oxidation.

The addition of an atom of oxygen to a molecule is not, of course, the only way of oxidising it. It may also be oxidised by having an atom of hydrogen removed from it. More generally, it may be oxidised by having an electron removed from it, as in the oxidation of ferrous ion to ferric ion.

$$Fe^{++} \rightleftharpoons Fe^{+++} + \epsilon.$$

Oxidising agents and reducing agents can be thought of in terms of the readiness with which they gain or lose electrons. Hydrogen atoms readily part with electrons and for this reason nascent hydrogen is a very powerful reducing agent. Oxygen atoms readily gain electrons, so oxygen is a powerful oxidising agent. We can generalise still further. Consider an electron in an electric field. To move the electron away from the positive pole in the direction of the negative pole we have to do work upon it and as a result it gains potential energy; if we allow it to move spontaneously towards the positive pole it loses energy which then becomes available for other purposes. Now consider an electron in orbit around the nucleus of an atom. In order to move it into an orbit of greater radius, i.e. away from the positively charged nucleus, energy must be supplied, and if enough energy is supplied the electron may be removed completely from the field of the atom. The atom is then oxidised. In aqueous solution relatively little energy is needed to remove an electron from a hydrogen atom, but much energy is needed to remove an electron from an oxygen atom. Conversely, if an electron finds its way from a hydrogen atom to an oxygen atom it will yield energy during the process; it is travelling, as it were, towards the positive pole of an electric field.

Oxidation and reduction can therefore be looked upon as the loss or gain of electrons; and this line of thought leads on to the possibility of following oxidative processes by means of electrical methods of some appropriate kind.

Suppose that we take a vessel containing a solution of $FeCl_2$ and $FeCl_3$ in equal concentrations and that we furnish it with two electrodes, one of which is a bright platinum wire and the other any kind of reference electrode, such as a calomel electrode—see Fig. 12.1 and legend. If we connect the electrodes to a potentiometer we will find that the

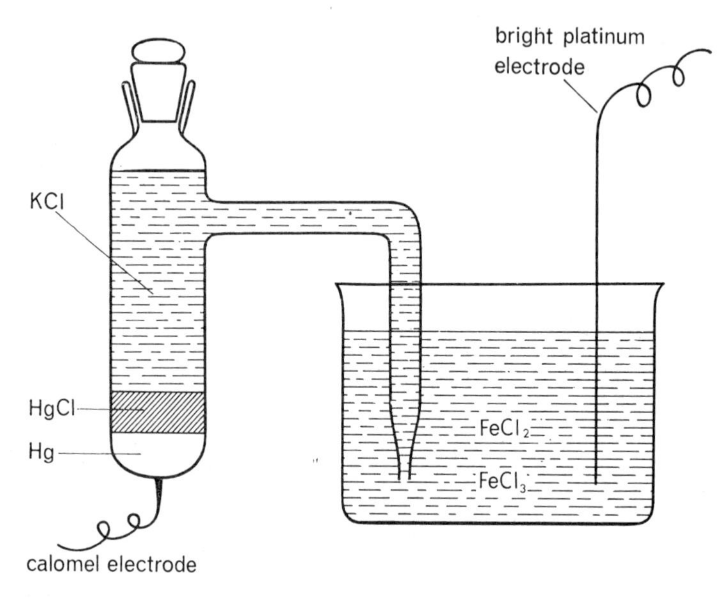

Fig. 12.1. Apparatus for measuring the oxidation-reduction potential of a mixture of $FeCl_2$ and $FeCl_3$.

We measure the potential between a bright platinum wire and some kind of reference electrode. As reference electrode we could use any electrode formed of a metal in contact with its own ions; the calomel electrode is basically mercury in contact with mercurous ions.

In physical chemistry potentials are quoted on the basis that the reference electrode is a standard hydrogen electrode, that is, a platinum wire coated with platinum black, saturated with hydrogen at 1 atm. pressure and in contact with a normal solution of hydrogen ion ($pH = 0$). But the hydrogen electrode is inconvenient to use. We therefore use a calomel electrode, whose potential with respect to the standard hydrogen electrode is $+0{\cdot}24$ volts, and correct our measurements accordingly.

platinum wire is positive with respect to the calomel electrode by about 0·5 volts. If we increase the concentration of $FeCl_3$ relative to $FeCl_2$ the platinum wire will become more positive; conversely, if we decrease the concentration of $FeCl_3$ it will become less positive. We can think of this in the following way: Fe^{+++}, being 3 electrons short, has a greater affinity for electrons than Fe^{++}, which is only 2 electrons short; conse-

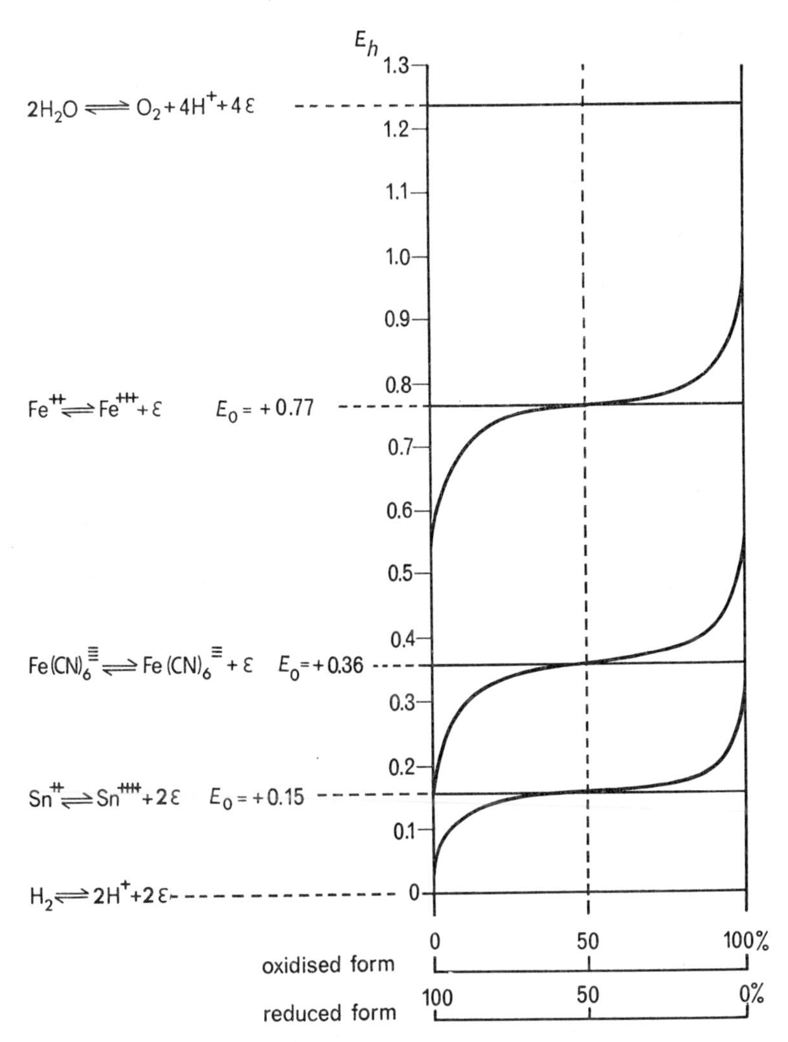

Fig. 12.2. The relation between oxidation-reduction potential and relative concentration of oxidised and reduced forms.

The figures relate to the following conditions of measurement: the reference electrode is a standard hydrogen electrode (see legend to Fig. 12.1) and the solutions are all at 25° C, 1 atm. pressure and pH 0. This is implicit in the use of the symbol E_h.

Each system may be characterised by quoting the value of the potential under conditions when the oxidised and reduced forms are present at equal concentrations; these values are designated E_0. A system having a more positive value of E_0 is capable of oxidising a system having a less positive value of E_0. Thus $K_3Fe(CN)_6$ is capable of oxidising $SnCl_2$ but will reduce $FeCl_3$.

quently the more Fe^{+++} relative to Fe^{++} in the solution the greater will be the tendency of the solution to take electrons from the platinum, leaving it with a positive charge. The important point for our present purposes is not the mechanism of the reaction but the fact that the relative concentrations of Fe^{++} and Fe^{+++}—that is, the 'oxidation status' of the solution—can be measured in terms of electrical potential difference. Potentials measured in this way are called oxidation-reduction potentials. The relationship between oxidation-reduction potential and relative concentrations of oxidised and reduced forms is shown in Fig. 12.2 for a few familiar inorganic reactions.

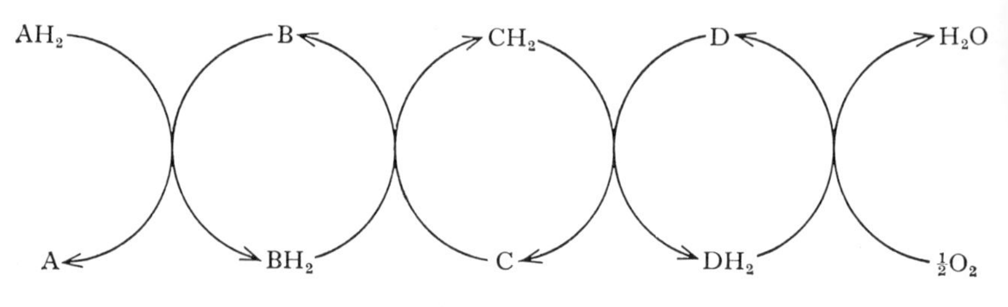

Fig. 12.3

Oxidation-reduction potentials are of great importance in biology for the following reasons. First, they provide a convenient way of defining the status of a system in the oxidation-reduction hierarchy, since from the value of E_0 (see legend to Fig. 12.2) for the system we know at once what it will oxidise and what it will reduce. Secondly, thermodynamics provides us with a relationship between E_0 and $\Delta F'$, whereby the free energy change of a reaction can be calculated from measurements of its oxidation-reduction potential. Thirdly, oxidation-reduction potential can be continuously recorded and we can thereby follow the progress of reactions. Fourthly, oxidation-reduction potentials can be measured on a very small scale, with wires down to 1 μ in diameter, making it possible to follow reactions in living tissues.

Most of the energy used by animals comes from the oxidation of hydrogen by oxygen to form water. The hydrogen enters the body incorporated in the organic matter of the food, from which it is split off by processes which we shall consider in detail at a later stage. Free hydrogen atoms, of course, are so reactive that they have virtually no free existence; having been split off from one molecule a hydrogen atom at once attaches itself to another, or alternatively it becomes a hydrogen ion and passes its electron to another molecule. What we find is that in organisms either hydrogen atoms or their electrons are passed from molecule to molecule until eventually they unite with oxygen. If AH_2 represents the molecule from which hydrogen atoms are split off and B, C, D, or X, Y, Z, represent the other molecules in the reaction chain, then we could represent the process either as Fig. 12.3 or Fig. 12.4.

In the first case the protons go along with electrons. In the second case the protons go into solution as hydrogen ions and the electrons travel down the chain alone; at the end they pick up two hydrogen ions out of the solution and join up with oxygen. We find both processes in

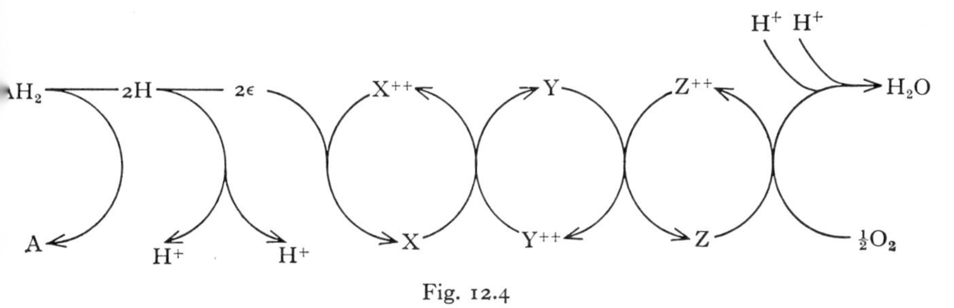

Fig. 12.4

organisms, and since the original reactants and the final products are the same the free energy change is the same for both processes.

$$AH_2 + \tfrac{1}{2}O_2 \rightleftharpoons A + H_2O,$$
$$\Delta F' \simeq -50{,}000 \text{ cal./mole of } H_2.$$

The actual value of the free energy change depends, of course, upon the nature of compound A, but what we usually find is that in organisms the oxidation of one mole of hydrogen results in the conversion of 3 moles of ADP into ATP, which at 8900 cal./mole each, accounts for 26,700 cal./mole of H_2.

Plants, on the other hand, derive their energy from the sun. Broadly speaking, what happens here is that the energy of sunlight is used to break up molecules of water,

$$H_2O \xrightarrow{\text{light}} \tfrac{1}{2}O_2 + 2H^+ + 2\epsilon.$$

The oxygen atoms combine to form molecular oxygen which is evolved, the hydrogen ions go into solution and the electrons are then captured by compounds such as X^{++} and brought into the energy-yielding sequence of reactions which result in the synthesis of ATP and which we shall consider in more detail later on.

13

ENZYMES AND CO-FACTORS

1. CATALYSIS AND ACTIVATION ENERGY

A high negative value of ΔF tells us that the reaction is likely to proceed spontaneously and that the products will greatly exceed the reactants at equilibrium, but it does not guarantee that the reaction will proceed with measurable speed. A well-known exergonic reaction is the combination of hydrogen and oxygen to form water, for which $\Delta F' = -120{,}000$ cal./mole of water; but this reaction does not proceed spontaneously at room temperature.

In order to understand why such a reaction fails to proceed we will make use of a hydraulic analogy. Let us suppose that we have a mass of water in a basin which is raised above the floor. By virtue of its height above the floor this mass of water has potential energy; but because the water is retained by the lip of the basin its potential energy is not immediately available. We must first raise the water so that it can cross over the lip—we must give it additional potential energy—before it can yield up the potential energy it already has; for example, we might take the basin and shake it so that some water would slop over the edge. In the same way molecular hydrogen and oxygen need energy before they will react. The two atoms of each molecule are held together by bonds of substantial strength and these have to be broken before other bonds can be established. To start the reaction we have to use an electric spark. This creates a very high temperature in its immediate vicinity and the violent thermal agitation thus engendered serves to break some of the bonds. The free atoms are then able to unite to form water molecules. This process is accompanied by the degradation of a large amount of free energy into heat, which raises the temperature further and serves to dissociate more molecules of reactants. Thus, once started, the reaction sweeps right through the system with explosive violence.

There is therefore a kind of energy barrier which has to be surmounted before the reaction can proceed. In the example just quoted we provided the necessary energy in the form of heat—this is very like shaking the basin; but in the biochemical context of constant temperature and pressure what we are interested in is the amount of free energy which is needed to get the reaction going. The important quantity is the free energy of activation, denoted by ΔF^*, and its relation to the net free energy change of the reaction, ΔF, may be represented as in Fig. 13.1, from which it can be seen that the free energy of activation is recovered when the reaction has taken place. The free energy of activation is difficult to measure directly, but once again thermodynamics comes to our aid with an equation which enables us to calculate the free energy of activation from the velocity constant of the reaction and its dependence upon temperature.

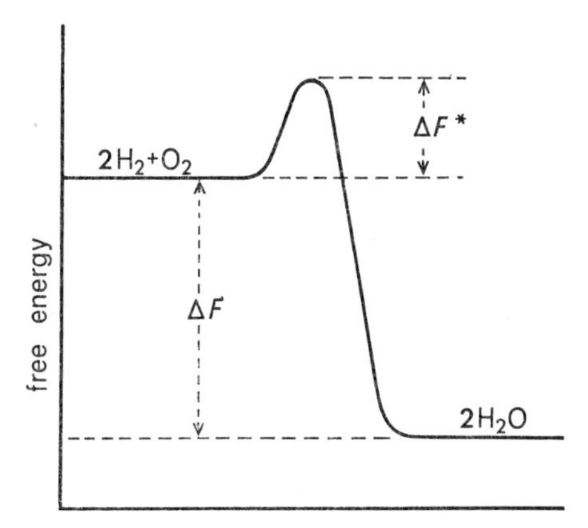

Fig. 13.1. To illustrate the relation between the free energy change of a reaction, ΔF, and the free energy of activation, ΔF^*.

Reactions which fail to proceed notwithstanding a high negative value of ΔF can often be persuaded to do so in the presence of a catalyst. From the thermodynamic point of view a catalyst is something which lowers the free energy of activation. Putting it another way, it lowers the energy barrier—or the lip of the basin—and so allows the reaction to proceed. From the physical point of view what probably happens is that the reactant combines temporarily with the catalyst and that as a result the energy of the reactant molecule is redistributed so that certain bonds become more liable to rupture by thermal agitation.

With this brief introduction to the nature of catalysis we will now go on to consider the catalysts found in living organisms.

2. ENZYMES

The word 'enzyme' was coined by Kühne in 1878 to replace the word 'ferment' about whose precise connotation much confusion had arisen.

'Enzyme', from Greek, means literally 'in yeast'. At the end of the nineteenth century the general consensus of opinion about the nature of the alcoholic fermentation of yeast was in support of the views of Louis Pasteur, who had stated his position in these words: 'I believe that alcoholic fermentation never occurs without simultaneous organisation, development, and multiplication of cells, or the continued life of cells already formed....' The subject of enzyme chemistry may be said to have been born in 1897, two years after Pasteur's death, when his views upon this subject were shown by Hans and Edouard Buchner to be wrong.

The Buchner brothers were not primarily interested in yeast from the point of view of its alcoholic fermentation. They were engaged in trying to prepare a cell-free protein for therapeutic purposes. Having obtained an extract of yeast cells, by grinding them with sand in a mortar and filtering off the debris, their next problem was how to preserve it; for since it was intended to be administered to patients the usual bactericidal agents were ruled out. An assistant suggested they should add large quantities of sucrose—make jam of it, in fact. This they did; but to their great surprise the jam quickly went bad, with the production of ethyl alcohol and carbon dioxide. In this way was opened the prospect of studying the chemistry of alcoholic fermentation in non-living, cell-free systems, and efforts were soon being made to isolate 'the enzyme' in pure form. It soon became apparent, however, that yeast juice did not contain one single enzyme, but many.

All the enzymes whose chemical composition has been investigated are proteins, and if anyone discovers an enzyme which is not a protein it will certainly create a stir. The methods which are used to separate and purify enzymes are the same methods as are used to separate and purify proteins. They are characterised by their mildness. Precipitation by concentrated ammonium sulphate ('salting-out'), or adsorption on to kaolin, all carried out under critical conditions of pH, are the sort of methods resorted to in the initial stages. Later on we might make use of the methods of chromatography and we might even attempt to obtain the enzyme in crystalline form. Crystallisation of protein, however, is by no means so good a test of purity as it is with substances of smaller molecular weight.

Being proteins, enzymes are susceptible to influences and agents which are known to affect proteins. They are rapidly denatured, and their catalytic properties destroyed, at temperatures of 50° C and over, and by the ions of heavy metals. Their catalytic activities are notably affected by pH. Enzymes differ from inorganic catalysts in several respects which are important to note.

First of all, enzymes are more efficient than inorganic catalysts such as platinum; that is to say, they effect a greater lowering of the free energy of activation with the result that the reaction goes more quickly. A classic example of this is the decomposition of hydrogen peroxide into water and oxygen. In the absence of a catalyst this reaction takes place very slowly; in the presence of platinum it goes faster and in the presence of the enzyme catalase it goes very fast indeed, as the figures in Table 13.1 show.

Table 13.1. *Decomposition of Hydrogen Peroxide*

	Free energy of activation (cal./mole)	Relative rate of reaction
No catalyst	18,000	1
Platinum	12,000	10,000
Catalase	2,000	10,000,000

Secondly, they show much greater specificity. There are many reactions which are known to be speeded up in the presence of platinum, but the catalytic activities of an enzyme are generally restricted to one particular reaction. This, however, is a matter of degree, for there is great variation between one enzyme and another in this respect. The lipase of pancreatic juice promotes hydrolysis of the glycerol esters of a wide range of fatty acids. At the other end of the scale is the enzyme urease which breaks down urea into carbon dioxide and ammonia; it is quite specific to this reaction and will not attack a urea molecule in which one of the hydrogen atoms has been replaced by a methyl group. Another example of high specificity is provided by the existence of two distinct classes of enzymes which deaminate (split off ammonia from) amino acids. Most of the amino acids can exist in two forms (stereoisomers), the one being the mirror image of the other.

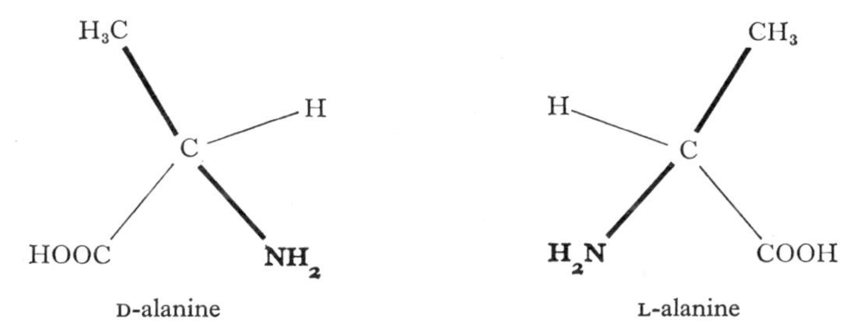

The enzyme which attacks D-alanine will not attack L-alanine.

Thirdly, they are less stable. A catalyst is supposed to be completely recoverable from the products of reaction. This is a criterion upon which enzymes fail to qualify, for their effectiveness slowly decreases with time, no doubt owing to denaturation in some form or other.

Much effort has been devoted to the study of enzyme action. It is generally agreed that the first stage must involve some kind of combination between the enzyme and the substrate molecule on which it acts. This is suggested by the effects of different concentrations of enzyme and substrate upon the velocity of the reaction. In a few cases in which the enzyme absorbs light of certain wavelengths it has been possible to show that in the presence of the substrate there is a change in the absorption spectrum of the enzyme, indicating a change of molecular structure as a result of the formation of a compound.

The catalytic activity of the enzyme resides in a part of the molecule known as an active centre. Information about the number of active centres in a single enzyme molecule can be obtained from the number of heavy metal ions required to put the enzyme molecule out of action. It is also possible, by carefully controlled use of digestive enzymes, to remove non-essential parts of the molecule and thus locate the active centres; but as yet no active centre has been fully described. Since the active centre is probably formed by amino acid residues on different parts of the polypeptide chain an adequate description would probably involve knowledge of the complete three-dimensional configuration of the protein.

The high specificity of enzymes to substrates, and their sensitivity to small changes in stereochemical configuration, suggest that in the union of enzyme and substrate there is some kind of point-to-point correspondence of structure, so that one particular group of atoms on the substrate molecule is held by a complementary group on the enzyme. High specificity could be achieved by having a large number of such points of contact irregularly spaced. The fact that enzyme activity is often inhibited by compounds which closely resemble the natural substrate may be quoted as circumstantial evidence. For example, the enzyme succinic dehydrogenase, which catalyses the conversion of succinic acid into fumaric acid (see Chapter 17) is inhibited by malonic acid.

This inhibition, unlike that due to heavy metals, is reversible and competitive; if we increase the concentration of succinic acid sufficiently the reaction will go forwards, even in the continued presence of malonic acid. The explanation put forward for this type of inhibition is that malonic acid is sufficiently similar to succinic acid to occupy the active site on the enzyme, excluding the rightful incumbent.

$$\begin{array}{c} COOH \\ | \\ CH_2 \\ | \\ CH_2 \\ | \\ COOH \end{array} \qquad\qquad \begin{array}{c} COOH \\ | \\ CH_2 \\ | \\ COOH \end{array}$$

succinic acid   malonic acid

This interpretation of enzyme-substrate specificity in terms of stereochemical configuration is entirely plausible, but it is important to bear in mind that it is speculation. No active centre is known in anything like the detail necessary to describe its stereochemical configuration. The lock-and-key diagrams, which one so frequently encounters, are symbolic only of our hopes for the future.

3. CO-FACTORS

One of the methods commonly used in the preparation of pure proteins is dialysis. The protein-containing solution is placed in a bag of some material—parchment, collodion, or nowadays usually cellophane—having submicroscopic pores of such a size that only relatively large molecules, of molecular weight 5000 or above, are retained by it. The outside of the bag is kept washed in running water. By this process all the substances of small molecular weight are leached out. In 1904 Harden and Young, investigating the enzymatic properties of yeast juice, found that after dialysis the juice lost its ability to ferment glucose. This ability could be restored by adding some boiled yeast juice which had not been dialysed. These observations showed that the ability of yeast juice to ferment glucose was dependent upon two factors: a heat-labile substance of high molecular weight, presumed to be a protein, and a heat-stable substance of low molecular weight which could be removed by dialysis. 'The enzyme' of yeast having by now received the name of zymase, the dialysable factor was referred to as co-zymase. This was the first indication of the need for something other than enzyme and substrate.

A study of the chemical nature of co-zymase was begun by von Euler in 1923. It soon became clear that whatever it was it was present in very small amounts. After some ten years of work it was established that adenine (a purine) and a pentose sugar formed part of the molecule. At about the same time Warburg, who was studying an enzyme present in mammalian red blood cells, found that his enzyme required some co-enzyme for its activity. In 1934 he and his associates isolated 20 mg of this co-enzyme from 250 litres of horse blood. It was found to be composed of 1 molecule of adenine, 2 molecules of pentose sugar, 3 equivalents of phosphoric acid and 1 molecule of nicotinamide, a derivative of pyridine. This announcement prompted von Euler to test for nicotinamide in his preparations of co-zymase, and sure enough the test was positive. The full structure of these co-enzymes took several more years to work out, but it was obvious from the start that they were not identical, for von Euler's co-enzyme contained only 2 equivalents of phosphoric acid. Both co-enzymes belong to a class of compound known as nucleotides (see Chapter 23 and Appendix, p. 318) which contain a purine, a pentose sugar and phosphate; and because the co-enzymes also contain pyridine they are called pyridine nucleotides. Von Euler's is diphospho-pyridine-nucleotide, generally known as DPN, and Warburg's is triphospho-pyridine-nucleotide, TPN.† Their full structural formulae are to be found in the Appendix, pp. 323–4.

Shortly after the completion of his first analysis Warburg further found that in the reaction which had originally attracted his attention, an oxidative reaction, the oxidation of the substrate was accompanied by the reduction of TPN. The reduced forms of DPN and TPN we shall denote by $DPNH_2$ and $TPNH_2$, representing reduction as the acceptance by the molecule of a pair of hydrogen atoms.

We will not pursue the history of these researches any further but will proceed at once to their outcome, which is that DPN and TPN are hydrogen carriers. As discussed on p. 150, many metabolic reactions, classed as oxidative, involve not the addition of an atom of oxygen but the removal of a pair of hydrogen atoms. The immediate fate of these hydrogen atoms is to be taken up by a hydrogen carrier; later, they may be used elsewhere in a reductive reaction or they may be oxidised to water. For each species of substrate molecule there seems to be a

† An international commission has recently decided that henceforward DPN will be known as 'nicotine adenine dinucleotide' (NAD) and TPN as 'nicotine adenine dinucleotide phosphate' (NADP).

specific dehydrogenase enzyme which can remove a pair of hydrogen atoms from it and pass them to a hydrogen carrier. Each dehydrogenase enzyme co-operates exclusively with one particular hydrogen carrier.† But since there are many more dehydrogenase enzymes than there are hydrogen carriers, each hydrogen carrier has a panel of dehydrogenase enzymes which it may be called upon to attend.

Out of this search for co-enzymes came the discovery of ATP. It was isolated from muscle by Lohmann in 1929, but at that time it was thought of not as an energy carrier but as a co-enzyme which participated in the production of lactic acid by muscle. To describe the steps by which its true significance came to be realised would involve an extensive digression into the chemistry of muscular contraction. The two most significant observations were, first, that a fibrous protein, actomyosin, prepared from muscle, would contract in the presence of ATP, and secondly, that purified actomyosin was also an enzyme which promoted the hydrolysis of ATP. Evidence then quickly accumulated to show that ATP was present in many other organisms—micro-organisms, plants and animals—and that it participated in reactions involving energy transfer, some of which will be described in later pages.

Before leaving the subject of co-factors in enzyme action two further points should be made. First, in addition to the organic co-factors already mentioned, a great many enzyme systems are only effective in the presence of traces of metallic cations—manganese, zinc, cobalt for example. Secondly, some enzymes, even when purified to the stage of crystallisation, are found to contain chemical combinations which are not characteristic of pure protein. This is also known to be the case for other physiologically important substances which are mainly protein in nature; in haemoglobin, for example, the molecule of haem (see Appendix, p. 329) is tightly bound to the protein globin. Such non-protein components are called prosthetic groups. The point to be made here is that the strength of the association is variable. Where the association is so strong that vigorous treatment is required to disrupt it we tend to speak of a prosthetic group; where the association is weak and the prosthetic group is detached by mild treatment we tend to speak of it as a co-factor. But whether the association is strong or weak we cannot but suppose that when the enzyme system is in action some very precise form of association is involved. This would apply equally to the metallic

† There are other hydrogen carriers besides DPN and TPN, notably the so-called flavin nucleotides (see Appendix, p. 325).

cations. The distinction between built-in prosthetic groups and free co-factors is probably not one that is worth pursuing.

In this book we shall not, in general, bother with the names of enzymes. They tend to be long and inelegant, referring to the reactants and products and ending in '-ase'. It may be assumed that all the reactions hereinafter mentioned are catalysed by specific enzymes unless the contrary is stated.

14

METABOLIC PATHWAYS

In Chapter 2 we discussed briefly the various ways in which organisms satisfy their requirements for energy. We saw that while some are able to capture the energy of sunlight or of simple inorganic reactions, others have to depend upon the energy locked up in complex organic compounds. This distinction, like so many others, is not absolute. For although in the long run plants depend upon the energy of sunlight, in the short run—during the hours of darkness—they depend, like animals, upon energy derived from the oxidation of the organic matter within their own bodies.

The food of a mammal consists of very large molecules—proteins, polysaccharides and neutral fats. In the alimentary canal these are broken down into amino acids, monosaccharides, fatty acids and glycerol, all of which are absorbed into the blood stream. But these processes of breakdown, though they are mildly exergonic, do not contribute any useful energy to the animal since they are not accompanied by any production of ATP; all the free energy is degraded into heat. Within the tissues of the body, notably in the liver, further breakdown occurs. Amino acids are deaminated, ammonia being split off, and the residues eventually become one or other of three carboxylic acids which we shall have more to say about later: oxaloacetic acid, α-ketoglutaric acid and acetic acid in the form of its compound with co-enzyme A (see Chapter 17, p. 184, and Appendix, p. 328). Long-chain fatty acids are oxidised at the β-carbon atom (the carbon atom next but one to the carboxyl group), a process which removes two carbon atoms at a time, again as acetyl co-enzyme A. Acetyl co-enzyme A may here be taken as the end-point of glycolysis, by which monosaccharides are broken down. Of these processes—deamination, glycolysis, β-oxidation—glycolysis is the only one which is directly concerned in the production of ATP. Further ATP is produced by oxidative decarboxylation, whereby carbon is oxidised to carbon dioxide. But by far the greatest production of ATP comes from the oxidation of the hydrogen atoms which are split off at various stages. These hydrogen atoms, in combination with DPN and other hydrogen carriers, enter into a complex reaction sequence involving the respiratory pigment cytochrome and are oxidised to water by atmospheric oxygen; this process is called oxidative phosphorylation.

A sequence of chemical reactions in organisms, connecting initial reactants with final products, is called a metabolic pathway. Metabolic

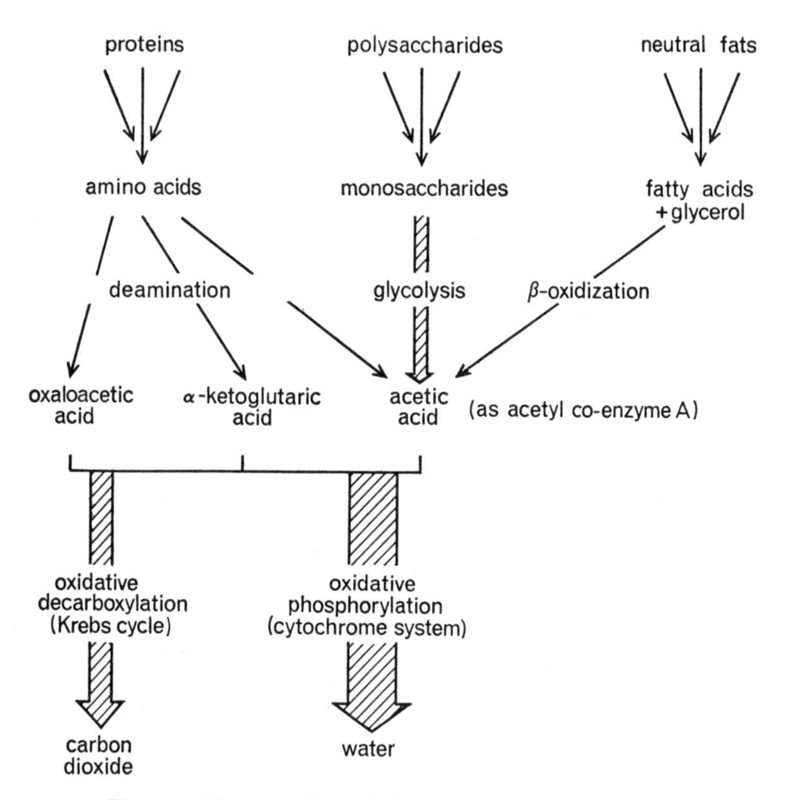

Fig. 14.1. The general metabolic pathways of the foodstuffs.

The heavy arrows indicate processes which are associated with the production of ATP. In deamination, glycolysis, β-oxidation and oxidative decarboxylation hydrogen atoms are removed from the substrates to DPN and are eventually oxidised in the cytochrome system. These subsidiary pathways have been omitted from the diagram to avoid undue complication.

pathways are usually very complicated. In a laboratory vessel we can bring about the oxidation of glucose into carbon dioxide and water merely by heating it, and no one feels it necessary to describe the process in more detail than to say simply that the hydrogen and oxygen come off as water and that the carbon is oxidised by atmospheric oxygen. But in organisms the oxidation of glucose, even by the most direct route, involves upwards of 20 separate reactions, each one of which is catalysed by a specific enzyme, and many of which involve the production of ATP.

Since metabolic pathways commonly start from diverse and complex molecules their initial stages show great variation. But as the large molecules are progressively broken down into smaller molecules the various pathways converge and their final common path is remarkable in being substantially the same, not only for all starting points but also for nearly all organisms. The metabolic pathways of oxidative decarboxylation (the Krebs cycle) and oxidative phosphorylation (the cytochrome system) have been demonstrated in organisms ranging from the higher animals and plants through protists to bacteria. What does this mean? Does it mean that these reactions occurring in mitochondria have been inherited from some common ancestor in the remote past? Or does it mean that for purposes of deriving useful energy from the oxidation of organic matter this is the most efficient system and so has been evolved many times independently? Questions like these are impossible to answer. Nevertheless, in view of the great resourcefulness of micro-organisms in exploiting chemical changes it is hard to believe that if a better method exists they would not have found it.

Not all organisms derive all their energy from oxidative processes involving atmospheric oxygen; and although strict anaerobes are not very common, a great many organisms including ourselves make use of energy derived from non-oxidative processes. In our muscles the conversion of glycogen to lactic acid

$$\underset{\text{glycogen}}{(C_6H_{10}O_5)_n} + nH_2O \longrightarrow \underset{\text{lactic acid}}{2nC_3H_6O_3}$$

is accompanied by the production of 3 molecules of ATP per 6-carbon unit of glycogen. This reaction enables us to engage in violent effort of short duration. The maximum sustained level of activity, as for example in a cross-country race, depends ultimately upon the ability of the heart and lungs to get oxygen to the muscles and to get carbon dioxide away. But in a 100 yards sprint the level of activity is much higher than could be sustained over a greater distance, and the energy for this extra activity comes from the breakdown of glycogen to lactic acid. The extra activity cannot be sustained because if it were the lactic acid would lower the pH of the blood to an intolerable level; and after the effort the runner breathes heavily for some minutes while the lactic acid is being removed. Following the same lines, the energy supply of yeast comes mainly from the breakdown of glucose into ethyl alcohol and carbon dioxide

$$C_6H_{12}O_6 \longrightarrow 2C_2H_5OH + 2CO_2,$$

for which atmospheric oxygen is unnecessary; and the various fermentative and putrefactive bacteria no doubt obtain energy from similar processes which do not involve the participation of molecular oxygen.

In the chapters to follow we shall be concerned with the tracing of metabolic pathways in more detail, but before we embark upon this it would not be out of place to take a glance at the methods of research which are used in this field of biochemistry. The biochemists of today mostly deal with reactions which can be followed in small glass vessels, but much of the early pioneer work involved experiments on whole animals, usually mammals. A typical experiment would be, for example, to feed a large amount of amino acid—glycine, say—to a dog, whereupon an increased urinary excretion of urea would be noted. The next step would be to discover what organ or organs in the body were responsible for the deamination of glycine and the synthesis of urea. We might remove the animal's liver and see if administration of glycine was still followed by the increased excretion of urea; or we might excise the kidney, keeping it alive for some hours under perfusion, and see what happened when glycine was added to the perfusing fluid. All experiments of this type are extremely laborious to conduct, and they often call for delicate surgery and skilled post-operative care. For these and for other reasons biochemists prefer to work not with whole animals or whole organs but with preparations of tissue which can be handled with less trouble and inconvenience.

When a piece of tissue is removed from the body its blood supply is abolished, and since it continues to use oxygen it will soon suffer from the lack of it. In making use of tissue preparations it is essential to use pieces of tissue which are so small that their oxygen requirements can be met by diffusion alone. This means that the tissue has to be cut up in some way, a process which inevitably destroys some of the cells. The least drastic method is to use tissue slices. The animal is killed by a blow on the head, the organ is removed as quickly as possible, and then with a sharp razor slices are cut about 1 cm square and 0·5 mm thick. They are immediately placed in oxygenated physiological saline

(Ringer's fluid). In tissue slices the great majority of the cells are intact and undamaged. Mincing is more destructive. Grinding with sand in a mortar produces what is called a *brei*, in which most of the cells are broken open. In homogenisation, as described on p. 73, the cells are completely disintegrated but their organelles, e.g. mitochondria, remain intact; a preparation made in this way is called a cell-free system. For the disintegration of subcellular particles and bacteria it is generally necessary to use ultrasonic vibrations.

Preparations made in these various ways are inevitably contaminated with bacteria and in the course of a few hours the bacteria may well have multiplied to such an extent that subsequent chemical changes are mainly due to their activities. Nowadays it is customary to add antibiotics such as penicillin; in the days before antibiotics it was usually considered risky to prolong an experiment beyond one or two hours.

What is now established as the independent discipline of biochemistry started life in the nineteenth century as physiological chemistry which was regarded as an appendage of medicine. The methods used were the methods of contemporary organic chemists. We tend perhaps to overlook the debt which we owe to the physiological chemists of the nineteenth century. It was they who established the chemical basis of living matter, extracting, precipitating, purifying, crystallising, analysing, listing the various chemical species which were to be found. It was slow work, not least because of the problem of separating single components out of highly complex mixtures, and biochemists continued to struggle with this difficulty up to the time of the Second World War. Then, almost overnight, the difficulty melted away with the introduction of chromatography. The principle of chromatography has already been described (pp. 27–9).

A second great technical advance which was made at about the same time was the introduction of the use of isotopes. The practicability of this method depends upon the fact that the isotopes of an element are chemically indistinguishable but can be detected in minute amount by physical methods. It therefore becomes possible to 'label' elements by means of isotopes; and such labelling does not affect the chemical reactions they enter into but enables us to trace their progress. Isotopes had indeed been used in experiments before the Second World War, so the idea was not entirely new; the new factor in the situation was that with the development of nuclear reactors it became possible to prepare a wide range of radioactive isotopes at relatively small cost.

Radioactive atoms, when they disintegrate, may produce radiation of three kinds: α-rays (nuclei of helium atoms), β-rays (high-energy electrons) and γ-rays (X-rays of short wavelength). For biochemical purposes we are mainly concerned with β-emitters, so we may reasonably restrict our consideration of instruments to the one which is most widely used for the assay of β-rays, namely the Geiger counter. The

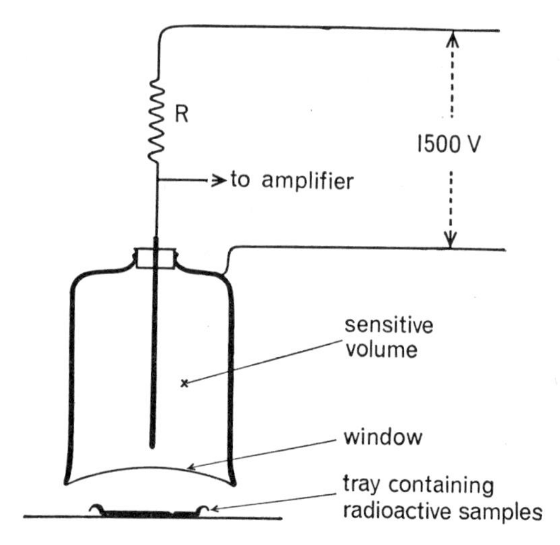

Fig. 14.2. The Geiger tube.

This diagram shows an end-window Geiger tube. The window is made of very thin and light material such as mica. The space (sensitive volume) contains gas at very low pressure. A potential of about 1500 volts is applied between the central electrode and the outer wall.

When a high-energy particle enters the sensitive volume it collides with molecules of gas and ionises them, more or less by knocking off electrons. The gaseous ions are at once accelerated towards the oppositely charged electrode and collide violently with other molecules, ionising them too. In this way the entry of the high-energy particle triggers a burst of ionisation. The current, however, has to flow through the resistance R and this causes the voltage across the electrodes to drop; the ions, no longer strongly accelerated, are able to regain their lost charges without further violence. The ionisation then dies away, the potential re-develops and the tube is ready to respond to another particle.

Geiger tube itself is shown in Fig. 14.2 and the principles of its operation are described in the legend to that figure. If radioactive material is placed close under the window of the Geiger tube, it emits radiation in all directions and some of this passes through the window into the sensitive volume. Each particle on passing through the gas causes a burst of ionisation and a brief voltage pulse is fed to the amplifier. After amplification the pulse is then made to operate some kind of electronic counting device which it would take too long to describe. The principle of the method is that the material to be assayed is placed on a tray in position close under the window and the number of pulses is counted over an interval of time, from which we obtain its count rate. A second identical tray containing a known amount of the same radioactive isotope is then placed in exactly the same position and its count rate is determined. After various corrections have been applied, which we need not go into, the amount of radioactive isotope in the experimental material is found by simple proportion.

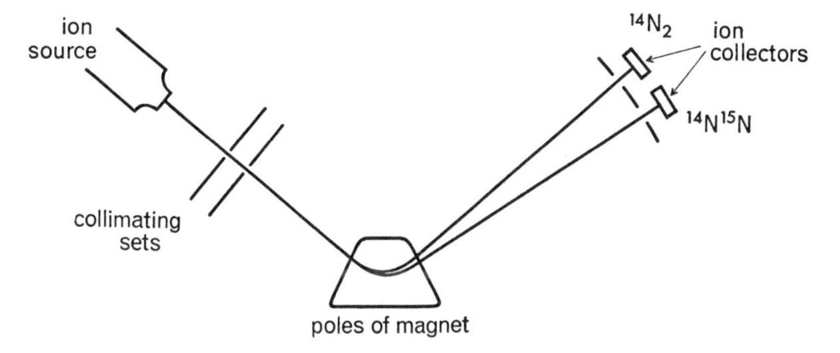

Fig. 14.3. The principle of the mass spectrograph.

The element—nitrogen, let us say—is provided in the form of a gas which contains the normal molecular species $^{14}N_2$ and the heavy molecular species $^{15}N^{14}N$ and $^{15}N_2$, though the last is so rare that it is generally disregarded. The ion source and all the various electrodes are built into a tube which is evacuated. As the gas is allowed to enter the tube at the ion source it is ionised by bombardment with a beam of electrons. The ions are then accelerated by an electrostatic field and made to pass between the poles of a magnet. The degree to which they are deflected depends upon their ratio of charge to mass, e/m; all three ionic species have the same charge, but their masses are respectively 28, 29 and 30. The different ionic species are thus separated and two of the ionic beams, one of $^{14}N_2$ and the other of $^{15}N^{14}N$, are brought to focus at different places where the ions are collected and assayed as ionic current. The ratio of the currents is the same as the ratio of the amounts of the two ionic species, whereby the amount of ^{15}N can be found as a percentage of the amount of ^{14}N.

The Geiger counter is relatively easy to use and forms part of the standard equipment of a biological research laboratory. By contrast, the mass spectrograph, which is the instrument mainly used for the assay of stable isotopes, is a more elaborate and expensive instrument and is much more difficult to use. Its principles are described in Fig. 14.3 and legend. The difficulties of operation stem from the fact that the gas is at appreciable pressure in the region where it is ionised and then has to traverse a highly evacuated region over which the accelerating and deflecting fields are applied. This calls for delicate adjustment of pumping rate—indeed, all the adjustments of slits and fields are highly critical.

Another method of separating stable isotopes by mass difference makes use of a very high-speed centrifuge; this method is described elsewhere (Chapter 23, p. 254).

When we speak of an atom as 'labelled' we mean that a certain proportion of the atoms, not necessarily all of them, are in the form of an isotope not found in nature. If these isotopes are radioactive they can be traced when present in exceedingly small amounts. The radioactive isotopes which are most used in biochemistry are ^{14}C, ^{3}H, ^{32}P and ^{35}S, these being among the most common elements in biological materials. These radioactive isotopes can be incorporated into organic molecules in various ways. For example, ^{14}C-glucose is made by allowing a culture of the unicellular alga *Chlorella* to carry out photosynthesis in an atmosphere of $^{14}CO_2$. If we wanted to discover the fate of glucose in the body of an animal we would inject it with radioactively labelled glucose; then, after a suitable time we would kill it, make extracts of its organs and by chromatographic separations and the like we would hope to be able to identify the compounds in which the radioactive label was present. In more refined experiments it is possible to synthesise organic compounds in such a way that the carbon atom at one particular position in the molecule is labelled; then if the radioactive carbon atom can be detected in another compound, and if its position in the molecule can be determined, we get a valuable indication of how the first compound is transformed into the second.

Unfortunately there are no useful radioactive isotopes of oxygen and nitrogen, and for these elements we have to make use of the stable isotopes ^{18}O and ^{15}N.

How then would we initiate a biochemical investigation? What tactics would we adopt? Let us suppose that we wish to trace the path of glucose oxidation by some micro-organism. To determine the progress of the reaction as a whole we would probably decide to measure the rate of oxygen consumption, and for this we might well use a manometric method such as is illustrated and described in Fig. 14.4. We might begin by trying to establish the expected relationship between the amount of glucose used up by a suspension of organisms and the amount of oxygen consumed; if substances other than carbohydrate are being oxidised at the same time, or if glucose is being metabolised by some non-oxidative pathway as well, this should show up in the results. Then we might try some simple experiments involving the addition of labelled glucose to the culture medium and see what labelled compounds we could isolate

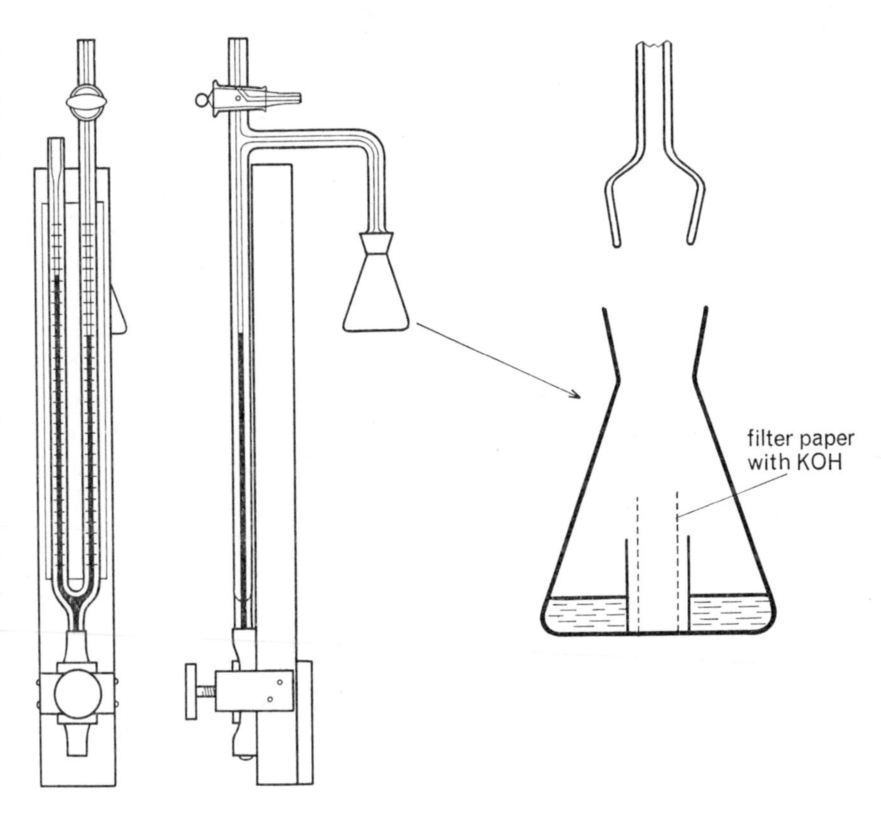

Fig. 14.4. The Warburg respirometer.

The respiring material, bathed in physiological saline, is placed in the respirometer vessel. The tube attached to the floor of the vessel holds a piece of filter paper soaked with KOH. This absorbs any carbon dioxide produced. Oxygen uptake can then be measured as reduction in volume.

The respirometer is set up with the vessel submerged in water in a thermostat and with the manometer outside. The respirometer is gently agitated. When sufficient time has been allowed for the vessel to take up the temperature of the bath the tap is closed and the manometer is read. After an appropriate time interval the fluid in the arm of the manometer to which the flask is attached is brought back to its original level (by decompression of a rubber tube at the base of the manometer) and the manometer is read again.

Calibration is necessary to enable readings of the manometer to be converted into change of volume. A similar respirometer, not containing any respiring material, is set up at the same time to enable correction to be made for change in atmospheric pressure during an experiment.

in an extract of the cells. In evaluating our results we would have to bear in mind, of course, the possibility that some of these compounds might not be in the main sequence of oxidation, but might lie on metabolic side lines. After that we might try the effect of various enzyme inhibitors and see whether any of the labelled products were accumulated or diminished in amount. If we found that some compound was accumulated under inhibition we would then see whether the normal organism could use it as a substitute for glucose and thereby maintain its rate of oxygen consumption. Having got some clues as to possible intermediates we would then try to fractionate the system for example by homogenisation and bulk separation (p. 73), so as to separate different parts of the reaction sequence. This would lead on to attempts to separate and purify enzymes, co-factors, etc. If these attempts were successful we would try to set up reconstituted systems, using crystalline enzymes and synthetic substrates so as to confirm that we had discovered and identified all the essential parts of the mechanism. Having pieced together what seemed to be a chemically plausible metabolic pathway, we might submit it to the test of thermodynamics and confirm that it involved a decrease of free energy. Finally, we would have to satisfy ourselves that the whole sequence of reactions would proceed fast enough to account for the observed rate of oxygen consumption.

For the task of tracing out metabolic pathways there is no single infallible method. Ideas suggested by one line of approach have to be confirmed by at least one other before they can look for general acceptance. As a consequence, some parts of the history of these investigations make dull reading and are not particularly instructive. The history of the tracing of metabolic pathways is a story of the laborious accumulation of a great wealth of fact, a story of false clues and rectified mistakes, out of which the true sequence of events was gradually discerned, and the importance of really significant discoveries was often not realised at the time they were made.

15

THE OXIDATIVE METABOLISM OF CARBOHYDRATE

To illustrate the flow of energy along a metabolic pathway we will take as an example the oxidation of carbohydrate by yeast. This differs only in certain minor respects from the oxidation of carbohydrate by muscle. Starting from glucose the main features of the process are as set out in Fig. 15.1. It may be regarded as made up of three main episodes, to each of which a subsequent chapter will be devoted.

(1) ***Glycolysis.*** The glucose molecule is broken down into 2 molecules of pyruvic acid with the transfer of 2 pairs of hydrogen atoms to DPN and the generation of 2 energy-rich bonds of ATP.

(2) ***Oxidative decarboxylation.*** Oxygen is taken from 6 molecules of water and is used to complete the oxidation of the 6 carbon atoms to carbon dioxide. 10 pairs of hydrogen atoms are transferred to hydrogen carriers and 2 more energy-rich bonds are generated. This involves a sequence of chemical changes commonly known as the Krebs cycle.

(3) ***Oxidative phosphorylation.*** All 12 pairs of hydrogen atoms are oxidised to water by molecular oxygen with the generation of no fewer than 34 energy-rich bonds. This is brought about by electron transfer through a sequence of respiratory pigments known as cytochromes.

The enzymes and carriers needed for episode (1) can be extracted from the cytoplasm in soluble form. The enzymes and carriers of episodes (2) and (3) are located in the mitochondria, from which they can be separated with varying degrees of difficulty, some not at all.

For the process as a whole

$$C_6H_{12}O_6 + 6O_2 \longrightarrow 6CO_2 + 6H_2O,$$

the standard free energy change is $\Delta F' = -685{,}600$ cal./mole of glucose. The yield is 38 energy-rich bonds of ATP, each worth 8900 cal./mole, making a total yield of 338,200 cal./mole of glucose and representing an efficiency of just under 50%.

We will now examine each episode in more detail.

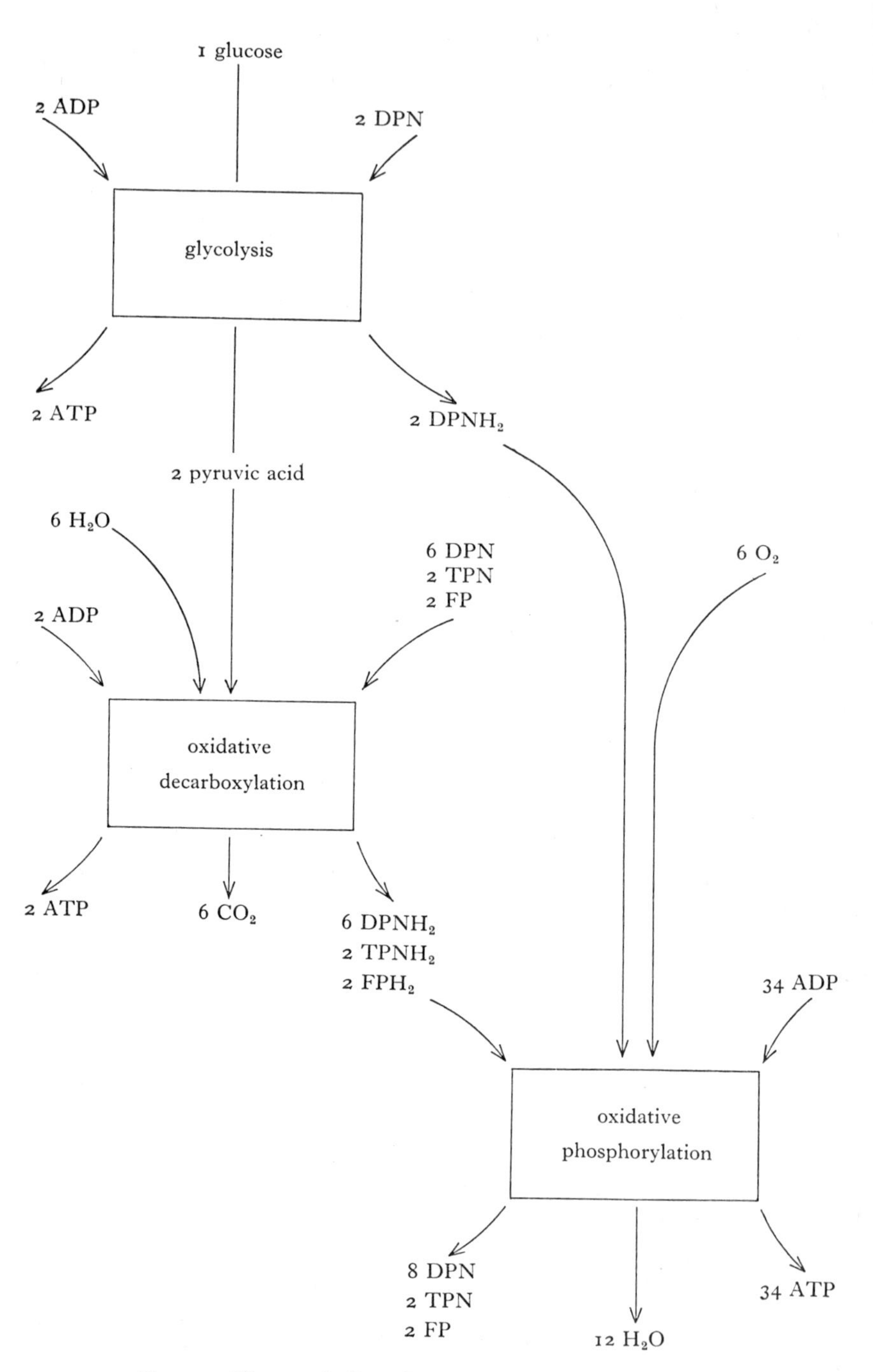

Fig. 15.1. The metabolic pathway of glucose oxidation in yeast.

16

GLYCOLYSIS

Glucose is fairly stable as organic molecules go and it has to be heated to about 250° C before it begins to break up. From the metabolic point of view it is quite refractory and indeed no enzyme is known which will break it up at one move. Nature's device is to attach phosphate groups to the molecule which seems to have the effect of redistributing the forces within the molecule so that the ring is weakened. But the reaction between glucose and phosphoric acid is endergonic, $\Delta F' = +3800$ cal./mole of glucose, so in order to attach a phosphate group to glucose we have to provide some free energy. This is arranged by bringing up the phosphate not as phosphoric acid but as ATP.

The first reaction of the sequence set out in Fig. 16.1 is the phosphorylation of glucose in the 6-position, for which purpose an energy-rich bond of ATP has to be sacrificed. The effect of introducing the phosphate group is to make possible an intramolecular rearrangement whereby the pyranose ring of glucose is contracted into the furanose ring of fructose.

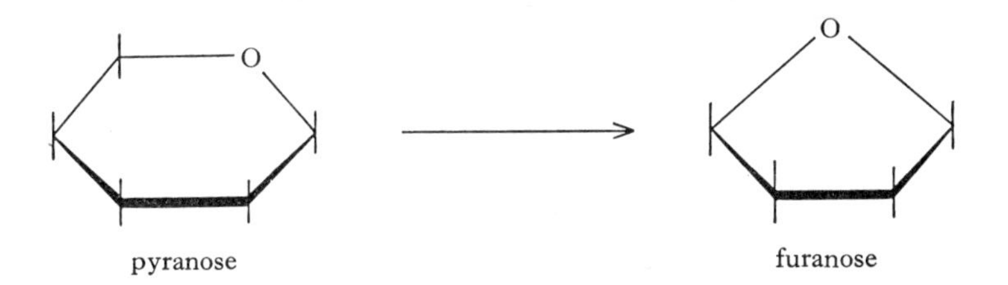

Another energy-rich bond of ATP is then expended in attaching a second phosphate group to the fructose, giving fructose 1,6-diphosphate, and this then falls apart into two more or less equal pieces, glyceraldehyde 3-phosphate and dihydroxy acetone phosphate, sometimes collectively known as 'triose phosphate'. Without their phosphate groups these two compounds have the same empirical formula, $C_3H_6O_3$, so in effect the glucose molecule has been split into two halves.

The next steps proceed from glyceraldehyde 3-phosphate into which dihydroxy acetone phosphate is readily converted. In the first of these

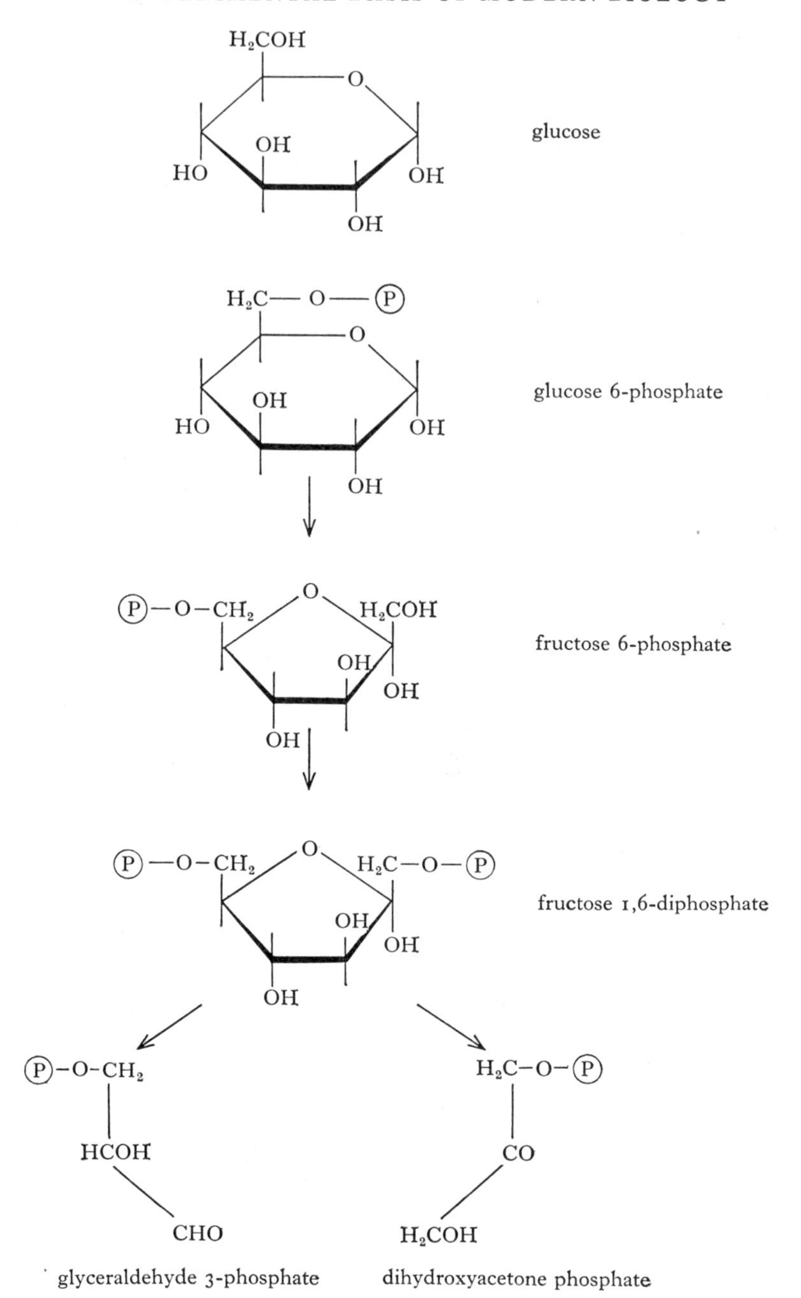

Fig. 16.1. Glycolysis in yeast, 1.

steps, as shown in Fig. 16.2, a second phosphate group is added to glyceraldehyde 3-phosphate to give glyceroyl 1,3-diphosphate. Both the phosphate groups are attached by energy-poor bonds. In the next

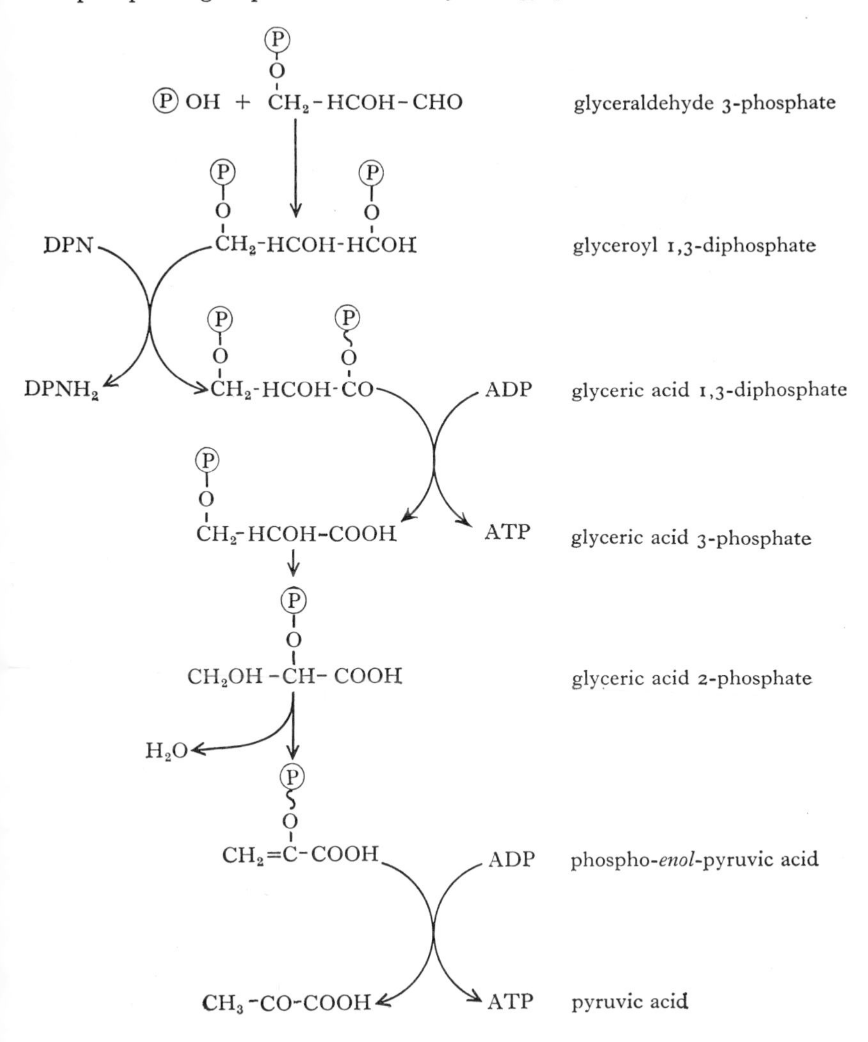

Fig. 16.2. Glycolysis in yeast, 2.

reaction two hydrogen atoms are removed from carbon atom (1) to give glyceric acid 1,3-diphosphate, and as a consequence of this change in its atomic environment the bond attaching the phosphate group to carbon atom (1) changes from the energy-poor to the energy-rich category.

This phosphate group is now transferred to ADP and we are left with glyceric acid 3-phosphate.

After this the remaining phosphate is transferred from the 3-position to the 2-position and the removal of water then gives phospho-*enol*-pyruvic acid. As already discussed on p. 145, as well as in the preceding paragraphs, the rearrangement of the molecule consequent upon the loss of the elements of water results in the change of the phosphate bond from the energy-poor to the energy-rich category. In the next reaction the phosphate group is transferred to ADP leaving pyruvic acid as the end product of the sequence.

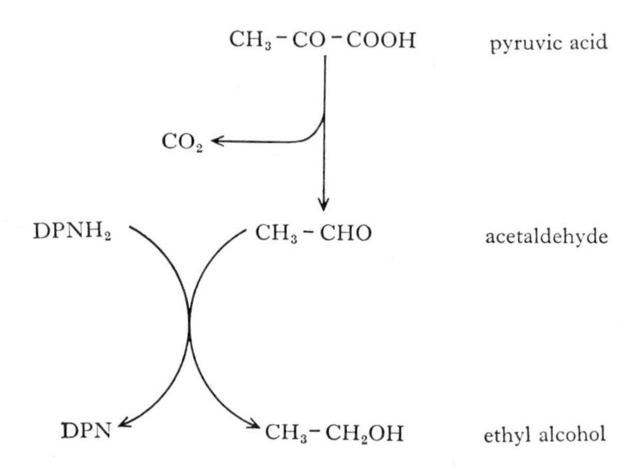

Fig. 16.3. Glycolysis in yeast, 3.

It is convenient to stop for a moment at this point. We are discussing the oxidation of glucose by yeast, and as far as oxidative metabolism is concerned this is where the process of glycolysis leaves off and the oxidation of carbon begins. But yeast is also able to carry glycolysis through two further stages which result in the production of ethyl alcohol, as shown in Fig. 16.3.

Now since these last two stages are not accompanied by the generation of any energy-rich bonds we may well ask why glycolysis does not stop at the stage of pyruvic acid. There appears to be a very good reason for going on with the process. At an earlier stage a pair of hydrogen atoms was passed to DPN. Now in general, carrier molecules are in limited supply, and if all the DPN has been converted to $DPNH_2$ the whole process of glycolysis is held up at the stage of glyceroyl 1,3-diphosphate.

By introducing a reductive process at a later stage, to balance the earlier oxidative process, the continued availability of DPN is assured.

In muscle events take a very similar course, in fact, over the middle part of the sequence the steps are identical in muscle and in yeast, deviations occurring only at the beginning and the end. In muscle, operations start not with glucose but with the polysaccharide glycogen. Glucose is readily obtained from glycogen by hydrolysis but in muscle glycogen is broken down not by hydrolysis but by phosphorolysis, a reaction in which H_3PO_4 replaces H_2O and in which the product is glucose 1-phosphate. Whereas the phosphorylation of glucose is endergonic, the phosphorolysis of glycogen is exergonic and ATP is not required. The phosphate group then passes from the 1-position to the 6-position with very little change of free energy and from then onwards muscle follows yeast as far as pyruvic acid. Beyond this stage the same problem arises again—how to restore DPN under anaerobic conditions. In muscle this is solved by reducing pyruvic acid to lactic acid, CH_3 CHOH COOH, for which a pair of hydrogen atoms is taken from $DPNH_2$. The most important difference, however, is that whereas 2 molecules of ATP have to be fed into the yeast sequence, only 1 is needed by muscle, so that the net yield of anaerobic glycolysis, per glucose unit, is 3 molecules of ATP for muscle as compared with 2 molecules of ATP for yeast.

Although the energy capture per glucose unit is only 3 molecules of ATP under anaerobic conditions as compared with 39 when oxygen is available, the lactic acid mechanism is of very great biological significance, making it possible for an animal to engage in violent effort of short duration, as described on p. 163.

The metabolic pathway illustrated in Figs. 16.1 and 16.2 is known as the Embden–Meyerhof pathway after two out of the many distinguished biochemists who contributed to its elucidation. The history of these investigations, however, is not such as can be summarised briefly, nor does it include any features which lend themselves to the illustration of general principles. The pathway has been described in some detail for two reasons: first, it provides a good example of the sort of chemical manipulation needed to bring a refractory molecule into a more reactive condition and to capture its energy; and second, it illustrates the relevance of thermodynamics to biochemistry, which we will now further examine.

Glycolysis affords us some insight into the tactics which Nature

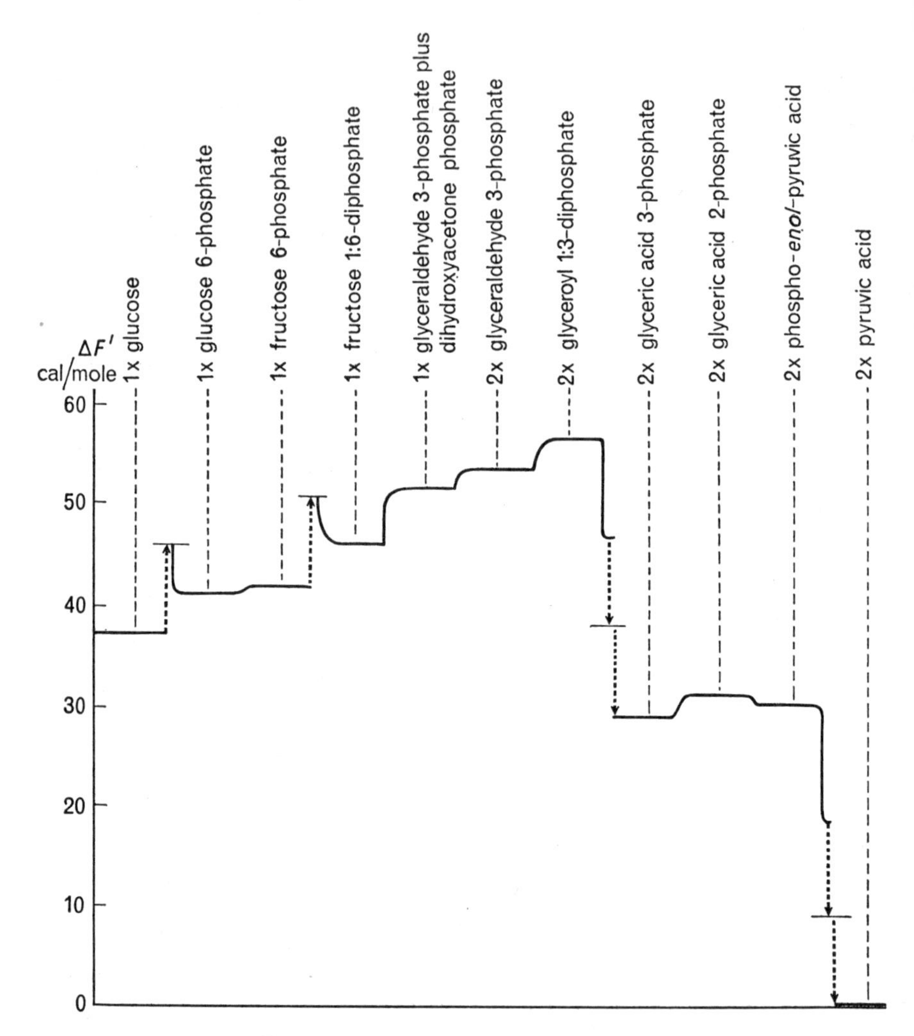

Fig. 16.4. The energy relations of glycolysis.

In this diagram the $\Delta F'$ values for the component reactions have been plotted cumulatively. When an energy-rich bond of ATP is used up in the reaction this is indicated by a broken line with an arrow pointing upwards; where an energy-rich bond is generated this is indicated by a broken line with an arrow pointing downwards. (Hydrogen atoms are assumed to be combined with DPN.)

This diagram should not be interpreted too literally. It is based upon standard free energy values for which the concentrations of reactants and products are specified as 1 mole/litre. We do not know what the actual concentrations are in living systems, but they are certainly much less than 1 mole/litre.

adopts in possessing herself of the energy of organic compounds. There are two steps in glycolysis at which energy-rich phosphate bonds are generated: the first is the oxidation of glyceroyl 1,3-diphosphate and the second is the dehydration of glyceric acid 2-phosphate. In both cases the tactic is the same. A phosphate group is already present in the molecule but its bond is of energy-poor status. The molecule enters a reaction which does not necessarily involve much change in free energy but which results in a change in the energy status of the phosphate bond. Taking the second case, the phosphate bond of phospho-*enol*-pyruvic acid is of great energy-richness; on hydrolysis it would yield 15,000 cal./mole of pyruvic acid. But there is no enzyme to catalyse this hydrolysis. Instead, there is an enzyme which catalyses the transfer of the phosphate group to ADP, a reaction for which $\Delta F' = -6100$ cal./mole. Thereby 8900 cal./mole are secured for the organism in the form of ATP.

In considering the energetics of glycolysis in yeast it is important to remember that from the stage of triose phosphate onwards everything has to be multiplied by two. Bearing this in mind we recollect that 2 molecules of ATP had to be fed in and that altogether 4 molecules of ATP were recovered, so the net yield is 2 molecules of ATP per molecule of glucose. For the process of glycolysis as far as pyruvic acid

$$C_6H_{12}O_6 + 2DPN \longrightarrow 2CH_3CO.COOH + 2DPNH_2,$$

$\Delta F' = -37,000$ cal./mole of glucose.

The 2 molecules of ATP represent 17,800 cal./mole of glucose, so the efficiency is about 48%.

The $\Delta F'$ values for all the component reactions are known, and it is instructive to set out the whole sequence of energy changes in graphical form as has been done in Fig. 16.4. Some of the reactions are seen to be endergonic and these would not take place in isolation. It is interesting to note that at the stage of $2\times$glyceroyl 1,3-diphosphate the energy level of the system is substantially higher than its original level, and that from fructose 1,6-diphosphate to $2\times$glyceroyl 1,3-diphosphate the sequence seems to run 'uphill'. Glycolysis has analogies with a siphon. In a siphon water will flow uphill provided that the level of the outflow is lower than the level of the inflow; in glycolysis, as in any sequence of coupled reactions, a negative ΔF for the sequence as a whole can take care of endergonic steps.

12-2

17

OXIDATIVE DECARBOXYLATION

Decarboxylation implies the removal of the carboxyl group of an organic acid with the production of carbon dioxide. In organisms this may take place in two ways: 'straight' decarboxylation and oxidative decarboxylation. 'Straight' decarboxylation is what happens to pyruvic acid in fermentation by yeast (p. 176). The carboxyl group breaks up giving carbon dioxide, the hydrogen atom attaching to the α-carbon atom.

$$CH_3.CO.COOH \longrightarrow CH_3.CHO + CO_2.$$

In oxidative decarboxylation oxygen participates, and in the present example would give acetic acid instead of acetaldehyde.

$$CH_3.CO.COOH + \tfrac{1}{2}O_2 \longrightarrow CH_3.COOH + CO_2.$$

In two of the reactions which we are now about to consider the substrates undergoing oxidative decarboxylation are α-keto acids and the oxygen comes from water;

$$R.CO.COOH + H_2O \longrightarrow R.COOH + CO_2 + 2H,$$

the hydrogen atoms being taken care of by various hydrogen carriers.

These reactions are organised in a cycle which is variously known as the Krebs cycle, the citric acid cycle and the tricarboxylic acid cycle. The operations of the cycle are summarised in Fig. 17.1. Briefly, the 3-carbon pyruvic acid undergoes decarboxylation and the 2-carbon residue joins up with the 4-carbon oxalo-acetic acid to give the 6-carbon citric acid, which then undergoes two successive decarboxylations in a complex series of changes whereby oxalo-acetic acid is reformed. Each step (and not all the steps are shown in Fig. 17.1) is catalysed by an enzyme and many of these enzymes have been isolated and purified.

The first steps towards the discovery of the Krebs cycle were taken in 1934 by Szent-Györgyi in the course of an investigation of the oxidative metabolism of muscle. For this investigation he chose to use pigeon breast muscle, which is known for its high rate of oxygen consumption. The oxygen consumption was measured by a manometric method, using a modification of the Warburg apparatus illustrated and described in Fig. 14.4. The muscle was finely minced so as to make it accessible to oxygen and to oxidisable substrates and was suspended in buffered saline. When this preparation was set up in the Warburg apparatus its oxygen consumption, initially high, fell off rapidly; and

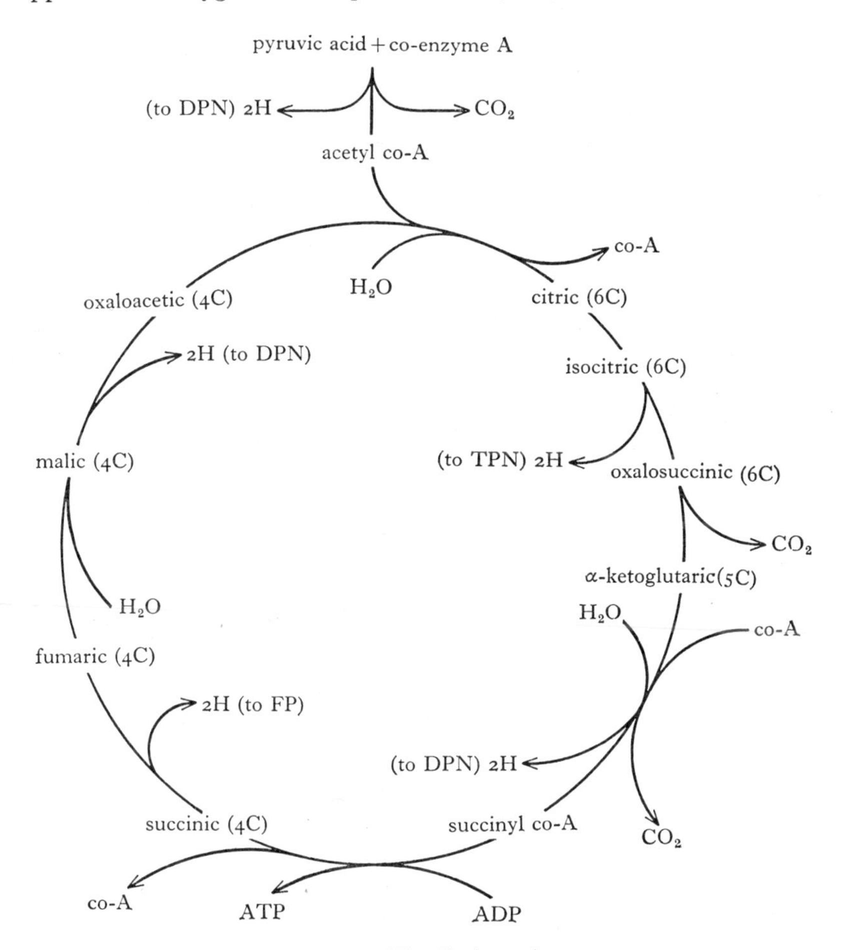

Fig. 17.1. The Krebs cycle.

the first aim of Szent-Györgyi's investigations was to discover what substrates, when added to the mince, would enable its oxygen consumption to be maintained.

One of the first substances he tested was succinic acid. With added succinic acid the muscle mince maintained its initial high rate of oxygen consumption for at least one hour.† But when the results came to be more closely examined it transpired that the extra oxygen consumed was much more than could be accounted for by the succinic acid used up; it was as if succinic acid acted as a catalyst in the oxidation of other substrates present in muscle mince. Following this up he showed that three other compounds—fumaric acid, malic acid and oxalo-acetic acid—had similar 'catalytic' action.

All these substances are 4-carbon dicarboxylic acids, and Szent-Györgyi guessed that they were probably interconvertible. To test this possibility it was necessary to develop methods of separating these acids from all the other soluble constituents of muscle mince—separations which today could so easily be effected by chromatography. When these methods had been perfected with the help of his colleagues, he was able to show that the following reversible transformations could take place in the presence of muscle mince.

$$
\begin{array}{ccccccc}
\mathrm{COOH} & & \mathrm{COOH} & & \mathrm{COOH} & & \mathrm{COOH} \\
| & & | & & | & & | \\
\mathrm{CH_2} & \underset{}{\overset{-2\mathrm{H}}{\rightleftharpoons}} & \mathrm{CH} & \overset{+\mathrm{H_2O}}{\rightleftharpoons} & \mathrm{CHOH} & \overset{-2\mathrm{H}}{\rightleftharpoons} & \mathrm{CO} \\
| & & \| & & | & & | \\
\mathrm{CH_2} & & \mathrm{CH} & & \mathrm{CH_2} & & \mathrm{CH_2} \\
| & & | & & | & & | \\
\mathrm{COOH} & & \mathrm{COOH} & & \mathrm{COOH} & & \mathrm{COOH} \\
\text{succinic} & & \text{fumaric} & & \text{malic} & & \text{oxalo-acetic}
\end{array}
$$

Shortly after this, in extension of Szent-Györgyi's original observations, Krebs showed that citric acid and α-ketoglutaric acid also had 'catalytic' action on the oxygen consumption of pigeon breast muscle; and Martius and Knoop, working on pigeon liver, showed that citric acid and α-ketoglutaric acid were interconvertible through *cis*-aconitic acid, isocitric acid and oxalosuccinic acid.

$$
\begin{array}{ccccccccc}
\mathrm{COOH} & & \mathrm{COOH} & & \mathrm{COOH} & & \mathrm{COOH} & & \mathrm{COOH} \\
| & & | & & | & & | & & | \\
\mathrm{CH_2} & & \mathrm{CH} & & \mathrm{CHOH} & & \mathrm{CO} & & \mathrm{CO} \\
| & & \| & & | & & | & & | \\
\mathrm{HOC.COOH} & \overset{-\mathrm{H_2O}}{\rightleftharpoons} & \mathrm{C.COOH} & \overset{+\mathrm{H_2O}}{\rightleftharpoons} & \mathrm{HC.COOH} & \overset{-2\mathrm{H}}{\rightleftharpoons} & \mathrm{HC.COOH} & \overset{-\mathrm{CO_2}}{\rightleftharpoons} & \mathrm{CH_2} \\
| & & | & & | & & | & & | \\
\mathrm{CH_2} & & \mathrm{CH_2} & & \mathrm{CH_2} & & \mathrm{CH_2} & & \mathrm{CH_2} \\
| & & | & & | & & | & & | \\
\mathrm{COOH} & & \mathrm{COOH} & & \mathrm{COOH} & & \mathrm{COOH} & & \mathrm{COOH} \\
\text{citric} & & \textit{cis}\text{-aconitic} & & \text{isocitric} & & \text{oxalosuccinic} & & \alpha\text{-ketoglutaric}
\end{array}
$$

† The experiments were not continued beyond one hour because of the risk of bacterial growth—these were the days before penicillin.

OXIDATIVE DECARBOXYLATION

It was already known that α-ketoglutaric acid underwent oxidative decarboxylation to give succinic acid.

The interrelationships of these acids, as they were understood at the beginning of 1937, were as shown in Fig. 17.2. In a paper published during that year Krebs and Johnson, having first confirmed the conclusions of Martius and Knoop upon pigeon breast muscle, put forward

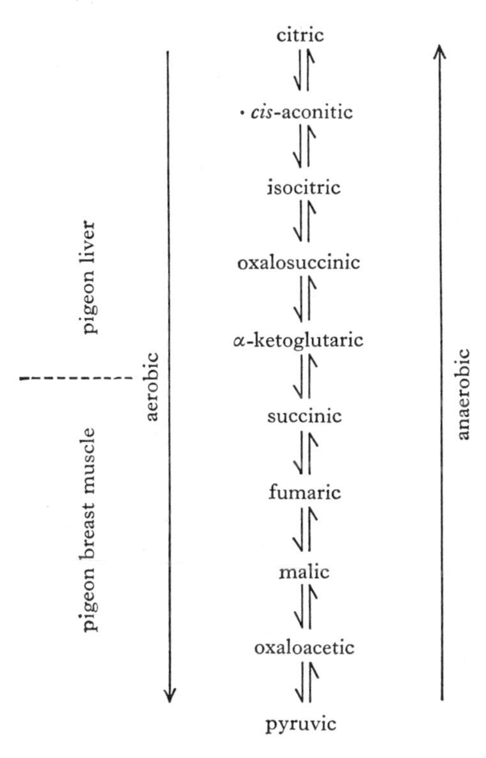

Fig. 17.2. Explanation in text.

the view that this chain of reactions was in fact joined up to form a cycle, whereby citric acid was formed from oxalo-acetic acid and some other as yet unknown compound. They presented two main lines of evidence:

(1) They showed that citric acid could be synthesised from oxalo-acetic acid. Excess of oxalo-acetic acid was added to minced pigeon breast in a small flask in the absence of oxygen; and after 20 minutes' incubation measurable quantities of citric acid were recovered.

(2) They showed that succinic acid could be synthesised from oxalo-acetic acid by two paths: via citric acid and α-ketoglutaric acid under

aerobic conditions, and via malic acid and fumaric acid under anaerobic conditions. This demonstration involved the use of malonic acid as an inhibitor; as mentioned on p. 156 malonic acid is a specific inhibitor of the enzyme succinic dehydrogenase which catalyses the interconversion of succinic and fumaric acids. Krebs and Johnson argued that under aerobic conditions malonic acid should increase the yield of succinic acid by preventing its breakdown to fumaric acid, while under anaerobic conditions it should decrease the yield of succinic acid by preventing its synthesis from fumaric acid. For the conception of a cycle this was the crucial experiment. Their results are reproduced in Table 17.1, exactly as they were presented in the original paper.

Table 17.1. *Effect of malonate on the anaerobic and aerobic conversion of oxalo-acetic acid into succinic acid*

0·75 gm wet muscle in 3 ccm. phosphate buffer. 40° C, pH 7·4.

Experimental conditions (final concentration of the substrate)	μl. succinic acid formed in 40 min.
1. O_2; 0·1 M oxaloacetate	1086
2. O_2; 0·1 M oxaloacetate; 0·06 M malonate	1410
3. N_2; 0·1 M oxaloacetate	1270
4. N_2; 0·1 M oxaloacetate; 0·06 M malonate	834

The citric acid cycle in the form in which it was originally proposed by Krebs and Johnson is reproduced in Fig. 17.3. It will be observed that at this stage Krebs had not identified the compound which condensed with oxalo-acetic acid. He had tested the effects of acetic acid and pyruvic acid upon the synthesis of citric acid but had found none. He thought it was likely to be a 3-carbon compound, which he referred to as 'triose', and supposed that a 7-carbon compound must precede citric acid in the cycle. No such 7-carbon compound was ever isolated, and eventually Stern and Ochoa, in 1950, showed that it was a 2-carbon compound, the acetyl radical in combination with co-enzyme A, which condensed with oxalo-acetic acid. The formulae of co-enzyme A and its acyl compounds are to be found in the Appendix, p. 328.

There certainly seems to be more chemistry than energy about the Krebs cycle. The free energy available from the conversion of pyruvic acid to carbon dioxide (the hydrogen atoms being taken up by hydrogen carriers) is $\Delta F' = -20{,}600$ cal./mole of pyruvic acid, of which 8900 is secured as ATP and 11,700 is left to drive the cycle. But by setting aside five pairs of hydrogen atoms the Krebs cycle has prepared the way for the great yield of energy which is realised when these hydrogen atoms are ultimately oxidised.

Since the idea was first put forward in 1937 the Krebs cycle has been tinkered with in various ways. At one time citric acid was placed outside the cycle, later to be reinstated. *Cis*-aconitic acid, once in, is now out. The step between pyruvic acid and acetyl co-enzyme A turns out to be

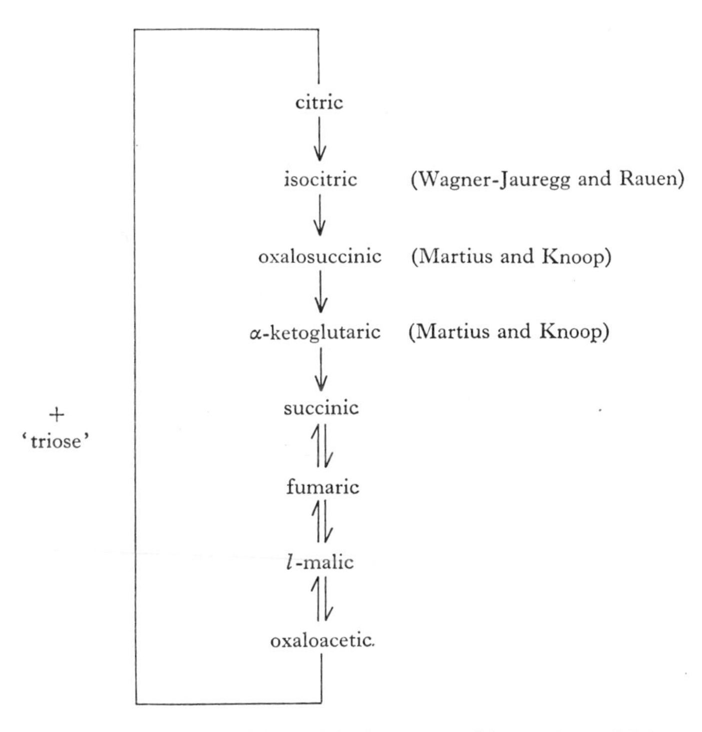

Fig. 17.3. The citric acid cycle as originally proposed by Krebs and Johnson in 1937. The names on the right refer to investigators recently concerned with the steps opposite.

horribly complicated, and co-enzyme A participates also at the stage of succinic acid. Undoubtedly the future holds further complications in store and no doubt some minor readjustments as well. But the main framework of the Krebs cycle has stood up to the strong currents of the last twenty years of biochemical research; powerful modern methods involving the use of ^{14}C have confirmed all its main features, so it looks as if it is here to stay, and as far as elementary textbooks are concerned it belongs among 'the facts of life'. But let Krebs himself speak to this:

'The schemes under consideration are of the nature of hypotheses. Like all hypotheses, they are suppositions which are made in order to account for experimental observations and to serve as a starting point for further investigations. For reasons to be stated later they belong to those types of hypotheses which are destined always to remain hypotheses. They can never become facts, i.e. something which can be observed, and therefore they cannot be proved (if to prove means to establish as a fact)—though they can be disproved. It is therefore beside the point if it is stated that such and such a scheme has not been "proved". The value of a reaction scheme should be assessed on the grounds of the observations which it "explains" and the discoveries to which it has given rise when used as a working hypothesis....'

And later—

'... it is not possible to investigate intermediary reactions under "physiological" conditions. Physiologically most intermediates exist only transitorily, i.e. in minute quantities. Moreover, they only occur intracellularly. These circumstances preclude their identification under "physiological" conditions. To investigate intermediary metabolism, the concentration of the metabolite must be artificially raised, or poisons must be added, and/or the tissue has to be removed from its normal site and to be perfused, or sliced, or minced, or extracted.

In short, while it can be proved that a tissue or a cell has the ability to perform certain reactions, the "physiological" occurrence of the reactions must remain an assumption.'

18

OXIDATIVE PHOSPHORYLATION

We now come to the terminal stages in the complete oxidation of carbohydrate. All the carbon atoms have now been oxidised to carbon dioxide, and for each molecule of glucose we are left with 12 pairs of hydrogen atoms in the care of DPN and other hydrogen carriers. Two of these pairs were removed during glycolysis, the remaining 10 pairs came from the Krebs cycle. These 12 pairs of hydrogen atoms now enter a reaction system which we will call the cytochrome system after its principal component, and are eventually oxidised by molecular oxygen to form water. The name 'oxidative phosphorylation' applied to the whole process registers the important feature that for every pair of hydrogen atoms from $DPNH_2$ thus oxidised 3 molecules of ADP are phosphorylated to give 3 molecules of ATP.

Ten pairs of hydrogen atoms come in combination with DPN or TPN. These are transferred to flavoprotein (FP) which appears to combine the properties of enzyme and hydrogen carrier. The remaining 2 pairs are transferred directly from succinic acid to FP. Next in sequence are the cytochrome pigments of which we are going to consider two only, cytochrome a_3 and cytochrome *c*. These pigments contain in the molecule an atom of iron which can exist in the oxidised (Fe^{+++}) and reduced (Fe^{++}) states. The hydrogen atoms from FP lose their electrons and become hydrogen ions. The electrons pass through cytochrome and eventually unite with hydrogen ions and oxygen in the presence of the enzyme cytochrome oxidase to form water.

The scheme which is set out in Fig. 18.1 is admittedly a simplified one, but it will serve to illustrate the principles which apply and the methods which have been used to investigate these processes.

It will be remembered that whereas the enzymes concerned in glycolysis are found free in the cytoplasm, the reactions of the Krebs cycle and the cytochrome system take place within the mitochondria. By disruption of the mitochondria, which can be brought about by ultrasonic vibrations, many of the enzymes of the Krebs cycle are liberated; but most of the enzymes and co-factors which comprise the cytochrome system are bound to the particulate debris of the mitochondria and efforts to separate them in pure form have met with limited success. This means that an approach based upon the use of reconstituted enzyme systems, which is possible in the case of glycolysis, is effectively barred in the case of oxidative phosphorylation. On the other hand, the study of the cytochrome system has prospered by virtue of the fact that so many of the substances which participate in the reactions of this system have characteristic absorption spectra.

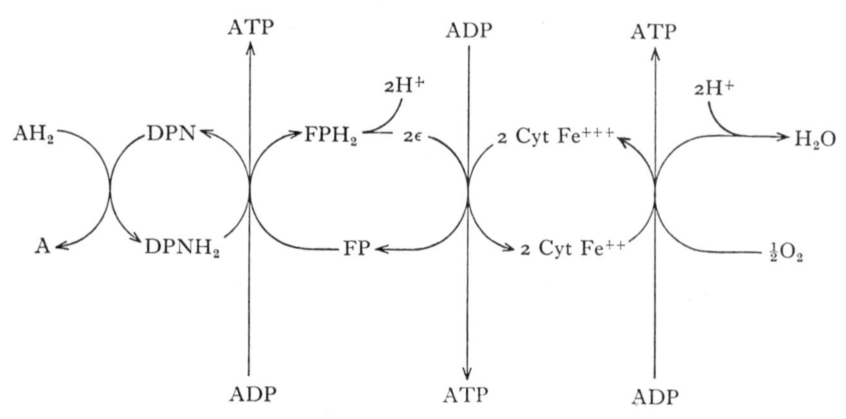

Fig. 18.1. Oxidative phosphorylation.

This scheme is simpler than any known to occur in nature. Some naturally occurring systems may comprise up to two FP steps and up to three cytochrome steps.

When viewed by transmitted light a substance appears coloured because some wavelengths of light are absorbed more than others. A solution of oxyhaemoglobin appears red because of greater absorption of other wavelengths, notably the yellow-green and deep violet. If a solution of oxyhaemoglobin is placed in the path of the light entering a spectroscope the spectrum is seen to have two dark bands, one in the yellow and one in the green, corresponding to wavelengths of 5790 Å and 5420 Å (Fig. 18.2). A more accurate representation of the absorp-

Fig. 18.2. The absorption spectrum of haemoglobin.

The upper figure shows the appearance of the spectrum of light which has passed through a solution of the pigment.

To obtain the quantitative data of the lower figure a spectrometer having a very narrow exit slit is used to select monochromatic light from various parts of the spectral range. The light is passed through a solution of the pigment and is registered by a photoelectric cell. The intensity of the transmitted light at some selected wavelength is taken as reference and the intensity at other wavelengths, relative to reference, is recorded and plotted as relative absorption. Since the range of wavelengths extends far into the UV, all the lenses, etc., of the spectrometer must be of quartz.

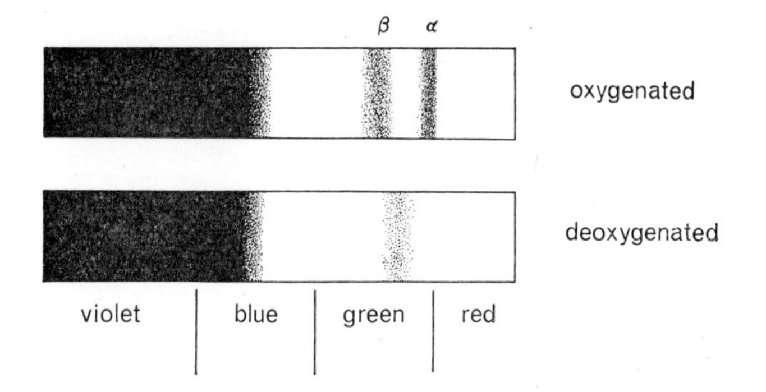
β
α
oxygenated
deoxygenated
violet
blue
green
red

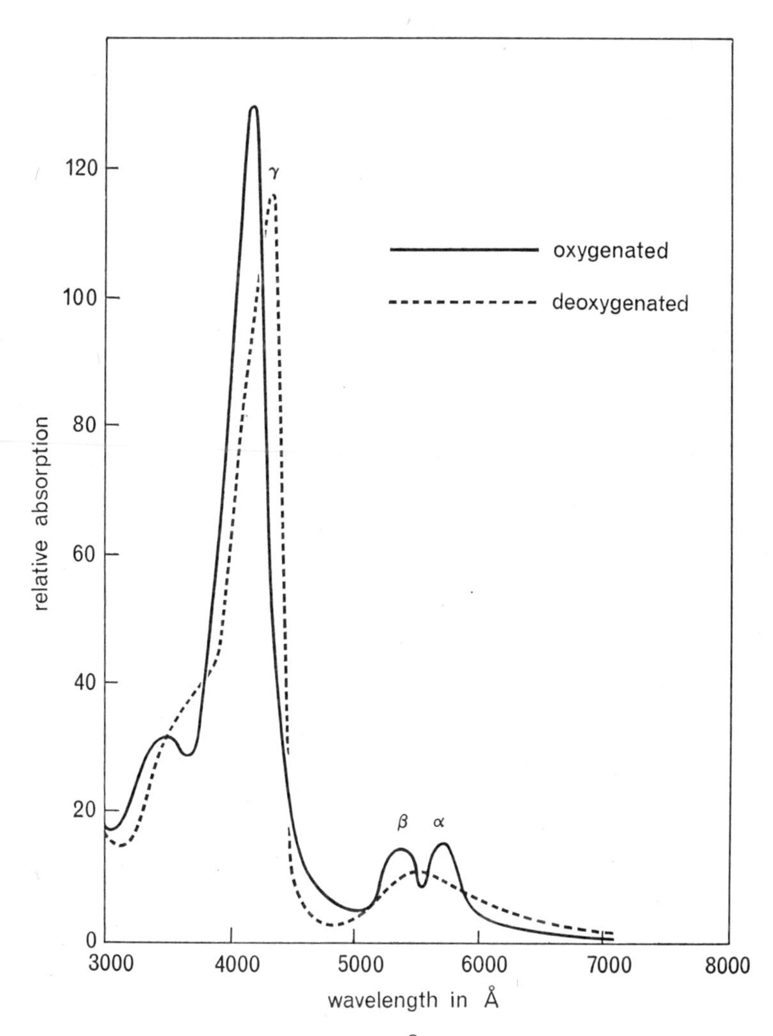
γ
oxygenated
deoxygenated
β
α
relative absorption
0
20
40
60
80
100
120
3000
4000
5000
6000
7000
8000
wavelength in Å

tion spectrum is obtained by measuring the relative absorption of light of different wavelengths and plotting relative absorption against wavelength. When this is done it can be seen that in addition to the bands at 5790 Å and 5420 Å, known respectively as the α and β bands, there is also a very much stronger band, the γ band, at 4320 Å in the deep violet. These three bands are characteristic of pigments which contain the prosthetic group known as haem (see Appendix, p. 329). Now as these figures show, if we convert oxyhaemoglobin into haemoglobin, by exposure to a vacuum, the α and β bands disappear. It is clear, then, that by following changes in the absorption of light of some suitably chosen wavelength we have in principle a method of measuring the state of oxygenation of the pigment—and, most important, a method which can be applied to a living system without interfering with its activities.

The first great name which we meet with in the history of the investigation of tissue oxidation is that of Warburg. Starting in 1918, and using the apparatus described and illustrated in Fig. 14.4, he showed that not only whole organisms but also their tissues were able to utilise molecular oxygen. Since this ability was rapidly lost at the comparatively low temperature of 50–60° C he attributed it to the presence of some enzyme, which he called the *Atmungsferment* or respiratory enzyme. This enzyme was recognisable by its property of being strongly inhibited by cyanide, by hydrogen sulphide and by carbon monoxide, and using these criteria Warburg established its presence in all aerobic organisms which he investigated.

One of the most important of Warburg's discoveries concerned the inhibition due to carbon monoxide. The action of enzyme inhibitors, as we discussed on p. 156, is often due to the formation of a stable compound between the inhibitor and the active centre of the enzyme. In the case of the inhibition of the *Atmungsferment* by carbon monoxide a compound is undoubtedly formed and this compound has the unusual property of being dissociated by light. Carbon monoxide is an inhibitor only in the dark; in strong light tissue respiration proceeds normally, even in the presence of carbon monoxide. Warburg set out to discover what wavelengths of light were effective in abolishing the inhibition caused by carbon monoxide. He provided himself with a set of light filters, each of which would pass only a narrow spectral band. He set up a suspension of yeast cells in a respirometer and recorded the rate of oxygen consumption under illumination with different wavelengths of light, ranging from the red to the ultraviolet. Now if light is to have any effect upon a chemical process it must be absorbed; so if the rate of oxygen consumption is plotted against wavelength a kind of absorption spectrum is obtained—called a photochemical absorption spectrum to distinguish it from an absorption spectrum obtained by direct measurement of the absorption of light. The photochemical absorption spectrum of the *Atmungsferment* is reproduced in Fig. 18.3. It shows α, β and γ bands, which at once suggests that the *Atmungsferment* is a respiratory pigment containing haem.

Warburg's contribution, then, was to demonstrate the existence of some enzyme system which was capable of promoting the oxidation of naturally occurring substrates by molecular oxygen, which was present in all aerobic organisms and which was probably a haem-containing pigment.

The pigment now known as cytochrome was first discovered by MacMunn in 1884. He found it to be present in the muscles and other tissues of a great variety of animals. The respiratory significance of the pigment was discovered by Keilin in 1925 in the course of an investigation of the respiration of parasitic insects. What he actually observed was the appearance of certain absorption bands while he was examining their tissues with a microspectroscope. Fig. 18.4, taken from his paper, shows the absorption spectrum of the thoracic muscles of a honey bee, as seen with the muscles compressed under a coverslip on a microscope slide. It shows four absorption bands. In fact, this spectrum turned out to be composite; this was indicated by the relative independence of the intensities of the bands as seen in different preparations and under different conditons. Keilin resolved this composite spectrum into three components, each having two bands, the α and β bands characteristic of haem pigments. Although he was unable to extract any of the components of cytochrome without their denaturation, he obtained enough indirect evidence to make it highly probable that they were indeed haem pigments as the evidence of the spectrum suggested.

The characteristic absorption spectrum of cytochrome is seen only if the tissue is deprived of oxygen, as when an actively respiring tissue is confined under a coverslip. As soon as oxygen is readmitted the bands fade. This observation suggests that the pigment is capable of undergoing reversible oxidation and reduction and that it plays some part in tissue respiration. Keilin also showed that in the presence of cyanide the pigment was permanently reduced, that is to say, the bands persisted even in the presence of abundant oxygen, whereas in the presence of

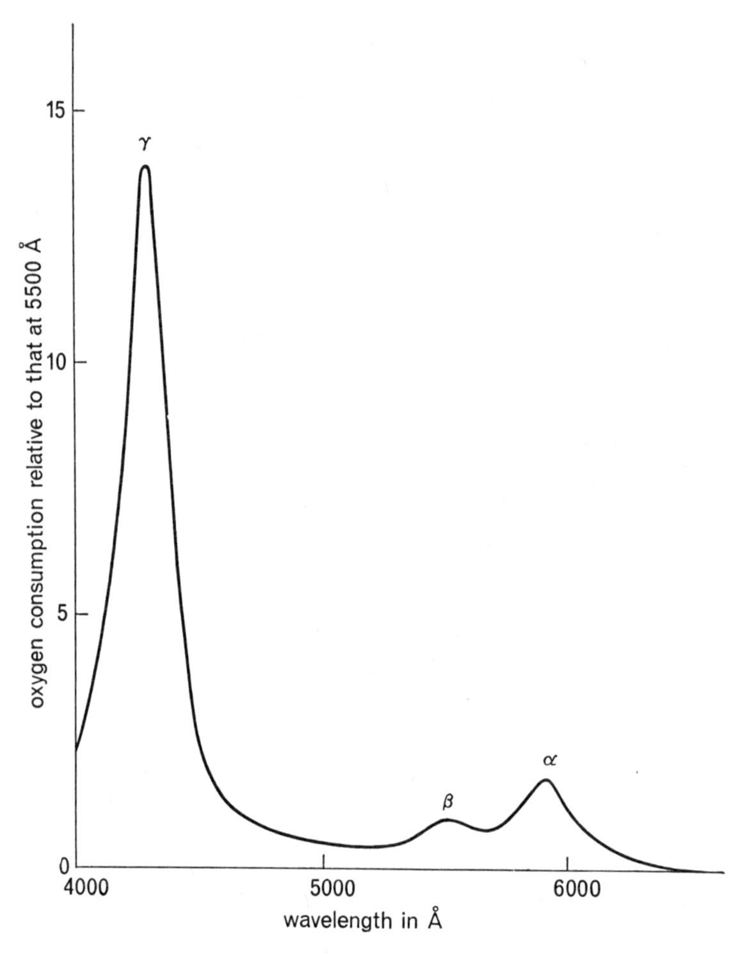

Fig. 18.3. The photochemical absorption spectrum of the *Atmungsferment* in yeast.

The photochemical absorption spectrum figured here is not the one originally published by Warburg in 1929, but is taken from a paper by Castor and Chance published in 1955. Warburg used light filters and the number of wavelengths he could obtain in this way was limited. Castor and Chance used a spectrometer with a narrow exit slit, which gave them an effectively unlimited choice of wavelengths, so they were able to work out the form of the graph in greater detail. They also used an electrical method of measuring the oxygen consumption which was much quicker than Warburg's respirometer method.

narcotics such as urethane it was difficult to get the bands to appear. At this stage Keilin pictured cytochrome as a respiratory carrier with relationships as below.

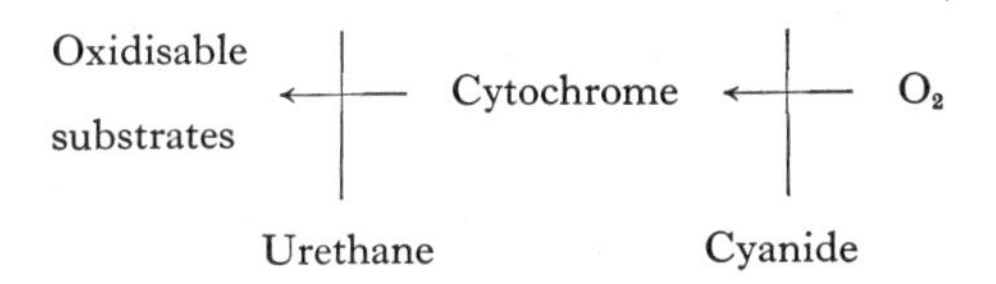

Finally, in this first paper, Keilin reported the presence of cytochrome not only in animals but also in some higher plants, in yeast and in some bacteria.

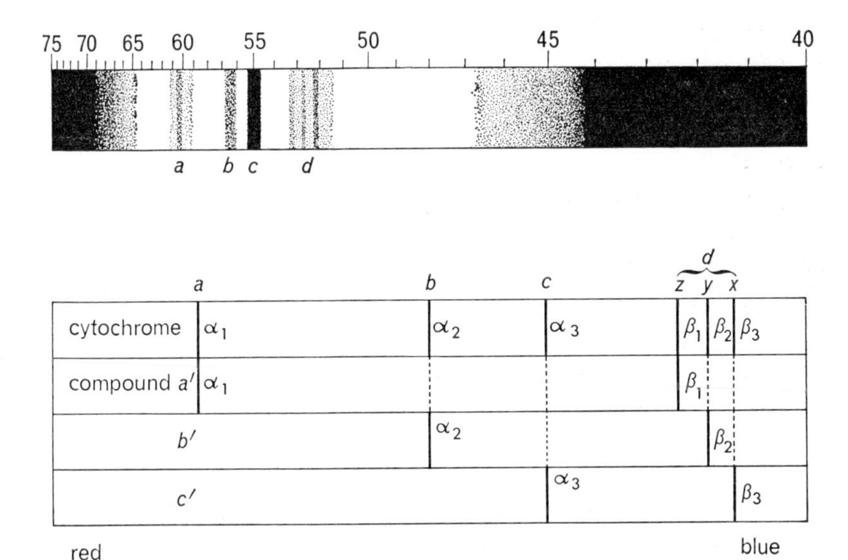

Fig. 18.4. The absorption spectrum of cytochrome.
Above, as seen in living tissue. Below, as resolved into its components.

Since the time of the publication of Keilin's paper much research activity has been concentrated upon cytochrome and has led to the identification of many variant forms of the three main components. Despite this great effort cytochrome *c* is the only form of cytochrome which has been isolated in the pure state. Much progress has been made with its chemistry, and its primary structure is now known. It is undoubtedly a haem pigment, showing a strong γ band as well as the α and β bands already described; but whereas haemoglobin enters into a loose combination with oxygen, the iron atom remaining in the ferrous state, cytochrome *c* reacts differently. The oxidation and reduction of cytochrome *c* are associated with a change in the valency of the iron atom, from ferrous to ferric. There is no indication of any other change in the molecule, such as the loss or addition of hydrogen atoms, and for this reason cytochrome *c* is to be regarded not as a hydrogen carrier like DPN but as an electron carrier. It is readily reduced in the presence of dehydrogenase systems, but it is not reoxidised by molecular oxygen alone; its oxidation by molecular oxygen requires the presence of some enzyme which was provisionally called cytochrome oxidase.

Cytochrome oxidase is associated with the particulate matter of mitochondrial debris and has never been separated and purified. Well-washed mitochondrial debris will catalyse the oxidation of cytochrome *c* by molecular oxygen; in the absence of cytochrome *c* it will not enable any tissue preparation to utilise molecular oxygen. We may therefore proceed on the assumption that the mitochondrial debris contains an enzyme which specifically catalyses the oxidation of cytochrome *c*. This enzyme can be shown to be inhibited by cyanide, by hydrogen sulphide and by carbon monoxide in the dark, as is Warburg's *Atmungsferment*; and the photochemical absorption spectra of the two enzymes are closely similar. There is, in fact, no good reason why we should not accept the identity of cytochrome oxidase and the *Atmungsferment*, and the latter term is no longer in current use.

Equally, there is no good reason why we should not believe that cytochrome oxidase is a haem pigment, and it is tempting to suppose that it may be one of the cytochromes. In 1939 Keilin and Hartree, having first established the separate identity of cytochrome a_3, went on to show that it formed a compound with carbon monoxide. In the absorption spectrum of the CO-compound of cytochrome a_3—not the photochemical absorption spectrum, but the absorption spectrum directly observed with the microspectroscope—they were able to see two absorption bands, one at 4300 Å and the other at 5900 Å (compare with Fig. 18.3) and this prompted them to suggest that cytochrome a_3 might be cytochrome oxidase; but they put this forward tentatively because of certain difficulties—for example, that they had not been able to demonstrate the dissociation of the CO-compound in strong light. It was later shown that they had been using carbon monoxide at too high a partial pressure, and when their experiments were repeated with less carbon monoxide in the gas mixture, photodissociation was easily observed. It is now generally accepted that cytochrome a_3, cytochrome oxidase and the *Atmungsferment* are all the same thing.

The position we have reached, then, is that cytochrome *c* is an electron carrier which can be reduced by the dehydrogenase systems of the cell *in vivo* and which can be oxidised by molecular oxygen through the offices of cytochrome a_3. This latter appears to be the terminal reaction of the cytochrome system.

Having seen how cytochrome *c* is oxidised we may now inquire into the conditions under which it is reduced. The first question we might ask is if it can be reduced by $DPNH_2$. The answer to this question is a qualified yes; $DPNH_2$ is capable of reducing cytochrome *c*, but if only the pure substances are present the reaction is very slow. Clearly if this reaction forms part of the normal chain of respiratory reactions there must be some enzyme to expedite its progress.

Careful search for this enzyme led eventually to the class of substances called flavoproteins. These had already been studied by Warburg and others and had for long been known collectively as 'yellow enzyme' before any functional connection between flavoproteins and cytochromes had been suspected. Flavoproteins, which we have denoted by FP, carry a prosthetic group containing the pigment riboflavin (see Appendix, p. 326). This prosthetic group functions as a hydrogen carrier, fairly firmly attached to the protein; indeed, the flavoprotein molecule can be regarded as a respiratory enzyme with a built-in hydrogen carrier. It seems, too, that iron is necessary for the effective co-operation of flavoprotein and cytochrome, but the iron is very loosely attached—in fact, the ease with which this iron is lost during the process of extracting and purifying the enzyme was of itself a major cause of delay in recognising the role of flavoproteins in the cytochrome system, as we shall see in a moment. Various flavoproteins were isolated from yeast, heart muscle, liver, etc., and were found to be readily reduced by $DPNH_2$ and $TPNH_2$, and those that reacted with $TPNH_2$ were able to react rapidly with cytochrome *c*. But those that reacted with $DPNH_2$ were slow to react with cytochrome *c*, too slow, as calculations showed, to handle the traffic which was known to be carried by DPN. Eventually iron was found to be the missing factor, and after the addition of a small amount of iron, either as Fe^{++} or Fe^{+++}, the reaction goes very much faster.

We may now turn to the energy relations of the system. For every pair of hydrogen atoms oxidised to water three molecules of ATP are formed from ADP. It is not altogether easy to demonstrate this quantitative relationship because the disruption of cells sets free all sorts of enzymes which break down ATP as fast as it is formed. A useful trick is to add glucose to the system. As we saw on p. 173 glucose reacts with ATP in the presence of the appropriate enzyme to give glucose 6-phosphate which is relatively stable; so if we determine the amount of glucose 6-phosphate produced we get a better estimate of the number of energy-rich phosphate bonds than if we try to determine the amount of ATP directly. The line of attack has been to relate the number of energy-rich phosphate bonds produced to the number of oxygen atoms consumed within a given time. The P/O ratio, as it is called, works out at something between 2 and 3. Now since stoichiometric considerations require us to deal with whole numbers only, and since for the reasons just given the estimate of energy-rich phosphate bonds is likely to err on the low side, it is generally accepted that for the process of oxidative phosphorylation in the cytochrome system the P/O ratio is 3. The standard free energy change for the reaction whereby two hydrogen atoms from $DPNH_2$ form one molecule of water is $\Delta F' = -52{,}400$ cal./mole. The generation of three molecules of ATP requires $3 \times 8900 = 26{,}700$ cal./mole, giving an efficiency of 51% which is of the order we might expect.

It is possible to analyse the process a little further. We can measure the oxidation-reduction potentials of the various carriers which take part, and thermodynamics provides us with a relationship whereby standard free energy changes can be calculated from oxidation-reduction potentials. These are as follows:

	$\Delta F'$ (cal./mole of hydrogen)
DPN	
	−11,100
FP	
	−16,100
Cytochrome	
	−25,200
Oxygen	
	−52,400

Ascorbic acid reduces cytochrome *c* directly and can be used as a substrate for the cytochrome (*sensu stricto*) system alone. For the oxidation of ascorbic acid we find a P/O ratio of just under 1. This suggests that one energy-rich phosphate bond is generated at the last step, although there is enough energy for 2. It is known that the two hydrogen atoms from succinic acid are not carried by DPN or TPN, but enter the cytochrome system at the level of FP; and for the oxidation of succinic acid the P/O ratio lies between 1 and 2. We therefore have some evidence for allocating the three energy-rich phosphate bonds to the three separate steps, as indicated above. But we have no information about the further details of this coupling, which presumably involves enzymes as yet unidentified.

So much, then, for the simplified scheme of the cytochrome system as presented in Fig. 18.1. Let it be said again that it is a highly simplified scheme, that whereas the chain of reactions therein displayed has a single cytochrome link, there may well be three cytochrome links in series and there may be two flavoprotein links in parallel. It is all very complicated; yet, surprisingly, with modern technique we can watch it in action.

The principle underlying this technique was first made use of by Millikan in 1937 and its modern refinements are largely due to Chance. As now developed it consists in taking simultaneous records of partial pressure of oxygen, of pH and of light absorption at two wavelengths. All these can be measured electrically and very rapid changes can be faithfully recorded. It would take too long to describe the methods of measuring oxygen and pH, even in outline, but these methods are not particularly difficult by comparison with the problem of making sense of the records of light absorption. All the participants in the cytochrome system, from $DPNH_2$ to cytochrome a_3, have absorption spectra of the same general level of complexity as those illustrated in Fig. 18.3; the problem is to find out what is happening to individual components when we are presented with all their absorption spectra superimposed. To begin with, it helps if we plot out the 'difference spectrum' of the system we are studying; we plot the absorption spectrum first in the oxidised condition, then (on the same graph) in the reduced condition; then we subtract one from the other, point by point, and plot the differences (having regard to sign) against wavelength. We then get a graph which wanders irregularly above and below the zero line, crossing the zero line at those wavelengths at which the oxidised and reduced forms have the same relative absorption. The principle of the method is that those wavelengths at which the difference spectrum is zero for pigment A can be used to follow changes in pigment B, since changes in pigment A will not affect the absorption of light of this wavelength. When we have also to take account of pigments C, D, E, ..., as well it becomes less and less easy to apply the principle with full rigour; but useful information can still be obtained from a complex system by judicious choice of wavelengths.

Fig. 18.5 presents the difference spectrum measured on a cell suspension. It shows peaks and troughs which are primarily associated with changes in the state of oxidation of one or other of the components. So if we wished to follow changes in, say, $FP \rightleftharpoons FPH_2$ we would choose to follow changes in absorption at 4560 Å. But in fact what the apparatus records is changes in light intensity which may be due to a variety

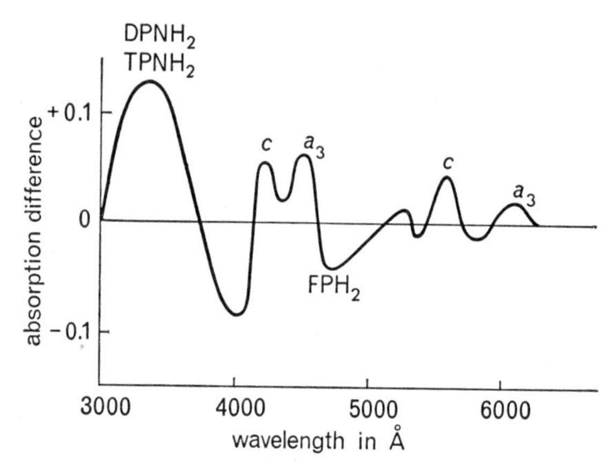

Fig. 18.5. The difference spectrum (oxidised *minus* reduced) of a suspension of cells (ascites tumour).

Peaks and troughs are indicated as being suitable for following changes in the various hydrogen and electron carriers: $DPNH_2$ and $TPNH_2$, FPH_2 and cytochromes c and a_3. Any of the wavelengths at which the graph cuts the zero line may be used as reference; thus for studying $FP \rightleftharpoons FPH_2$ records were taken at 4650 Å and the reference wavelength was 5100 Å.

of causes such as variations in turbidity of the suspension or variations in the strength of the light source. To eliminate these sources of error and to obtain a true record of changes in absorption at 4650 Å we have to compare the intensity of light at 4650 Å with the intensity at some reference wavelength, and for this we would choose a wavelength at which the difference spectrum is zero.

This technique can be applied, for example, to suspensions of mitochondria and we can see how, as the availability of oxygen is decreased, the various carriers become reduced in a definite order. Thus we obtain a most valuable check upon conclusions which have been reached by studying the various links of the chain in partial isolation. The technique can also be applied to whole tissues and organs. Living muscles can be investigated while they are stimulated to contract, and the time course of the oxidative processes can be followed.

However, the main difficulty which further progress encounters is that of separating the components of the system. Mitochondria have been broken up, by ultrasonic waves and other means, and quite small pieces of mitochondria are found to be able to carry on all the operations of the cytochrome system. Disruption and fractionation can be carried to a point at which some particle fractions are deficient. It is possible to separate a fraction which has no cytochrome oxidase activity from a fraction which has no succinic dehydrogenase activity—and indeed disruption and fractionation can be carried some way beyond this. But if mitochondria are subjected to such treatment as renders the characteristic cristae with their double membranes no longer visible in the electron microscope then the enzymatic activity is effectively abolished. There is therefore little prospect of being able to obtain enzymes which are both functional and in a state of crystalline purity.

19

PHOTOSYNTHESIS

The whole living world as we know it depends upon the energy of sunlight. The activities of organisms and the operations of the carbon and nitrogen cycles in nature are necessarily accompanied by the degradation of free energy into heat. Left to itself the whole living world would ultimately grind to a halt and this is avoided only because the sources of free energy are continually replenished from the energy of sunlight through the process of photosynthesis. The investigation of photosynthesis is inspired not only by its fundamental importance in nature but also by its inherent biochemical and biophysical interest.

The key substance in the vitally important capture of the energy of sunlight is the pigment chlorophyll. Its structure was worked out by Wilstätter and by Fischer over a period of 30 years, from 1910 to 1940, and is set out in the Appendix (p. 330). It is clearly a substance of the same nature as the haem pigments, iron being replaced by magnesium. The absorption spectrum of chlorophyll is shown in Fig. 19.1 and the green colour is clearly due to absorption in the red and in the deep violet. Chlorophyll is found within plant cells in special bodies known as chloroplasts (see p. 116) from which it can be readily extracted with alcohol. But extracted chlorophyll cannot carry out photosynthesis.

Within the framework of all the synthetic activities of plants any distinction as to what is or what is not photosynthesis can only be arbitrary; so we may conveniently restrict our field by deciding to treat only of the synthesis of carbohydrate. Photosynthesis is the reverse of respiration. This was established by Joseph Priestley who showed as long ago as 1778 that green plants removed carbon dioxide from the atmosphere and replenished the oxygen. The importance of light for this process was recognised by Ingenhousz at about the same time. Although the green pigment was long suspected of being concerned in the process, this was finally demonstrated a hundred years later, by Engelmann in 1880.

The earliest method of measuring photosynthesis was the dry weight method of Sachs, in which the dry weight of a piece of leaf which had been exposed to the light was compared with the dry weight of a similar piece which had been kept in the dark. Much use has been made of aquatic plants since the oxygen which they evolve is readily collected and is a measure of the amount of photosynthesis. But since multicellular plants introduce the complications arising from diffusion through tissue spaces most investigators have favoured unicellular algae such as *Chlorella*. Photosynthesis is measured in the Warburg apparatus (Fig. 14.4). The alga is suspended in a medium containing

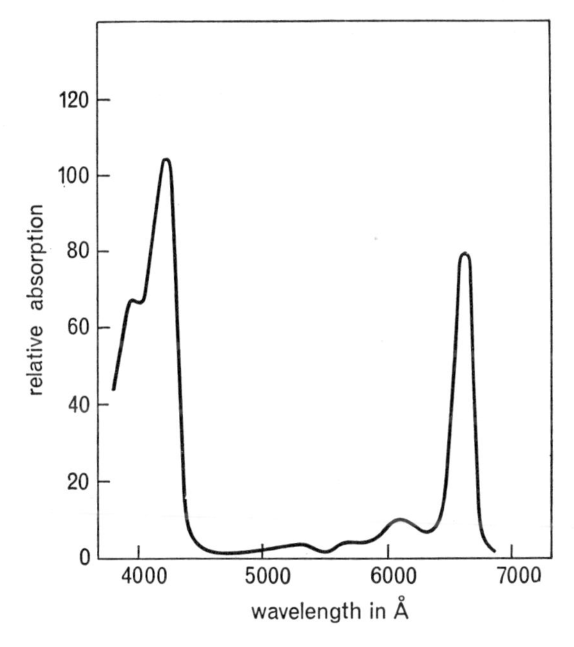

Fig. 19.1. The absorption spectrum of chlorophyll.

bicarbonate as a source of carbon dioxide and the production of oxygen is measured as increase in volume.† More recently suspensions of chloroplasts prepared from spinach leaves have begun to rival *Chlorella* in popularity. All these methods, of course, simply measure one of the end products of photosynthesis and by themselves give no clue as to its chemical mechanism; indeed, the chemical changes involved in the fixation of carbon dioxide were virtually unknown before isotopes became available.

† The filter paper soaked in KOH, which is used to absorb carbon dioxide in respiration experiments, is here omitted.

The first important step towards the analysis of photosynthesis was taken in 1905 when Blackman showed that it was possible to distinguish between a 'light reaction' and a 'dark reaction' in photosynthesis. He studied the effect of temperature upon the rate of photosynthesis of the moss *Fontinalis* under two different sets of conditions: (*a*) in weak light and high concentration of carbon dioxide, and (*b*) in strong light and low concentration of carbon dioxide. He made the reasonable assumption that under (*a*) the factor limiting the rate would be light intensity, and under (*b*) concentration of carbon dioxide. He found that the rate of photosynthesis under (*b*) was doubled for a 10° C rise in temperature, as the rate of any chemical reaction might be, whereas under (*a*) it was independent of temperature as would be expected for a photochemical reaction.

The distinction between a 'light reaction' and a 'dark reaction' was further underlined by an observation made by Warburg in 1919. Warburg found, using *Chlorella*, that when photosynthesis was carried out in intermittent light, produced by a rotating shutter in front of the light source, the rate of photosynthesis *per unit of light energy* could be as much as 400 % greater than in continuous light of the same intensity. This result he interpreted as meaning that in continuous strong light the 'light reaction' outpaces the 'dark reaction' and the products of the 'light reaction' accumulate, whereas in intermittent light the products of the 'light reaction' can be disposed of during the intervening dark periods.

In the absence of any definite information to the contrary most people inclined to the view that carbon dioxide formed a compound with chlorophyll and that the energy of light enabled it to lose its oxygen and enter into some further reaction with water. Other possibilities were suggested, notably by Kluyver and van Niel; but up to the outbreak of the Second World War, it had not been possible to separate any of the component reactions from the complete photosynthetic reaction, and in consequence current ideas were little more than speculation. In 1939 Hill, investigating the reactions of chloroplasts isolated from the leaves of flowering plants, discovered that an illuminated suspension of chloroplasts could reduce ferric oxalate with the evolution of oxygen, and in the stoichiometric relation of four atoms of iron reduced per molecule of oxygen evolved; there was no evidence that carbon dioxide participated in this reaction. This was a most important advance. First, a separation had been effected between the evolution of oxygen and the fixation of carbon dioxide; and secondly, it was now clear that, whatever might be the source of the oxygen evolved in photosynthesis, it was unlikely to be carbon dioxide.

At about the same time van Niel was making a study of the photosynthetic bacteria and he showed that in some of them the photosynthetic reaction could be represented by the equation

$$2H_2S + CO_2 \longrightarrow (CH_2O) + 2S + H_2O.$$

No production of oxygen could be detected in this reaction, which took place under strictly anaerobic conditions. It appeared that hydrogen sulphide acted as a hydrogen donor, being oxidised to free sulphur. By analogy, he suggested that in the photosynthesis of green plants it was possible that water acted as a hydrogen donor, and that the initial reaction was the photolysis of water.

Telling evidence in favour of this view was provided at about the same time by Ruben. Using the stable isotope ^{18}O he first prepared $H_2{}^{18}O$ and $C^{18}O_2$. He set up cultures of *Chlorella* in media containing these compounds,† collected the oxygen evolved and analysed it with the mass spectrometer. The results of these experiments are reproduced in Table 19.1.

Table 19.1

% ^{18}O in carbon dioxide	% ^{18}O in water	% ^{18}O in oxygen evolved
0·61	0·85	0·86
0·50	0·20	0·20
0·40	0·20	0·20

It seems clear that the oxygen evolved is derived from the water and not from the carbon dioxide. Later it was shown by Vishniac and Ochoa in 1951 that the hydrogen atoms from the water are taken up by the hydrogen carrier TPN. On this basis we might provisionally write for the 'light reaction':

$$2H_2O + 2TPN \xrightarrow[\text{chlorophyll}]{\text{light}} 2TPNH_2 + O_2.$$

Having acquired a molecule of $TPNH_2$ the plant would be in a position to begin chemical operations. It might either use the $TPNH_2$

† The isotopes of oxygen undergo slow interchange between water and carbon dioxide in solution. Thus even if one starts with $H_2{}^{18}O$ and $C^{16}O_2$, there will be some $C^{18}O_2$ present at the end of the experiment.

to insert hydrogen into carbon dioxide or it might present the $TPNH_2$ to a mitochondrion and receive 3 molecules of ATP in exchange. In principle both these paths are followed, but events do not take just this course.

For some twenty-five years fruitless efforts were made to show that isolated chloroplasts were capable of carrying out the whole process of photosynthesis. Success came in 1954 to Arnon who devised a method of preparing chloroplasts from spinach leaves without damage to them. Illuminated suspensions of isolated chloroplasts were shown to be capable of carrying out all the essential processes of photosynthesis. The suggestion advanced in the preceding paragraph that photosynthesis involves co-operation between chloroplasts and mitochondria is therefore superfluous, and indeed the relative scarcity of mitochondria in the mesophyll cells of plants could have been advanced as an argument against it. It becomes necessary to think again about how the energy potentially available as $TPNH_2$ can be realised in the form of ATP. A partial answer to this question is provided by the discovery that cytochrome is present in chloroplasts. This might suggest that the chloroplast has a built-in mitochondrion; but again the suggestion is not borne out by the evidence, for isolated chloroplasts are incapable of oxidising $TPNH_2$. But isolated chloroplasts, when illuminated, are able to produce ATP, a process which we will call photophosphorylation to distinguish it from the comparable process of oxidative phosphorylation which goes on in mitochondria.

We have excellent evidence that both $TPNH_2$ and ATP become available as a result of the 'light reaction' and we may now consider how they are used in the process of carbon dioxide fixation, which is the 'dark reaction'.

Carbon dioxide fixation in plants does not seem to follow any single metabolic pathway, and as the subject is one of some complexity we will only touch upon it very briefly. Our present knowledge stems from the work of Calvin and Benson who, beginning in 1948, introduced a method based upon the use of ^{14}C. A suspension of *Chlorella* is set up in a long narrow tube under strong illumination, cooled so as to slow down the reaction, and kept stirred by a stream of bubbles of an air-CO_2 mixture. When the system has had time to settle down to a steady rate of photosynthesis some $^{14}CO_2$ is injected into the air stream, whereupon it is quickly dispersed throughout the suspension and is taken up by the cells. Then after a definite time interval, which may lie between

5 and 30 seconds, the whole suspension is allowed to fall into a large volume of boiling alcohol. This rapidly and effectively puts a stop to whatever reactions have been going on in the cells. The alcohol is evaporated off, the residue is taken up in water and the compounds it contains are separated by chromatography. The positions of compounds containing ^{14}C are ascertained by the method of autoradiography, that is, by placing the chromatogram in contact with a photographic plate which is then locally darkened by the radiation. After 5 seconds' exposure to $^{14}CO_2$ the main radioactive spot is found to occupy the same position upon the chromatogram as would be occupied by a spot of glyceric acid 3-phosphate. After longer exposures the radioactivity spreads to an ever-increasing number of compounds among which sugar phosphates are conspicuous.

Thus it appears that the pathway of carbon dioxide fixation leads into the pathway of glycolysis; given $TPNH_2$ and ATP, synthesis of glucose from glyceric acid 3-phosphate is known to be possible. But there is more to it than this, for the radioactivity also appears at an early stage in various compounds having 5 or 7 carbon atoms.

The early stages were difficult to work out, and as has been mentioned already various pathways are involved. What seems to happen is that the synthesis of carbohydrate does not build up directly upon carbon dioxide, but instead carbon dioxide is combined with a 5-carbon compound ribulose 1,5-diphosphate (see Appendix, p. 318) and enters a cycle of reactions which is illustrated in Fig. 19.2. The 6-carbon atom compound formed by the combination of carbon dioxide with ribulose 1,5-diphosphate at once breaks down into 2 molecules of glyceric acid 3-phosphate; that is to say, starting with 3 molecules of carbon dioxide (as we must do to get the equations to balance) 6 molecules of glyceric acid 3-phosphate are produced. By the reverse of the corresponding steps in glycolysis these give, with 6 ATP and 6 $TPNH_2$, 6 molecules of glyceraldehyde 3-phosphate. One of these eventually goes on to become part of a glucose molecule; the other 5 undergo a very complex rearrangement, which we shall not go into, to give 3 molecules of ribulose 5-phosphate. The 3 molecules of ribulose 5-phosphate are further phosphorylated by ATP to restore the original 3 molecules of ribulose 1,5-diphosphate which re-enter the cycle.

It is known that glucose can be formed from glyceraldehyde 3-phosphate by the reversal of glycolysis without any further help from ATP, and this is readily understandable since glyceraldehyde 3-phosphate is on the 'energy hump' of glycolysis (see Fig. 16.4). From Fig. 19.2 we see that in order to produce one molecule of glyceraldehyde 3-phosphate, available for further synthesis to glucose, we need 9 ATP + 6 $TPNH_2$. So for the synthesis of one molecule of glucose from water and carbon dioxide we need altogether 18 ATP + 12 $TPNH_2$.

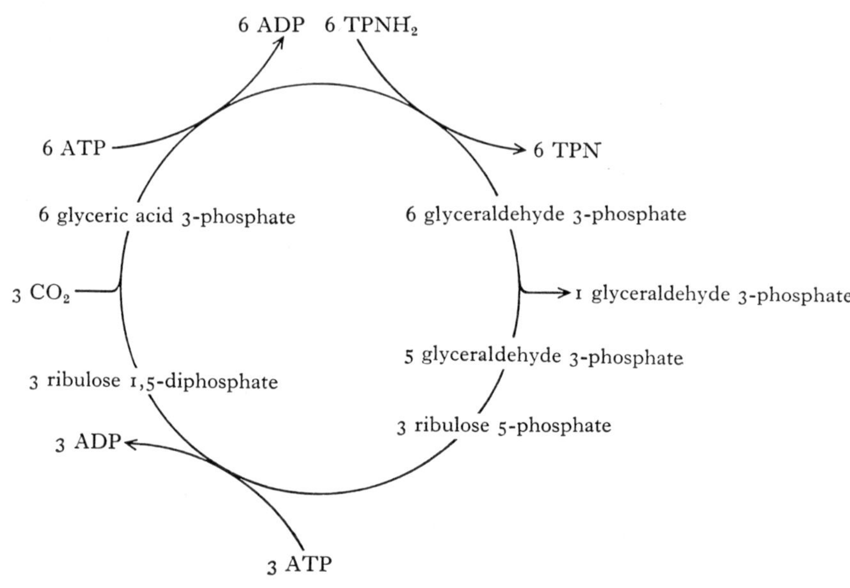

Fig. 19.2. One of the pathways of carbon dioxide fixation in photosynthesis.

Is this a reasonable conclusion from the point of view of energy relations? The reactions we have to consider are the following:

			$\Delta F'$
$6CO_2 + 6H_2O$	$\longrightarrow$	$C_6H_{12}O_6 + 6O_2$	+685,600
$6O_2 + 12TPNH_2$	$\longrightarrow$	$12TPN + 12H_2O$ ($-52,600 \times 12$)	−631,200
18ATP	$\longrightarrow$	18ADP (-8900×18)	−160,200
			−105,800

The process as a whole has a negative value for $\Delta F'$, so we can be confident that it will tend to go in the right direction.

The foregoing calculation, of course, refers only to the 'dark reaction'. When we come to examine the energy relations of the 'light reaction' we immediately encounter problems which demand a new approach.

Although for purposes of describing the workings of optical instruments we can think of light as a continuous wavemotion, this conception is inadequate to handle problems concerning the energy of light. For the past half-century physicists have known that the energy of light comes not in a continuous stream but in little packets, known as quanta. Just as an electric current in a wire, originally thought of as the continuous flow of 'electric fluid', is now known to consist in the flow of charged particles, the electrons, so now a beam of light is known to be made up of units which are called photons, and with each photon there is associated a quantum of energy. The amount of energy in a quantum is not a universal constant (as is the charge of an electron) but is inversely proportional to the wavelength, λ, of the light—or, alternatively, directly proportional to its frequency, ν. The energy of a photon is given by the product $h\nu$, where h is Planck's constant, one of the fundamental quantities of physics, having a value of $6{\cdot}62 \times 10^{-27}$ erg-sec. For red light of wavelength 6500 Å, which is near one of the absorption peaks of chlorophyll, we have

$$\lambda = 6{\cdot}5 \times 10^{-5} \text{ cm},$$

$$\nu = c/\lambda, \quad \text{where } c \text{ is the velocity of light, } 2{\cdot}99 \times 10^{10} \text{ cm/sec.,}$$

$$= 4{\cdot}6 \times 10^{14}/\text{sec.}$$

Then

$$h\nu = 6{\cdot}62 \times 10^{-27} \times 4{\cdot}6 \times 10^{14}$$

$$= 3{\cdot}05 \times 10^{-12} \text{ ergs.}$$

This is the energy in a quantum of red light of wavelength λ = 6500 Å.

Quanta are indivisible and the reactions in which they take part are all or nothing. When a quantum is taken up by an atom the consequence is that an electron transfers to a more distant orbit. This in effect involves moving the negative charge of the electron away from the positive charge of the nucleus, a process requiring energy, and all the energy of the photon is used up in this process. The energised electron may then fall back to some closer orbit, some energy being lost by re-radiation of light. But it can happen that the energised electron slips over, as it were, into the electron circulation of a neighbouring atom of some different element, and this move is potentially the first step of a photochemical reaction. According to the Stark–Einstein law of photochemical equivalence each molecule taking part in a chemical reaction induced by exposure to light absorbs one quantum of the radiation causing the reaction.

The object of the exercise in which we are about to engage is to assess the quantum efficiency of photosynthesis. By convention, and for convenience of measurement, photosynthesis is reckoned in terms of the amount of oxygen produced. We are therefore asking how many quanta are required for the liberation of one molecule of oxygen in the photosynthetic reaction.† This implies that we are going to take cognisance of what happens at the level of individual atoms and molecules; we are going to have to come up for a brief moment out of our comfortable dugout in classical thermodynamics and join action in the field of molecular biophysics.

The free energy required for the synthesis of carbohydrate is 685,600 cal./mole of glucose, or 114,300 cal./mole of oxygen. It will be convenient to recalculate this figure in terms of ergs/molecule. There are $4{\cdot}187 \times 10^7$ ergs in a calorie (Joule's mechanical equivalent of heat) and $6{\cdot}024 \times 10^{23}$ molecules in a mole (Avogadro's number). So to convert cal./mole into ergs/molecule we have

$$114{,}300 \times \frac{4{\cdot}184 \times 10^7}{6{\cdot}024 \times 10^{23}} = 7{\cdot}95 \times 10^{-12} \text{ ergs/molecule of oxygen.}$$

When we come to compare this figure with the energy of a quantum of red light, $3{\cdot}05 \times 10^{-12}$ ergs, we encounter the difficulty that whereas the energy requirement for photosynthesis is reckoned as free energy, the energy of the quantum is reckoned as total energy, and we do not know how much of this is 'free', the classical concepts of temperature and pressure being inapplicable to single molecules and to quanta. But let us assume that we may treat it as free energy and see where we get to. As a first approximation we may divide 7·97 by 3·05 and obtain the figure of 2·6 quanta/molecule of oxygen. Since we are obliged to deal in whole numbers this would suggest that the lowest conceivable figure for quantum efficiency is 3. But this approximation is based upon unrealistic assumptions as to the mechanism of photochemical reactions. If we are to take the view, as we have done hitherto, that the first reaction in photosynthesis is the splitting of water molecules with the liberation of molecular oxygen, then we must envisage something like the following:

$$2H_2O \rightleftharpoons O_2 + 4H^+ + 4\epsilon.$$

Thus to liberate 1 molecule of oxygen we have to take away 4 electrons. We have no grounds for believing that in photochemical reactions

† Strictly speaking, if we are assessing quantum efficiency we should ask how many molecules of oxygen are liberated per quantum.

the energy of photons can be saved up as though in a bank and then paid out later in amounts appropriate to the energy requirements. This is what the calculation in the previous paragraph implies. On the contrary, the evidence is quite clear that in photochemical reactions one quantum has to be used for each electron which is energised, even though the energy of the quantum may be greatly in excess of the energy needed by the electron for its subsequent operations; any energy left over simply goes to waste, i.e. is degraded into heat or re-radiated as light. We must therefore accept the proposition that the lowest conceivable figure for quantum efficiency is 4.

We may now ask whether the energy of the quantum, when transferred to the electron, is sufficient to enable it to enter into subsequent reactions. The next reaction of which we have knowledge is the capture by TPN of the hydrogen atoms split off from water, which we may write as

$$\text{TPN} + 2\text{H}^+ + 2\epsilon \longrightarrow \text{TPNH}_2.$$

We can tackle this problem in terms of oxidation-reduction potential. The oxidation-reduction potential of a system obeying the equation

$$2\text{H}_2\text{O} \rightleftharpoons \text{O}_2 + 4\text{H}^+ + 4\epsilon$$

at 25° C, pH 7 and oxygen at 1 atm. pressure is known to be +0·82 volts. For the system TPN ⇌ TPNH_2, under the same conditions, $E_0' =$ −0·28 volts. This means that the electrons have to be moved from +0·82 volts to −0·28 volts, i.e. through a potential difference of 1·1 volts. The energy required per electron is therefore 1·1 electron-volts. Knowing that 1 electron-volt = $1{\cdot}601 \times 10^{-12}$ ergs we can convert the energy of the quantum into electron-volts for comparison. It works out at 1·9 electron-volts. We may therefore conclude that the energy of a quantum of red light is amply sufficient to transfer an electron (or a hydrogen atom) from water to TPN. *A fortiori*, there is sufficient energy to transfer electrons to cytochrome. In view of the assumption we had to make earlier on, namely that all the energy of the quantum is 'free', there is some comfort in this ample margin.

Looking back to p. 206 we see that for the synthesis of 1 mole of glucose we need 12 TPNH_2 + 18 ATP. For every molecule of oxygen liberated in photosynthesis we must therefore provide 2 TPNH_2 + 3 ATP. For the 2 TPNH_2 we must allocate 4 quanta. Not a great deal is known about the details of photophosphorylation in chloroplasts, but there is some evidence that, as with mitochondria, for every pair of electrons passing through cytochrome 1 molecule of ADP is phosphorylated. At the worst we must allocate a further 6 quanta to provide the 3 molecules of ATP. We thus reach the conclusion that for every molecule of oxygen liberated in the photosynthesis of glucose at least 4

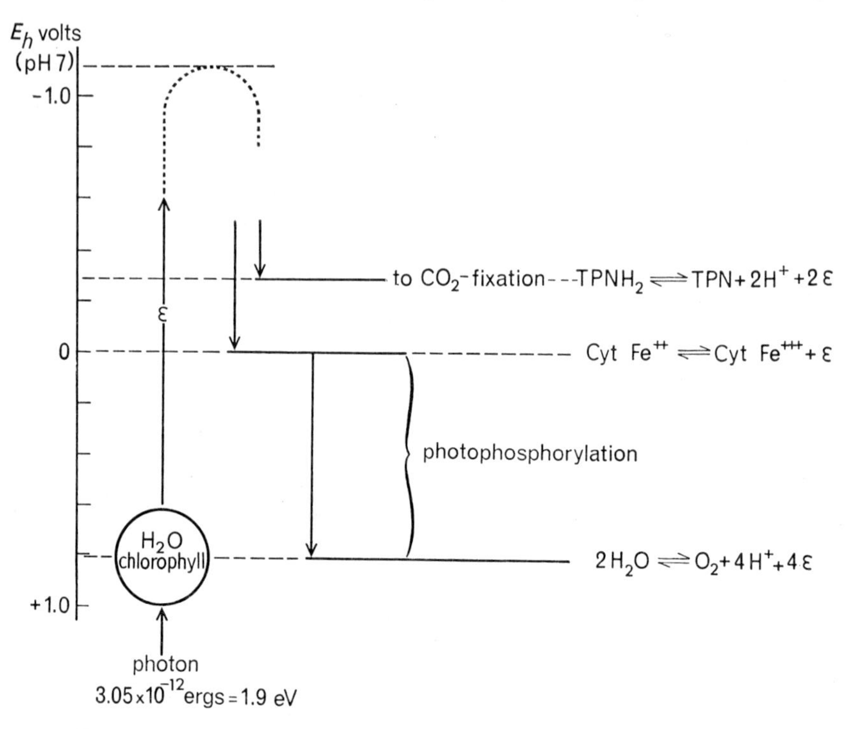

Fig. 19.3. The energy relations of the 'light reaction' expressed in terms of oxidation-reduction potential.

The energy of the photon, 1·9 eV, drives the electron far up the scale, making it a powerful reductant. The energised electron, joining up with a hydrogen ion, may either be captured by TPN and used in CO_2-fixation or it may be captured by cytochrome where its energy is used to produce ATP. For every 4 electrons captured by TPN a molecule of oxygen is liberated. But the electrons which pass through cytochrome eventually reduce oxygen to water so there is no net liberation of oxygen in respect of electrons which traverse this path.

This diagram is not altogether respectable since it implies that the energy of the photon can be treated as free energy.

No attempt has been made to incorporate modern views of the mechanism of the 'light reaction'. The diagram is intended only to illustrate how a photon of red light can give sufficient energy to an electron to enable it to be captured either by TPN or by cytochrome.

Note. The baseline conditions here specified are: 25° C, pH 7 and oxygen at 1 atm. pressure, for which the potential is +0·82 volts. Figure 12.2, in which the 'oxygen potential' is given as +1·24 volts, assumes pH 0.

and probably not more than 10 quanta will be required. The higher figure implies that there is no coupling between the reactions leading to $TPNH_2$ and the reactions leading to ATP. But experiment shows that there is a stoichiometric relationship as follows,

$$2TPN + 2H_2O + 2ADP + 2(P) \longrightarrow 2TPNH_2 + 2ATP + O_2,$$

so presumably there is a 'dark reaction' coupling of some kind, with the possibility of greater efficiency. Clearly our estimate of quantum efficiency will depend upon the mechanism we propose for the 'light reaction'. The currently favoured concept is that it involves two quantum steps in series; but the evidence for this concept is too difficult and complex to be gone into here.

The experimental determination of quantum efficiency is not altogether an easy matter. The difficulty is to decide how much light is actually absorbed. One can readily measure the intensity of a beam of light falling upon a suspension of *Chlorella* and one can readily measure the intensity of the beam after it has passed through the suspension, but the difference between the two values does not enable us to calculate the amount of light absorbed because some of the light is scattered and the amount scattered is difficult to measure. By a variety of methods, applied to a variety of photosynthetic systems ranging from suspensions of *Chlorella* to whole leaves, figures ranging from 8 to 12 quanta per molecule of oxygen have been obtained, and the general consensus of opinion is that the quantum efficiency of photosynthesis lies within these limits. In view of the uncertainties associated with the theoretical estimate of 4–10 quanta and in view of the difficulties of measurement this may be regarded as reasonable agreement.

Taking 10 quanta as an average figure and relating the energy of 10 quanta of red light at $3{\cdot}05 \times 10^{-12}$ ergs/quantum to the figure of $7{\cdot}95 \times 10^{-12}$ ergs/molecule of oxygen the overall efficiency of photosynthesis works out at about 40%. The quantum efficiency is more or less constant over the range of wavelengths from 4000 to 6500 Å. This is as we might expect in view of the fact that a single quantum can only energise a single electron and any excess energy given to the electron simply goes to waste. But of course the overall efficiency must be less with light of shorter wavelengths.

When we extend our considerations to include photosynthetic bacteria it appears that the essential feature of photosynthesis is the energising of electrons by light in the presence of chlorophyll and the subsequent capture of some part of their energy in the form of ATP. The production of $TPNH_2$ by the splitting of water is a device favoured by plants. As already mentioned, some photosynthetic bacteria obtain hydrogen for carbohydrate from H_2S with liberation of sulphur. Others can make use of molecular hydrogen to reduce TPN, having enzymes which catalyse the reaction $H_2 \rightleftharpoons 2H^+ + 2\epsilon$. In these bacteria it appears that the electrons travel in a closed circuit, being energised in chlorophyll and de-energised in cytochrome with the production of ATP, a process which has been called 'cyclic photophosphorylation'.

20

REFLECTIONS UPON THE USE OF ENERGY BY ORGANISMS

The writers of popular science never tire of telling us how wonderful Nature is and with what effortless superiority she conducts her operations of energy transformation as compared with the derisory attempts of human engineers; how the energy of glucose is transformed into the energy of ATP with an efficiency of 50–60% as compared with a mere 30–40% for a modern power station—and so on.

But let us be fair to our engineers. The power station is a heat engine and starts with the disadvantage that heat is not a 'useful' form of energy. When we remember that even a perfect heat engine cannot convert all the heat energy of its fuel into electrical energy, but only some part of it depending on the temperatures between which it operates (see p. 135), we see that the power station does in fact reach about 50% of its possible maximum efficiency, which certainly does not put it in a class below the yeast cell. Furthermore, since the calculated efficiency of the cell is based upon the conversion of 'useful' energy from one molecular form into another, it would be reasonable to compare its efficiency with the efficiency of some man-made device which converts one form of 'useful' energy into another—an electric motor, for example. An engineering firm which marketed a large electric motor having less than 90% efficiency would soon be out of business.

The power station takes in chemical energy in the form of coal, degrades it into heat, converts part of the heat into mechanical energy and finally converts this into electrical energy. Figures not being available for the performance of electric fish, the best comparison we can make is with a man who takes in chemical energy in the form of food and converts it into mechanical energy in his muscles. How well does he do?

In attempting to answer this question we have to remember that man is not different from other organisms in that he needs energy not only for his manifest activities, but also just to keep alive. If in reckoning the thermodynamic efficiency of a manual labourer we had to take account of the energy he dissipated during his hours of rest and recreation his efficiency might not look too good. A reasonable basis of reckoning would be to measure the work done and the fuel consumed during a limited period of effort, and then deduct the fuel which would have been consumed during the same period of rest. Measurements of this sort have been made from time to time by physiologists using human subjects. The device commonly used to measure the work done is the bicycle ergometer, which is no more than a fixed bicycle having the back wheel replaced by some means of measuring horsepower. While pedalling the bicycle the subject breathes into a respirometer and his oxygen consumption is registered. The results of an experiment of this type were as shown in Table 20.1.

Table 20.1

Rate of oxygen consumption	
(*a*) Pedalling ergometer under load	1·338 litres/min.
(*b*) Sitting still on ergometer	0·278
Difference	1·060
Rate of doing work, expressed as rate of heat production	1080 cal./min.

We will assume that the subject uses glycogen as fuel.† The free energy change for the complete oxidation of one glucose unit of glycogen is $-690,700$ cal./molar unit—just a little more than the corresponding figure of $-685,600$ cal./mole of glucose. In terms of oxygen consumed this is $690,700 \div 6 = 115,100$ cal./mole of oxygen. One mole of oxygen at N.T.P. occupies 22·4 litres, so the excess oxygen consumed, 1·06 litres, was 0·0473 moles. Thus in 1 minute the free energy available from the oxidation of carbohydrate was $0·0473 \times 115,100 = 5450$ cal. The amount of work done on the ergometer in a minute was 1080 cal., which means that the overall efficiency of the human machine is $\frac{1080}{5450} = 19·8\%$—not so very impressive after all.

Certain bacteria apart, it does seem that most organisms are reasonably thrifty in their use of free energy, and do not allow it to be degraded into heat without having something to show for it. In our own bodies the consumption of oxygen—as good a measure as any of energy utilisation—is obviously related to the amount of work we do. But in our muscles the enzymes and co-factors necessary for the oxidation of glycogen are always present. Oxygen is available. How is it that the

† This assumption is not strictly true. The respiratory quotient, i.e. the ratio (carbon dioxide produced)/(oxygen consumed) is about 0·9 instead of 1·0 as it would be if carbohydrate alone was used as fuel.

enzyme systems of muscle are prevented from consuming the stores of glycogen while the muscle is at rest? To understand the means whereby restraint is imposed we have to remember that the reactions which these enzymes carry out are coupled into a sequence which may be likened to a train of gears. No one wheel can move without all the others moving, and if one wheel is locked the whole train comes to rest. We can readily appreciate that any step in the sequence suffering from shortage of reactant can hold up the whole sequence, and that the smaller the amount of the reactant the more quickly is its shortage likely to make itself felt. Thus it is the carrier molecules, and ATP in particular, which, being present in amounts of the order of 10 millimoles per kilogram of tissue, are in the best position to control the progress of the whole sequence. In muscle the oxidation of glucose—or its conversion to lactic acid—involves the net conversion of ADP to ATP. This conversion can continue only for so long as there remains some ADP as yet unphosphorylated; so when all the ADP has been converted to ATP this reaction, and with it the whole sequence of reactions, comes to a stop. As soon as we start to use our muscles ATP is broken down, ADP becomes available and the metabolic pathway becomes active again.

But although there is usually a very close coupling within any one metabolic pathway it is evident that the many and various biochemical reactions which go on within the cell cannot all be in complete functional continuity. All cells seem to contain enzymes which hydrolyse ATP. Yeast juice contains such an enzyme, and the continued fermentative activity of yeast juice depends upon the fact that ATP is hydrolysed as fast as it is formed, with degradation of all its precious free energy into heat. In intact yeast cells it is hard to believe that this completely wasteful process finds any place; somehow or other enzymes and substrate must be kept apart until they can meet in the appropriate situation.

In so far as the glycolytic enzymes are free in the cytoplasm whereas the enzymes of the Krebs cycle and the cytochrome system are restricted to the mitochondria we already have a structural basis for functional separation. It seems likely, however, that the really significant structural units are nothing like so large as a mitochondrion. As described on p. 199 the activity of the cytochrome system survives disruption of the mitochondria, which indicates that a mitochondrion comprises several functional units. Equally, the fact that activity disappears with disappearance of the characteristic double membrane structure is added justification for believing that in the mitochondrion the various enzymes and carriers are not just packed all together in a bag but are arranged in some sort of ordered pattern. It may be that the structural proteins of the mitochondrial membranes are the very enzymes themselves, with their active centres exposed to one side or the other. But to say this is to go far in advance of any evidence which is yet to hand. The idea of some spatial organisation of the enzymes and carriers is compelling, but in our present state of knowledge diagrams illustrating arrangements of enzymes in the mitochondrial wall or of chlorophyll in the chloroplast are no more than *possible* arrangements, based upon what we know of the shapes and sizes of some of the molecules and of the spaces into which they have to fit.

The task of the biochemist in discovering how the energy of a complex molecule is mobilised in the energy-rich bonds of ATP has been to trace the molecule's progress through successive stages of breakdown and to ascertain at what stages the energy-rich bonds of ATP are generated. In the preceding chapters the impression has perhaps been given that in tracing metabolic pathways biochemists are accustomed to test their conclusions against the requirements of thermodynamics at every stage in this process. In principle this is perhaps how it should be done, but in practice there are various reasons why it would serve no useful purpose to go about the problem in this way. As has been mentioned (p. 140) it is not easy to measure $\Delta F'$, and where $\Delta F'$ values are not available it may well take longer to obtain them than to set up a reconstituted system and see if it works. Many of the listed values of $\Delta F'$ may be in error by as much as 10%. In any case we do not know the concentrations of reactants and products within the cell, so even if we have accurate values for $\Delta F'$, the standard free energy change, we are not able to calculate ΔF, the actual free energy change. For the biochemist thermodynamics is not so much a useful tool as it is a convenient theoretical framework upon which to display his wares.

The rate at which a reaction takes place is often of greater practical importance than its free energy change. It is the relative rates of reactions which largely determine whether one metabolic pathway rather than another is followed, and in their turn the rates of reactions are determined by the presence and activities of enzymes. Thus it is that the biochemist is preoccupied with enzymes rather than with energy.

The biochemist can describe with some confidence the metabolic pathways along which energy flows into the energy-rich bonds of ATP, and we are apt to be left with the impression that the problem of energy transfer in organisms is thereby solved. But it is important to realise how limited our knowledge really is and how its limitations are mainly those of the thermodynamic framework upon which it rests. We may ask—and be told—how much ATP can be had for the oxidation of a gram of glycogen; it would also be in order to ask how many ergs of mechanical work can be produced when this amount of ATP is used up in a muscle. But if we ask how ATP makes a muscle contract, that is a different matter. In asking this question we at once transfer discussion from the plane of thermodynamics to the plane of molecular biophysics. We cannot say how the energy of ATP is converted into work by a muscle without reference to the molecular basis of muscular contraction. We do not understand the molecular basis of muscular contraction any more than we understand the molecular basis of enzyme action because in neither case is the molecular structure known in sufficient detail. These are problems for the future. In the meantime we would do well not to allow our extensive knowledge of the metabolic pathways leading to ATP to blind us to our continued ignorance of how ATP is used.

PART III

INHERITANCE

21

THE GENETICS OF HIGHER ORGANISMS

Genetics, *stricto sensu*, is the science of inheritance and its study may be pursued without any reference to the structure of the germ cells or to the details of mitosis and meiosis. But we will here interpret the word in its widest sense and we shall be particularly concerned with that aspect of the subject known as cytogenetics. Cytogenetics is a union and integration of two lines of investigation which were originally entirely separate, namely the study of inheritance and the study of the microscopic structure of cells, and from this parentage there sprang a new branch of biological science of exceptional vigour and fruitfulness.

If cytogenetics is indeed a union of two disciplines it cannot be said to have come into existence until these separate disciplines had each a firm scientific basis. The publication of Mendel's paper in 1866 is universally acknowledged as marking the birth of the science of inheritance; but the paper was published in a little known scientific journal where it remained unnoticed until discovered by de Vries in 1900. Cytology got off to a slightly earlier start. Hofmeister in 1848 observed that at cell division the nucleus was resolved into smaller bodies, now called chromosomes. The essential facts of mitosis and meiosis, though not all the details given in Chapter 6, were established in the decade 1880–90, following improvements in the techniques of fixing and staining and above all of section cutting, as described in Chapter 4.

Thus when Mendel's work was discovered in 1900 the stage was already set for the emergence of the new subject of cytogenetics, and in 1902 Sutton was the first of the many, including de Vries himself, who drew attention to the parallel between the behaviour of the chromosomes and the behaviour of Mendel's independently segregating characters.

Although it is assumed that the reader is familiar with the basic facts and theory of Mendelian inheritance, a brief account will be given to illustrate the way in which Mendel worked and to clear up certain possible misconceptions. Among the latter is the widely held belief that Mendel formulated certain laws of inheritance as a result of his experiments. In fact, in Mendel's original paper the principles which he discovered were never formally stated as laws; his successors have therefore been free to state 'Mendel's Laws' in various terms, often incorporating such words as 'gene' which were unknown to Mendel.

Numerous investigators prior to Mendel had studied hybridisation in plants without discovering any principles of general applicability. The offspring of a cross in some cases resembled one or other of the parents, in other cases they were intermediate. When these hybrids were bred together some of their progeny were occasionally noted as resembling one or other of the grandparents rather than the parents themselves. Mendel was the first to apply quantitative methods to the study of inheritance. Not content with describing the progeny of hybrids and placing them in categories, he set out to determine the relative numbers in each of the categories he recognised; and it was to his quantitative approach that his success was largely due. He also chose his material, the garden pea (*Pisum sativum*), very wisely, having regard to its floral structure, and he carried out careful preliminary tests to establish the purity of the stocks from which he intended to breed. He selected seven pairs of differentiating characters for study, of which we shall mention only two: (i) the form of the dried seed, either round or wrinkled, and (ii) the colour of the cotyledons (visible through the transparent seed coat), either yellow or green.

In the first experiments he studied single pairs of differentiating characters. He applied pollen from a 'round' plant to the stigmata of a 'wrinkled' plant and vice versa. All the seeds produced by these plants were round, irrespective of the direction in which the cross was made. These hybrid seeds were then planted out and 253 hybrid plants were reared from them. The hybrid plants were self-fertilised and produced 7324 seeds of which 5474 were round and 1850 were wrinkled. The ratio of round to wrinkled was 2·96 : 1, which Mendel noted as a close approximation to 3 : 1.

The experiment was carried further. Plants grown from the wrinkled seeds and self-fertilised produced only wrinkled seeds—they were, in Mendel's terminology, 'constant', i.e. they bred true like the original 'wrinkled' stock. Of 565 plants grown from the round seeds and self-fertilised 193 were constant, giving round seeds only, while 372 gave both round and wrinkled seeds approximately in the ratio of 3 : 1. Among the round seeds the ratio of non-constants to constants was 1·93 : 1, a close approximation to 2 : 1.

Thus of the 7324 seeds produced by the first generation of hybrid plants 25% were constant wrinkled, 25% were constant round and 50% resembled their parents in giving round and wrinkled seeds in the ratio 3:1. This experiment was carried on as far as the sixth generation; and substantially the same results were obtained with the other six pairs of differentiating characters which had been chosen for study.

As stated above, Mendel and his contemporaries were familiar with the fact that in a hybrid one or other of the parental characters might be suppressed, and in designating 'round' as dominant and 'wrinkled' as recessive Mendel was not announcing a discovery but simply inventing useful descriptive terms. The decisive step which he took was to suppose that the characters 'round' and 'wrinkled' were not merged or blended together in the hybrid, but retained their individuality and could be distributed separately among the egg cells and pollen cells of the hybrids. Two further assumptions were needed to account for the numerical ratios which had been established by experiment: (i) that the hybrid produces 'round' egg cells and 'wrinkled' egg cells in equal numbers—and pollen cells likewise—and (ii) that it is a matter of chance which of the two sorts of pollen will unite with an egg cell. Expressed in symbolic form, A denoting 'round' and a denoting 'wrinkled',

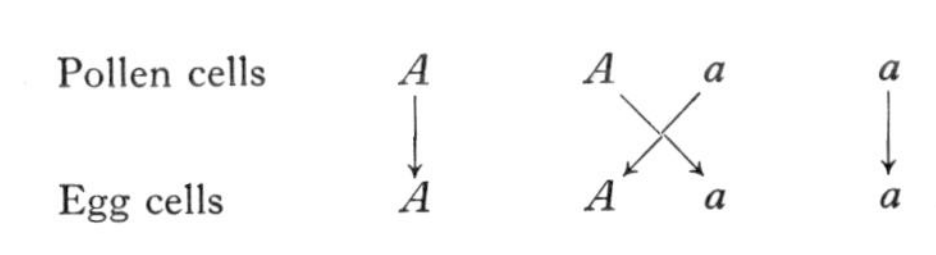

give

$$\frac{A}{A} + \frac{A}{a} + \frac{a}{A} + \frac{a}{a},$$

which represents (in Mendel's symbolism)

A	+	$2Aa$	+	a
constant 'round'		hybrid		constant 'wrinkled'

According to the modern convention the last line would be written

$$AA \quad + \quad 2Aa \quad + \quad aa,$$

and we will adhere to the modern convention in the rest of this description of Mendel's work.

In the second series of experiments Mendel investigated the results of crosses involving more than one pair of differentiating characters.

Seed plants from a stock having round seeds and yellow cotyledons were fertilised with pollen from a stock having wrinkled seeds and green cotyledons. The seeds so produced (the F_1 generation as we shall now call it) were all round and yellow. From these seeds 15 plants were raised, which on being self-fertilised yielded 556 seeds of the F_2 generation. These were:

yellow, round	315
yellow, wrinkled	101
green, round	108
green, wrinkled	32

This is of course the familiar 9:3:3:1 ratio, in an approximate form as might be expected from the small numbers involved. Mendel did not, in fact, draw attention to the existence of this particular ratio. He was more concerned to show, by further breeding, that there were not four categories but nine. Denoting 'round', 'wrinkled' by A, a as before, and 'yellow', 'green' by B, b, these categories are

$$AABB,\quad AABb\quad AAbb,\quad AaBB,\quad AaBb,\quad aaBB,\quad aaBb,\quad Aabb,\quad aabb,$$

from which it may be deduced that the differentiating characters are independently assorted in the reproductive cells.

It would be impossible, without taking up more space than can be spared, to describe the other experiments which Mendel carried out and to follow his penetrating analysis of their results. His outstanding contributions are those which have been summarised above—the idea that the characters retain their individuality from generation to generation. Although he wrote elsewhere of '...elements...which determine opposite characters...' he preferred to discuss his results in terms of the visible characters rather than in terms of the genetic elements which controlled them, and this is a little surprising since the principles he discovered can be formulated so much more precisely when expressed in the latter form.

Although Mendel's main contribution to genetics has stood the test of time it was only to be expected that some reinterpretation would become necessary in the light of further study. One of his generalisations, that of independent assortment of two pairs of characters, became a casualty in 1904. Bateson and Punnett, working with sweet peas, investigated the inheritance of two pairs of differentiating characters: purple flower, dominant to white flower, and long pollen grain dominant to round pollen grain. The parent plants were 'purple, long' and 'white, round'. According to the principle of independent assortment a 9:3:3:1

ratio was to be expected in the F_2 generation; in fact, Bateson and Punnett found as follows:

purple, long	583
purple, round	26
white, long	24
white, round	170

The observed ratio 'purple, long':'white, round' is 3·4:1, instead of the expected 9:1, for these combinations of characters. The characters were therefore not assorted independently; according to Bateson and Punnett 'purple' was in some way coupled to 'long' and 'white' to 'round'. We now speak of this coupling as linkage.

Although the suggestion was already current that the determinants of the Mendelian characters were carried on the chromosomes, Bateson and Punnett did not seek to interpret their experiment on this basis, no doubt because the obvious explanation—that both pairs of characters were associated with the same pair of chromosomes—would fail to account for the combinations 'purple, round' and 'white, long' which appeared in small numbers.

Five years later Janssens published a description of the pairing of homologous chromosomes during meiosis in the amphibian *Batrachoseps*. His figures show the chromosomes twisted together and that at certain places, the chiasmata, one lies across the other. Janssens suggested that where chiasma occurs there could be breakage and reunion whereby the terminal portions of the chromosomes could be exchanged. This has already been discussed elsewhere (p. 88). The suggestion went very far beyond any evidence he could adduce to support it; and indeed clear cytological evidence of crossing-over was not obtained until 1937. Nevertheless, the idea of crossing-over offered a possible explanation of incomplete linkage and these observations and suggestions by Janssens were the starting point for the remarkable work of Morgan and his associates on *Drosophila*.

There are obvious advantages in choosing as material for genetic study a small insect which breeds rapidly and whose external appearance provides a wealth of structural detail in which variation can readily be detected. In addition to these advantages *Drosophila* offers another which is of strong appeal to the cytogeneticist—it has only four pairs of chromosomes and these pairs are individually recognisable under the microscope. The diploid set of chromosomes in *Drosophila* is shown in Fig. 21.1.

Mendel had availed himself of different stocks of garden pea which he obtained from seedsmen. Morgan had to depend mainly upon variant forms which appeared spontaneously in the stocks which he kept in his laboratory. The reader will be aware that a spontaneous genetic change was first observed by de Vries in the evening primrose, *Oenothera lamarckiana*, and was called by him a mutation. From now on we shall be concerned with various mutant forms of *Drosophila* and this is a convenient point at which to introduce the symbolism which is conventionally used in the genetics of *Drosophila*. When a fly of abnormal

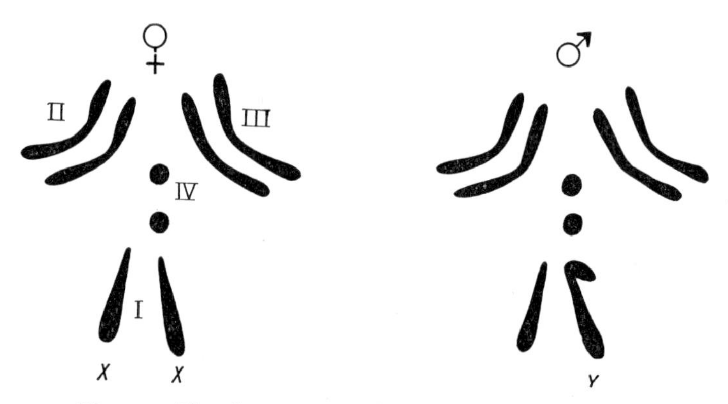

Fig. 21.1. The chromosomes of male and female *Drosophila*.

The pairs labelled I are the sex chromosomes, the other pairs being collectively known as autosomes. In the female the sex chromosomes are similar and are known as the *X*-chromosomes. In the male there is one *X*-chromosome; its partner, the *Y*-chromosome, is of different appearance and is genetically almost empty. In the male there is no crossing-over between the *X*-chromosome and the *Y*-chromosome, as is perhaps understandable; but it is further found that in the male there is no crossing-over between members of any pair of chromosomes, and for this no explanation is forthcoming.

appearance is found in the stock it is isolated and subjected to breeding tests; and if the abnormality passes these breeding tests it is concluded that a mutation has occurred and the mutant form is given a name which describes its appearance, e.g. *vestigial W*, denoting vestigial wings. The genetic element, or gene as we shall now call it, which determines the appearance of this character, is designated *vg*. The corresponding gene† which determines the appearance of normal wings in the normal or wild-type fly is designated vg^+, or, where there can be no ambiguity,

† More correctly, there is only one gene, which can exist in various states known as alleles (see p. 301); vg^+ is the dominant allele found in the wild type and *vg* is the mutant allele.

simply +. The words '*purple E, vestigial W, humpy B*' describe the appearance (the phenotype) of a fly which has purple eyes, vestigial wings and a humpy body. These mutant characters being recessive, such a fly would be homozygous for the genes concerned, designated respectively *pr*, *vg*, *hy*, and its genetic constitution (the genotype) would be written

$$\frac{pr \quad vg \quad hy}{pr \quad vg \quad hy}.$$

The offspring of a triple mutant mated to a wild-type fly would be written

$$\frac{pr \quad vg \quad hy}{+ \quad + \quad +},$$

and this heterozygous fly would be wild-type in appearance.

The decisive step which Morgan took was to suppose that the genes are linearly arranged in paired linkage groups and that closeness of linkage is a measure of their distance apart.

Let us consider three linked genes arranged in the order *A*, *B*, *C* upon a chromosome pair.

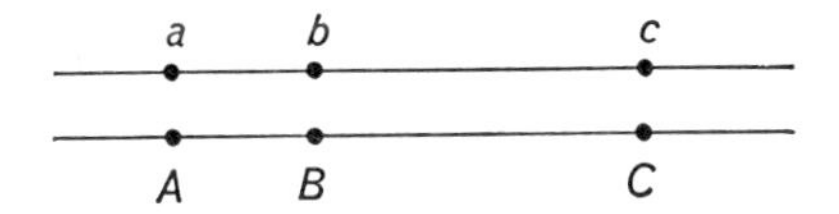

Let us suppose that crossing-over occurs with equal readiness at all points on the chromosome pair. From this it follows that the percentage of crossing-over between any two pairs of genes will be directly proportional to the distance between them. To test this hypothesis that the genes are arranged in a linear order we proceed as follows. We determine the percentages of crossing-over between *Aa* and *Bb*, between *Bb* and *Cc*, and between *Aa* and *Cc*. Denoting these percentages respectively as *AB*, *BC* and *AC*, it must follow, if our theory is correct, that

$$AB+BC = AC.$$

Now let us take the actual case of three linked mutant characters, all recessive: *yellow B*, *bifid W* and *white E*. When *yellow* and *white* are crossed recombination occurs with the appearance of the double mutant *yellow white* and the wild type, and these recombinants constitute 1·5 % of the offspring. The percentage of recombination between *yellow* and *bifid* is found to be 6·9 % and that between *bifid* and *white* is 5·4 %. The greatest percentage of recombination, 6·9 %, being between *yellow* and *bifid*, we conclude that these genes are the furthest apart, with *white* occupying the middle position. The ordering of the genes is then

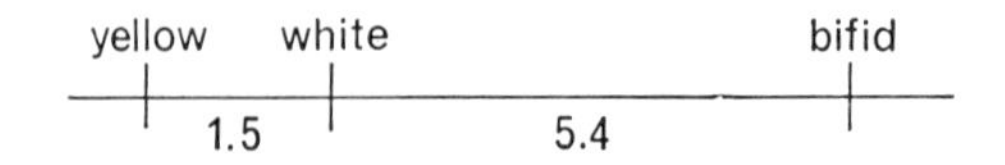

and since $1 \cdot 5 + 5 \cdot 4 = 6 \cdot 9$ additivity is preserved.

This test of additivity is crucial. If we decide to represent percentage of crossing-over as proportional to distance, then the distances between *yellow*, *white* and *bifid*, being additive, can be measured along one and the same line.† We then have to extend the data to include all the known genes; we have to determine the percentages of crossing-over between all the known genes within a linkage group and show that they are additive.

It might be thought that the percentage of crossing-over between genes could be obtained directly from the results of breeding experiments such as that of Bateson and Punnett. But although this is true, and the percentage of crossing-over can be ascertained from the extent to which the F_2 generation departs from a 3:1 ratio, the calculation is complicated. For this reason it is found more convenient to use not the F_2 generation, but the offspring of a 'recessive back-cross', for which the calculation is simple. It is also convenient to determine the percentages of crossing-over from a single breeding experiment in which all three pairs of genes are involved rather than to conduct three separate breeding experiments as envisaged in the hypothetical case considered above. Such a breeding experiment is known as a 'three-point cross'.

As an example we will take the mutations: *yellow B* (the wild-type body being grey), *cut W* (the wild-type wings being of normal shape), and *white E* (the wild-type eyes being red). If we mate a *yellow cut white* female with a wild-type male the female offspring are found to be of the wild type, as might be expected, but the male offspring are *yellow cut white*. This result tells us at once that these mutant characters are recessive and that they are sex-linked; that is to say, the mutant genes must lie on the *X*-chromosome (see legend to Fig. 21.1). If they were to lie on any of the autosomes both sexes would be wild-type since the

† Our assumption that crossing-over occurs with equal readiness at all points on the chromosome is of secondary importance. If it is true, then the distances between the genes, measured as percentages of crossing-over, will be directly proportional to the actual distances between the positions of the genes on the chromosome—otherwise they will not; but the question of linear order is not affected by this issue.

autosomes from the wild-type male would carry the dominant alleles; but if they lie on the X-chromosome then the male cannot supply any wild-type alleles to his sons, since the Y-chromosome of the male is genetically empty. The genotype of the mutant female parent may be provisionally written (i.e. without prejudice to the actual order of the genes) as

$$\frac{y \quad ct \quad w}{y \quad ct \quad w}.$$

The genotype of the wild-type male parent may be written as

$$\frac{+ \quad + \quad +}{\text{o}}.$$

The F_1 females are then

$$\frac{y \quad ct \quad w}{+ \quad + \quad +}$$

and the F_1 males are

$$\frac{y \quad ct \quad w}{\text{o}}.$$

In a normal recessive back-cross an F_1 heterozygote is mated with the recessive parent. In the present case, since the characters are sex-linked, we have to modify the normal procedure. On account of the sex-linkage only the females of the F_1 generation are properly heterozygous, i.e. containing both dominant and recessive alleles, so it is impossible to mate the F_1 heterozygote with the recessive parent, since this is also female. But because the genes in question are totally absent from the Y-chromosome, the F_1 males are, in effect, triple recessives. So we perform what is equivalent to a recessive back-cross by mating an F_1 female with an F_1 male.

If the three mutant characters are recombined in all possible ways by crossing-over we may expect to find 8 different phenotypes among the offspring. These phenotypes are listed in Table 21.1 together with the actual numbers found in an experiment in which 1000 offspring were examined.

Table 21.1

	Phenotypes			Numbers found
(1)	*grey*	*normal*	*red*	404
(2)	*grey*	*normal*	*white*	2
(3)	*grey*	*cut*	*red*	93
(4)	*grey*	*cut*	*white*	5
(5)	*yellow*	*normal*	*red*	7
(6)	*yellow*	*normal*	*white*	89
(7)	*yellow*	*cut*	*red*	1
(8)	*yellow*	*cut*	*white*	399

Of the 8 possible phenotypes, categories (1) and (8) are the parental combinations; the others show recombination of the parental characters. These recombinants are further analysed in Table 21.2.

Table 21.2

Phenotypes involving recombination between	Categories as in Table 21.1	Numbers as in Table 21.1	Total
grey/*yellow* and *normal*/*cut*	(3), (4), (5), (6)	93+5+7+89	194
grey/*yellow* and *red*/*white*	(2), (4), (5), (7)	2+5+7+1	15
normal/*cut* and *red*/*white*	(2), (3), (6), (7)	2+93+89+1	185

From a superficial inspection of the figures in Table 21.2 we may conclude that the provisional order we adopted, *y ct w*, is incorrect, and that the true order is *y w ct*, since the greatest number of recombinants is between *grey*/*yellow* and *normal*/*cut*. The test fails, however, on the criterion of additivity, since the number of recombinants between *grey*/*yellow* and *red*/*white* is 15, and that between *red*/*white* and *normal*/*cut* is 185, making a total of 200, whereas the observed figure for recombinants between *grey*/*yellow* and *normal*/*cut* is 194. But we have overlooked the fact that, since *w* occupies the middle position, each of the recombinants *yellow red cut* (of which there was one) and *grey white normal* (of which there were two) must have involved two cross-overs, one on either side of *w*, i.e.

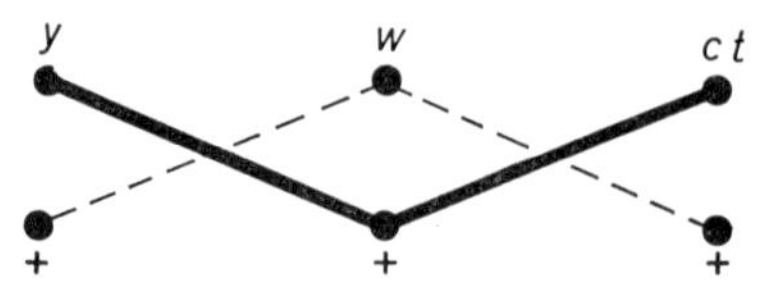

In order to correct the observed number of *recombinants* between *grey*/*yellow* and *normal*/*cut* so as to give the number of *cross-overs* between *y* and *ct* we have to add $2(1+2) = 6$, and this brings the figure of 194 to 200 as it should be. In any three-point cross, corrected for double crossing-over, additivity is strictly preserved.

The outcome of such investigations as these finds expression in the form of linkage maps (Fig. 21.2). The scale of a linkage map is conventionally that one unit of map length corresponds to 1% of crossing-over. On the linkage map of chromosome 2 the distance between *arista-less*

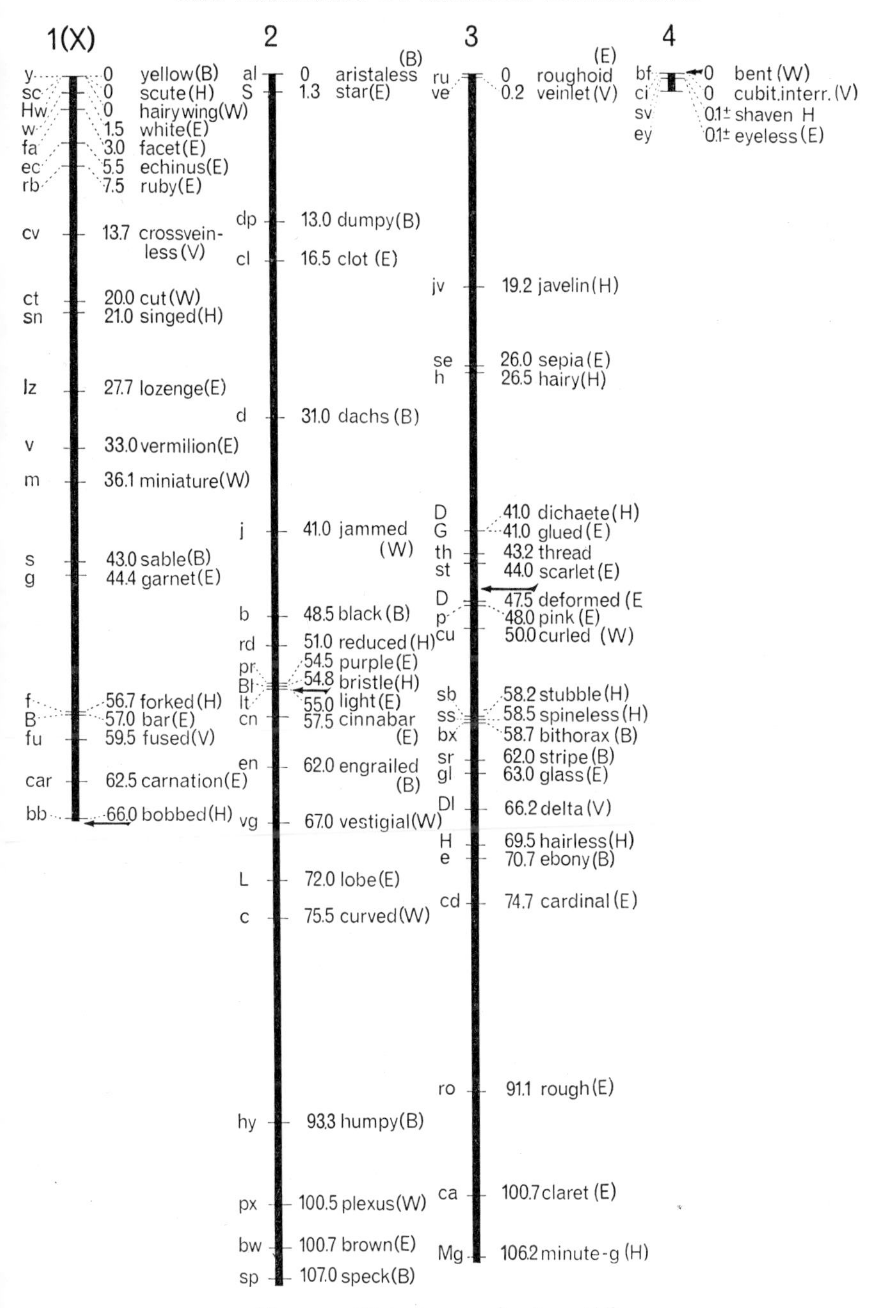

Fig. 21.2. Linkage maps for *Drosophila.*

B and *speck B* is 107 units. How can we observe a percentage of recombination which is greater than 100%? In point of fact the maximum percentage of recombination which can be observed is 50%, which is the percentage of recombination observed between unlinked genes. Within a linkage group, when the map distance between two genes is 40 units or more, there is so much multiple crossing-over between them that the percentage of recombination tends towards 50% and the accuracy of the method decreases sharply. Obviously then, a map distance of 107 units can only be arrived at as the sum of many shorter distances.

Morgan summarised his views in *The Theory of the Gene*, a book published in 1926, and the following is a paraphrased rendering of his formulation:

(i) the characters of the individual are referable to paired elements (genes) in the germinal material, held together in a definite number of linkage groups;

(ii) the members of each pair of genes separate in the formation of the germ cells;

(iii) the genes belonging to different linkage groups are assorted independently;

(iv) the genes belonging to the same linkage group undergo an orderly interchange (crossing-over);

(v) the frequency of crossing-over furnishes evidence of the linear order of the genes within each linkage group and of the relative position of the genes with respect to one another.

And in his own words: 'These principles, which, taken together, I have ventured to call the theory of the gene, enable us to handle problems of genetics on a strictly numerical basis and allow us to predict with a great deal of precision, what will occur in any given situation. In these respects the theory fulfils the requirements of a scientific theory in the fullest sense.'

Perhaps the most remarkable feature of this testament is the absence of any reference to chromosomes. Morgan of course believed that the genes were carried on the chromosomes, but he was scrupulously careful not to overstate his case. The data upon which Morgan's theory of the gene was based were derived from breeding experiments in which the numbers of different phenotypes were recorded and analysed; cytological observations were not essential to their interpretation. And if we re-examine the working hypothesis with which we entered upon this study (p. 227) we see that it is nowhere necessary to suppose that there is any connection between genes and chromosomes. The essential features of the hypothesis were: (i) that the genes were arranged in a linear order in paired linkage groups; (ii) that sections of the linear order could be exchanged between members of the pair in a process called crossing-over (which has not necessarily anything to do with the phenomenon which Janssens observed in *Batrachoseps*); and (iii) that the frequency of crossing-over is proportional to the relative distance between the genes.

In 1926 the strongest evidence in support of the view that the genes lie on the chromosomes was the fact that in *Drosophila* there are four linkage groups and four pairs of chromosomes. Similar evidence comes from the study of maize in which there are ten linkage groups and ten pairs of chromosomes, and many other examples of the same kind could be quoted. Stronger evidence came later, in the 1930's, after the rediscovery of the giant salivary chromosomes (p. 89). It then became possible to correlate observed cytological abnormalities with abnormalities of inheritance. This is further discussed in Chapter 26, p. 303. The evidence that the genes lie on the chromosomes, although very strong, is circumstantial and must always remain so until genes can be recognised under the microscope and can be seen to reside on the chromosomes. This is not intended to throw doubt upon the generally accepted view. But it is intended to point out that the generally accepted proposition that the genes are disposed in a linear order upon the chromosomes really consists of two distinctly different propositions based upon entirely different lines of evidence.

We will not pursue the study of classical genetics any further here, but will go on to consider in the next chapter the outstanding modern development in genetics which attributes to nucleic acids the role of being the material basis of inheritance.

22

NUCLEIC ACIDS AS THE MATERIAL BASIS OF INHERITANCE

In this chapter we shall be concerned with the evidence that the genetic information† which is necessary for the manifestation of the inherited characters of organisms is carried in the form of nucleic acids. In the next chapter we shall be concerned with the chemical nature of nucleic acids and its relation to their genetic properties. For our immediate

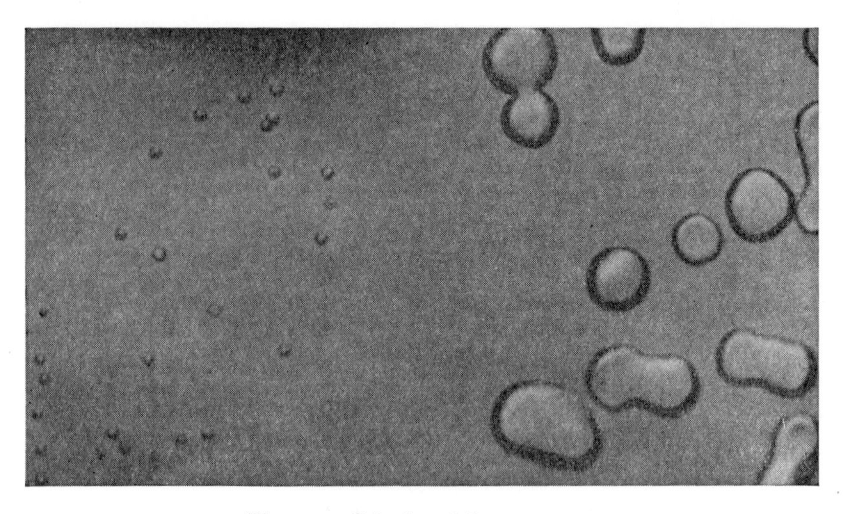

Fig. 22.1. Colonies of *Pneumococcus*.
Left, rough, non-virulent strain. Right, smooth, virulent strain.

purposes it is sufficient to know that there are two kinds of nucleic acid, deoxyribonucleic acid (DNA) and ribonucleic acid (RNA). In most organisms the genetic information is carried in DNA, and only in certain viruses does RNA assume the role of carrier.

The origins of this outstanding development in biology can be traced to a paper published by Griffith in 1928. Griffith was a medical bacterio-

† It is perhaps necessary to draw attention to the special meaning which the word 'information' has acquired when used in this context. Some writers prefer to put it that DNA carries a 'set of instructions' or a 'blueprint'.

logist. He was engaged in a study of the differences between various strains of *Pneumococcus*. The strain normally met with is virulent, and when mice are infected with it they invariably die of septicaemia. From this strain, by a series of passages from one mouse to another, Griffith isolated a non-virulent strain which was not fatal to mice. The two strains could further be distinguished by the form of the colonies which they produced when grown on agar plates. The non-virulent strain produced small colonies of rough texture and was designated strain R, whereas the virulent strain produced large smooth glistening colonies and was designated strain S—see Fig. 22.1. This difference is due to the fact that the bacteria of strain S have each a thick surface coating or capsule formed of polysaccharide, while those of strain R lack this capsule.

Griffith found much to his surprise that if he injected mice with a mixture of live strain R and heat-killed strain S the mice died; and from their bodies he could recover live bacteria of strain S. The results of his experiments may be summarised as follows:

Injected with live strain S, the mice die.
Injected with live strain R, the mice survive.
Injected with heat-killed strain S, the mice survive.
Injected with live strain R and heat-killed strain S, the mice die, and live strain S can be recovered.

Thus it would appear that there is some principle present in the heat-killed strain S which can transform live bacteria of strain R to live bacteria of strain S.

These results were repeated and confirmed by other workers, and transformation *in vitro* was achieved by 1931. But the significance of this phenomenon was not appreciated at the time, and this is not surprising in view of the bewildering way in which all bacteria seemed able to change their properties for no apparent reason. The publication of the work of Avery, McLeod and McCarty in 1944 put the whole matter on a different footing.

The object of this work was to establish the chemical nature of the transforming principle, at least sufficiently to place it in a general group of known chemical substances. Extracts of heat-killed strain S were prepared in accordance with standard procedures for isolating carbohydrate, protein, etc. In any one experiment the extract was set up at dilutions of 1:1, 1:5, 1:25, etc., in a series of test-tubes containing sterile broth, and each test-tube was seeded with a small inoculum from an actively growing culture of strain R. Even in successful experiments the percentage of bacteria transformed was very low, less than 1%, and it was therefore advantageous to make use of some treatment by which the two strains could be separated. For this purpose an antiserum against strain R was added to the sterile broth in the test-tubes; this caused the strain R bacteria to clump together and sink to the bottom, while any bacteria transformed to strain S were able to grow diffusely throughout the medium in the test-tube. The ultimate criterion of transformation was the appearance of the colonies when bacteria taken from the test-tubes were allowed to grow on agar plates.

The experiments were beset by many technical difficulties. Transformation is a capricious phenomenon, and a great deal of preliminary work had to be done before reproducible results could be obtained. But these details cannot be gone into here; the outcome was that Avery and his colleagues traced the transforming principle to DNA. Chemical purification by fractional precipitation and like processes was followed by treatment with enzymes known to break down proteins, polysaccharides, etc., and contaminants were reduced below the minimum detectable level. Nothing was left undone which could have been done at that time to establish the purity of the preparation. Moreover, from the transformed bacteria extracts could be made which had transforming activity far in excess of the amount used to bring about the original transformation; evidently the transforming principle itself grew during the growth of the transformed bacteria.

In retrospect this evidence, that DNA is a self-replicating agent which determines the character of the phenotype, seems decisive, but its publication was not immediately followed by a general recognition of the genetic role of DNA. The idea was altogether too shocking. One must remember that biologists had long been conditioned to the view that only in protein could be found that diversity of chemical form which the genetic material must undoubtedly possess. The genetics of bacteria were as yet unexplored. Other explanations were therefore canvassed; it was suggested, for example, that DNA might simply have stimulated variation and that the correspondence between the direction of transformation and the character of the donor strain was coincidental. But whatever may have been the general consensus of opinion the world of biology was alerted and a rapidly increasing research effort was brought to bear upon nucleic acids.

The following decade saw the confirmation of the work of Avery,

McLeod and McCarty and its extension to other transformations. As will be described in Chapter 25, a start had been made with the study of bacterial genetics. It was now known that mutation occurred in bacteria and many mutant strains showing resistance to particular drugs had been isolated; these provided new material for the study of transformation. The general line which the experiments followed is illustrated in Fig. 22.2, from which it may be noted that the frequency of *mutation*

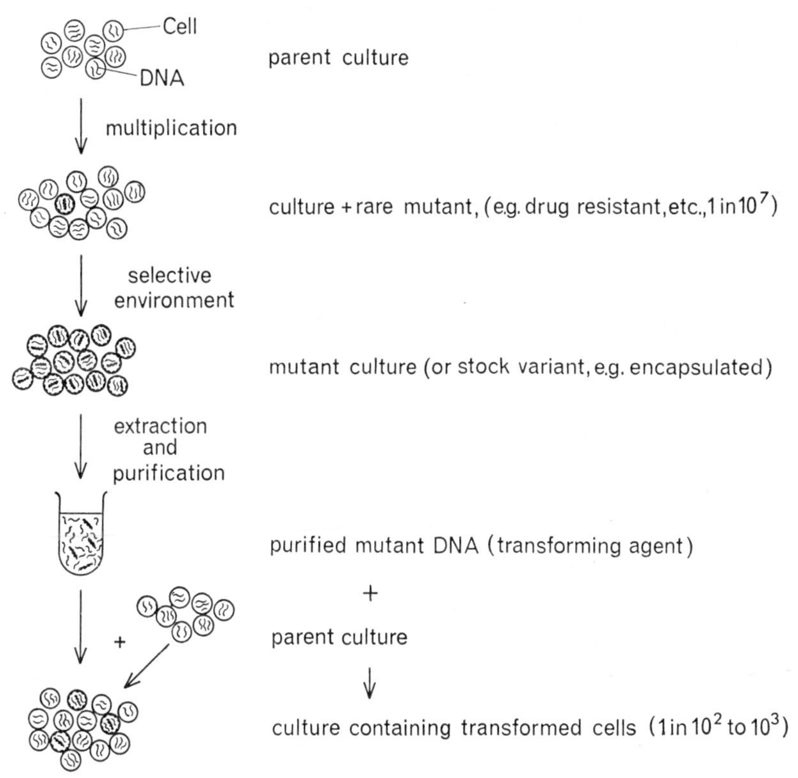

Fig. 22.2. Procedure for bacterial transformation.

to drug-resistance is of the order of 1 in 10^7, whereas the frequency of *transformation* to drug-resistance can be of the order of 1 in 10^2. DNA could now be purified until it contained less than 0·02% of protein. By 1955 some 30 different cases of bacterial transformation had been brought about by purified preparations of DNA. In all cases the transformation was in the direction which characterised the donor strain from which the DNA originated. In the face of this evidence the argument that the correspondence was coincidental could no longer be sustained and the reality of genetic transformation by DNA was generally conceded.

Most organisms contain both DNA and RNA. Most viruses which attack animals and viruses which attack bacteria—bacteriophage—contain DNA alone. Viruses which attack plants contain RNA alone. We must therefore suppose that if genetic information is carried in nucleic acids then RNA as well as DNA must have a genetic role, at least in plant viruses.

The first evidence for this came from experiments carried out on turnip yellow mosaic virus by Markham and Smith in 1949. From a suspension of virus they separated two fractions by centrifugation; both fractions contained protein, apparently identical, but only one contained RNA. The RNA-containing fraction was capable of infecting host plants, the other was not. These experiments suggested that RNA was the vehicle of inheritance but of course did not establish it, since protein was present as well. Stronger evidence was obtained in 1956 by Gierer and Schramm. Working with tobacco mosaic virus they were able to separate the RNA from the protein, by mild treatment with phenol, and to purify it until less than 0·02% of protein remained; and they were able to show that this purified RNA could infect host plants, albeit less readily than when it was combined with protein.

This work of Gierer and Schramm provides the critical evidence, comparable with the evidence of bacterial transformation in the case of DNA, that RNA can carry genetic information. It would be a pity, however, to leave this subject without mention of the developments which it has undergone in the hands of Fraenkel-Conrat. Working likewise with tobacco mosaic virus Fraenkel-Conrat and Williams prepared RNA-free protein by extracting virus with dilute NaOH, and protein-free RNA by removing protein with detergent. When these preparations were mixed together infective particles were reconstituted. In 1956 Fraenkel-Conrat, following up this discovery, was able to reconstitute infective virus particles from the RNA of one strain and the protein of another strain. The proteins of the two strains were readily distinguishable, both in amino acid composition and by serological reaction; and the two strains could also be distinguished by the type of lesion produced on the host plant. The reconstituted virus particles produced lesions which were characteristic of the strain from which the RNA originated, and the virus particles which were recovered from the infected plants were found to have protein characteristic of the strain from which the RNA originated. Thus the RNA not only reproduced itself but also formed its own characteristic protein in the progeny.

The evidence that nucleic acids, both DNA and RNA, are able to transmit and replicate genetic information is very strong, and in this context we may recall the experiment of Hershey and Chase, described on p. 126. But this evidence relates only to bacteria and viruses. Acceptance of the genetic role of DNA in these organisms does not automatically oblige us to accept that it has the same role in other organisms. The evidence that it does is circumstantial and on the whole rather weak. DNA can be demonstrated to be present on the chromosomes and nowhere else. Other evidence, such as that the chemical composition† of DNA is species-specific, is not convincing; proteins are also species-specific. It would be fair to say that such evidence as we have is not in conflict with the idea that DNA is the material basis of inheritance in higher organisms as in bacteria and viruses, but it cannot be said to establish this position in any positive sense. In the case of higher organisms strong evidence is lacking, and for this the reason is not far to seek. The genetic apparatus of bacteria and viruses is relatively simple. There are no chromosomes in the classical sense and nothing of the elaborate ritual of mitosis and meiosis; and the whole organism is very much less complex and less sensitive to disturbance of its functions. Attempts have already been made to effect genetic transformation in higher organisms, by injecting foreign DNA, and some success has been claimed. But in experiments with higher organisms, besides the unlikelihood of injected DNA finding its way to the chromosomes of the germ cells, we have also to reckon with the unlikelihood that the offspring would be viable after the great disturbance of their delicately balanced genetic system which this type of transformation would be likely to entail.

Notwithstanding that the evidence for the genetic role of nucleic acids is virtually restricted to bacteria and viruses, there are very few biologists who do not accept, at least provisionally, the idea that nucleic acids—mainly DNA—are the material basis of inheritance in all forms of life. Biologists accept the theory of evolution although the direct evidence for the occurrence of evolution in nature within the span of human

† By chemical composition in this context is meant the relative proportions of bases—see next chapter.

observation is very slight; the theory of evolution is accepted because through it we can perceive relations between observations which otherwise appear to be unconnected. In much the same way the proposition that DNA is the material basis of inheritance has a very strong appeal because we can see in the molecular configuration of DNA an obvious basis for the replication of genetic information, which on other grounds we know must take place

23

THE STRUCTURE OF NUCLEIC ACIDS

The story of the nucleic acids starts with the work of Miescher in 1868. By tryptic digestion of the pus cells from discarded bandages he obtained naked nuclei; and from these he isolated a substance which he called nuclein and which we now call nucleoprotein. Later, he showed that nuclein could be split into an acid component, which we now call nucleic acid, and a basic component of protein nature.

The two chief sources used by later workers for the preparation of protein-free nucleic acid were yeast and mammalian thymus gland. Certain differences, to be referred to in more detail later, were noticed in the nucleic acids prepared from these two sources, and it seemed that nucleic acids prepared from plants resembled yeast nucleic acid whereas those prepared from animal tissues resembled thymus nucleic acid. From this arose the idea that the two types could be designated plant nucleic acid and animal nucleic acid; but we now know that there is no such exclusiveness about their distribution, both types being found both in plants and in animals. What was called plant nucleic acid is now known as ribonucleic acid (RNA) and what was called animal nucleic acid is now known as deoxyribonucleic acid (DNA).

Without going into the details of what was a purely chemical investigation we will proceed forthwith to its outcome, which was to show that both types of nucleic acid are made up of units, called nucleotides, each of which consists of phosphoric acid, a pentose sugar and an organic base which may be either a purine or a pyrimidine. Since these latter compounds may be unfamiliar to the reader it will be useful to indicate their basic structure here in the text. The full structural formulae are given in the Appendix, p. 318.

There are two purine bases which are found in nucleic acids: adenine and guanine. The fundamental configuration of a purine is a double ring structure as shown on p. 242.

Purines differ one from another in the side groups attached to the ring in positions 2, 6, 8 and 9.

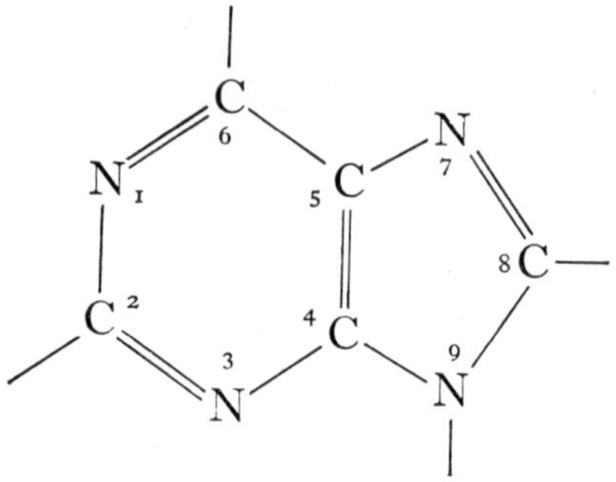

The fundamental configuration of a pyrimidine is a single ring which corresponds to the larger ring of purine.

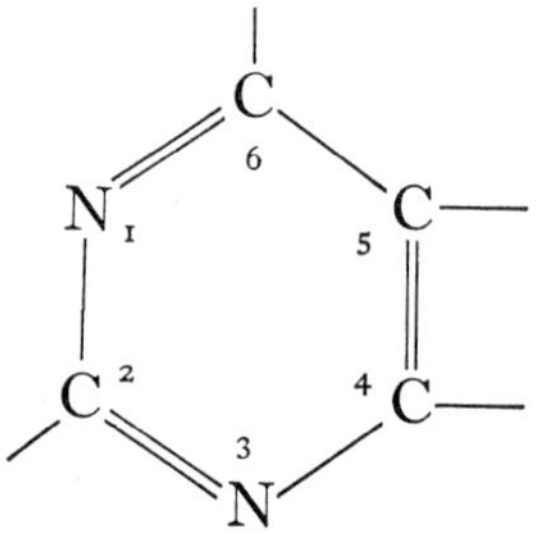

Pyrimidines differ in the side groups attached at positions (2), (4) and (5). There are three principal pyrimidine bases which are found in nucleic acids: thymine, cytosine and uracil. RNA contains adenine, guanine, cytosine and uracil; DNA contains adenine, guanine, cytosine and thymine.

The pentose sugar is either D-ribose (in RNA) or 2-deoxy-D-ribose (in DNA).

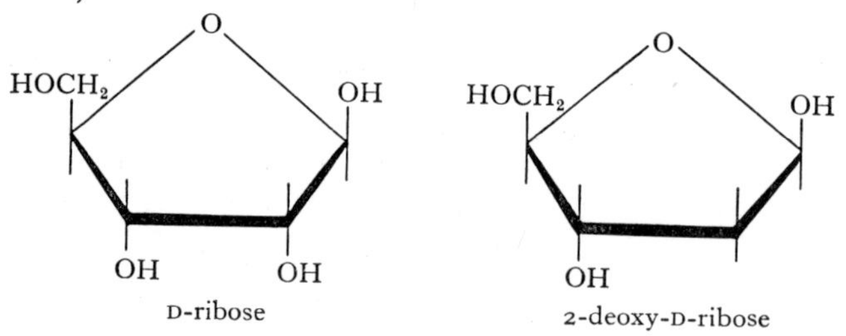

D-ribose 2-deoxy-D-ribose

In a nucleotide, such as adenylic acid, the nitrogen atom at position (9) in adenine is attached to carbon atom (1) of the sugar, and the phosphate group is attached to carbon atom (5) of the sugar. The full structure of adenylic acid is given in the Appendix, p. 320. We have met it before under another name—adenosine monophosphate or AMP (p. 144).

Nucleotides are linked together to form nucleic acids by the formation of a phosphate di-ester link between the phosphate group of one nucleotide and carbon atom (3) on the sugar of another.

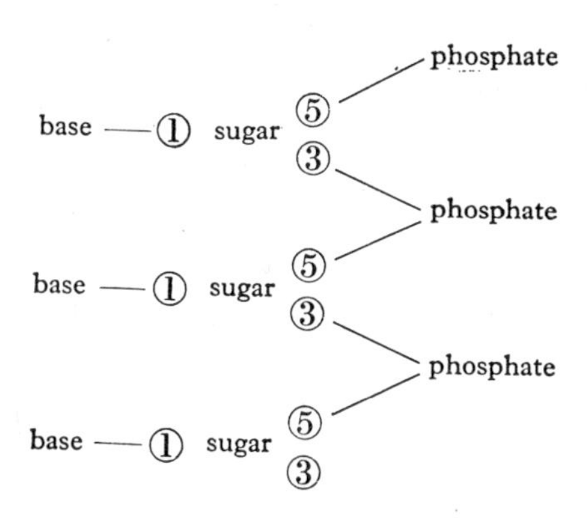

Thus far had our knowledge of the chemistry of nucleic acids progressed up to the time of the Second World War. It had been a leisurely progress but a solid one, each step being established in accordance with the exacting criteria of the organic chemist. Up to this time nucleic acid seemed to be just one more of Nature's many products, of academic interest, to which no special importance was attached. But with the evidence of bacterial transformation by DNA a new chapter was opened.

In 1950 Chargaff, having set on foot extensive analyses of DNA from a variety of different sources, drew attention to certain regularities which emerged from these analyses:

(i) the sum of the purine nucleotides = the sum of the pyrimidine nucleotides;

(ii) the molar ratio adenine/thymine = 1;

(iii) the molar ratio guanine/cytosine = 1;

(iv) the number of 6-amino groups of adenine + cytosine = the number of 6-keto groups of guanine + thymine.

The precise significance of these regularities could not be perceived at that time; but, as we shall see presently, they were of decisive importance in the elucidation of the structure of the molecule.

While these chemical investigations were in progress attempts were also being made to bring physical methods to bear upon the problem. The most powerful of these methods is the method of X-ray diffraction analysis. An outline of the principles of X-ray diffraction analysis has already been given (Chapter 3, p. 32) and it will be remembered that the method depends for its applicability upon the use of crystalline material. In general, the larger the molecule the more difficult it is to get it to crystallise. The molecule of DNA is very large. Solutions of

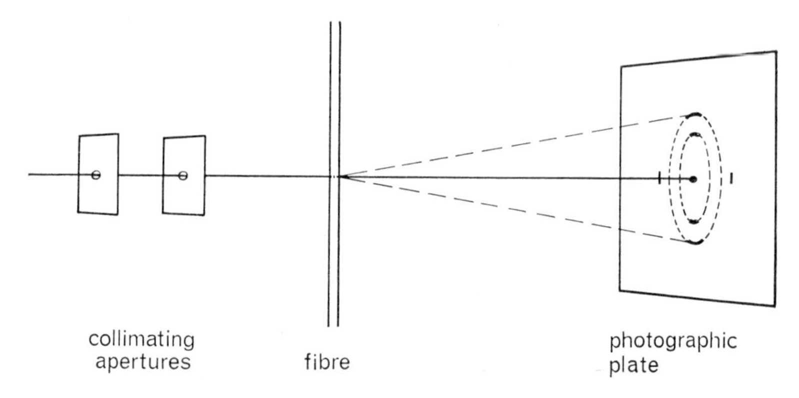

Fig. 23.1. Arrangement for obtaining the X-ray diffraction pattern of a fibre.

A narrow beam of monochromatic X-rays is allowed to fall on the fibre at right angles to its long axis and the diffracted beams produce diffuse arcs upon a photographic plate beyond the fibre.

DNA are viscous, almost glue-like in consistency, and a less promising material from the crystallographer's point of view can hardly be imagined.

Not surprisingly, the earlier attempts to obtain X-ray diffraction patterns from imperfectly crystalline preparations of DNA were disappointing. It could be seen, however, that there was indication of a periodicity of 3·4 Å which would correspond to the expected distance between adjacent nucleotides. There was also a suggestion of a helical arrangement, which is in any case a likely arrangement, *a priori*, for a long molecular chain. Pauling and Corey advanced a tentative suggestion that the core of the helix was formed of the linked phosphate groups, with the bases directed outwards and radially arranged on a helical pattern. Sharper X-ray diffraction patterns were obtained by Wilkins in 1953. He prepared fibres by slowly withdrawing a needle point from a solution of the sodium salt of DNA. Such fibres probably consist of small crystallites with their long axes parallel but otherwise disposed at random. In view of the random orientation of the crystallites in relation to the long axis of the fibre there is no need to rotate the fibre, as is done in the case of fully crystalline material (see Fig. 3.7). The fibre is suspended vertically at right angles to a horizontal beam of X-rays and the diffraction pattern is recorded on a flat plate at some distance beyond the fibre, as shown in Fig. 23.1. By reason of the poor ordering of the material the diffraction pattern lacks precision; instead of a large number of sharp spots we obtain a smaller number of diffuse arcs (Fig. 23.3).

Notwithstanding the imperfections of the material it was now possible to obtain more precise information. In addition to the previously observed periodicity of 3·4 Å a new periodicity of 34 Å could be detected, and at right angles to the length of the fibre a periodicity of 20 Å appeared.

In a letter to *Nature*, published in April 1953—a landmark in the history of biology—Watson and Crick proposed a structure for the DNA molecule. Having before them the evidence of Wilkins' diffraction patterns, they supposed that the DNA chain was disposed in the form of a helix, the 3·4 Å periodicity corresponding to the distance between successive nucleotides and the 34 Å periodicity corresponding to the pitch of the helix, there being 10 nucleotides per turn. They further supposed that the equatorial periodicity of 20 Å corresponded to the width of the chain. On this basis, there being 10 nucleotides within a space measuring 34 Å long and 20 Å in diameter, it was possible to calculate the density of the material. It worked out to be roughly half the known density of DNA. This suggested that there must be not one chain, but two. Watson and Crick next took account of the chemical regularities discovered by Chargaff. These regularities suggested some kind of pairing between purine nucleotides and pyrimidine nucleotides, and this in turn led them to consideration of a possible structure which would give some point to this pairing, namely a structure in which the bases were in proximity and the phosphates widely separated. They then constructed accurate models of the four nucleotides of DNA and tried to see how they might fit together. This is a more difficult and elaborate process than might be supposed. Account has to be taken not only of the size of the atoms and the lengths of the bonds connecting them, but also of the possibilities of rotation about some of the bonds and of the forces of attraction between different parts of the nucleotide molecules. From these studies it emerged that a satisfactory structure could be devised if

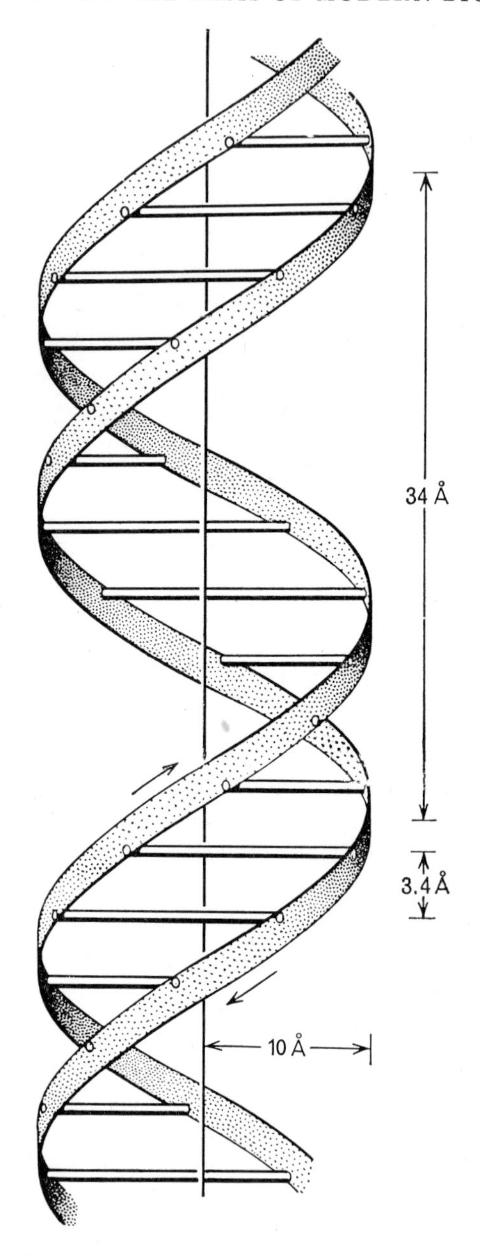

Fig. 23.2. The Watson–Crick model of DNA.

an adenine nucleotide of one chain paired with a thymine nucleotide of the other chain, and correspondingly if guanine paired with cytosine, the two members of each pair being held together by hydrogen bonds. From this proposed structure it was then possible to work out the expected X-ray diffraction pattern and compare it with the pattern observed by Wilkins. The agreement was good, and appeared to justify the whole conception. Watson and Crick summarised their proposed structure in a simple diagram which is reproduced in Fig. 23.2. In presenting this structure the authors observed: 'So far as we can tell it is roughly comparable with the experimental data, but it must be regarded as unproved until it can be checked against more accurate results.'

Not long afterwards Wilkins and his associates obtained even better X-ray diffraction patterns. These necessitated some slight revisions of the original spacings, but not so great that the Watson–Crick structure could not be adjusted to meet them. Greater precision was given to the pairing of the purine and pyrimidine bases (Fig. 23.5). By 1956 an agreed solution had been reached. Fig. 23.6 is a photograph of a model constructed in Wilkins' laboratory. It has received general acceptance and no observations have been made since that time to suggest that it is incorrect.

Given this more clearly defined structure for DNA, let us now consider how suitable it may be to play its part as the material basis of inheritance. With charming restraint Watson and Crick wrote: 'It has not escaped our notice that the specific pairing we have postulated immediately suggests a possible copying mechanism for the genetic material.' That is to say, if we were to separate the two chains of a DNA molecule, by rupturing the hydrogen bonds between the base pairs, and if we were to provide new partners for each chain out of a pool of nucleotides, we would end up with two molecules of DNA each identical with the original molecule. This is an inevitable consequence of the requirement that adenine must pair with thymine and guanine with cytosine; each chain can be regarded as a template upon which only its specific counterpart can be built up.

The attractiveness of this idea must not blind us to the realisation that it lacks any direct evidence to support it; and clearly such evidence must be sought by an investigation into the manner in which DNA is replicated.

So long as replication can be studied only as it occurs in the intact living cell, progress will be slow; what the biochemist aims to do in a

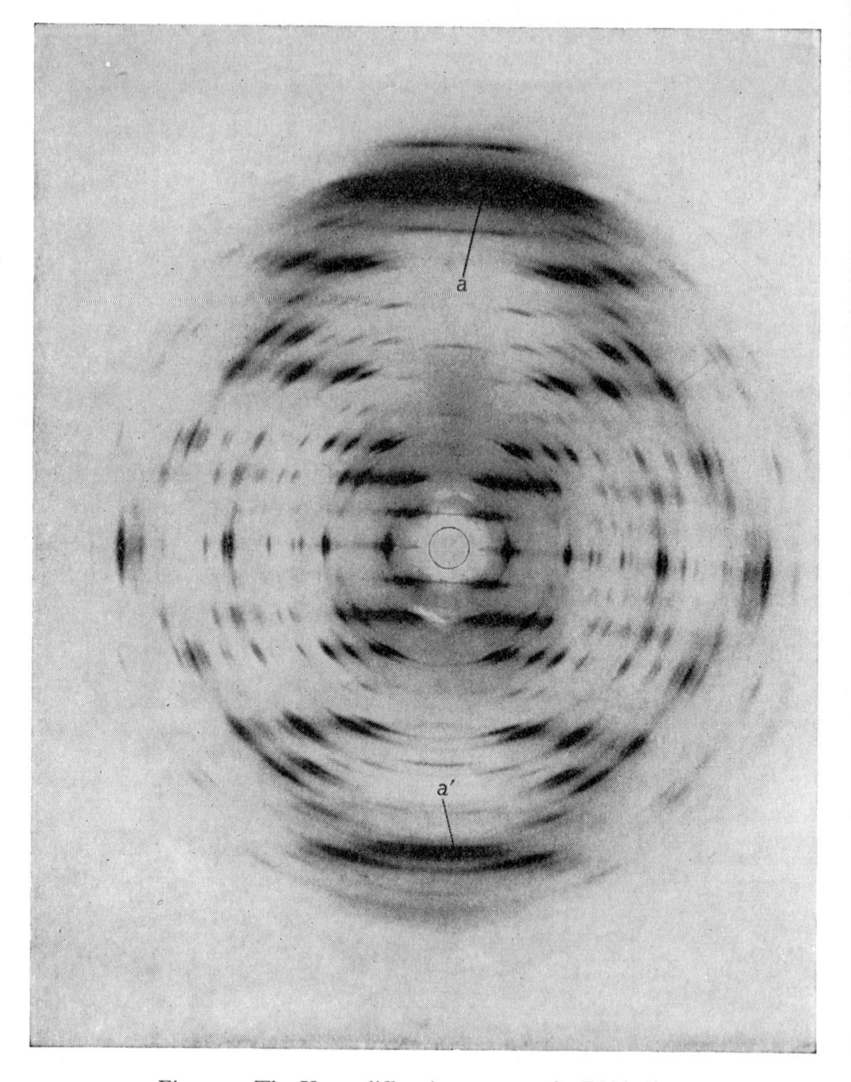

Fig. 23.3. The X-ray diffraction pattern of a DNA fibre.

The two large spots marked *a* and *a′* on the meridian are due to the first-order diffracted beams arising from the periodicity of 3·4 Å between neighbouring nucleotide pairs. From *a* to the equator the pattern is traversed by 10 horizontal lines upon which smaller spots are variously disposed; these may be related to the periodicity of 34 Å, the pitch of the helix, and correspond to diffracted beams of up to tenth-order. It will be observed that no spots of this series lie actually on the meridian; to the initiated this suggests a helical structure.

In the taking of this photograph the fibre was slightly inclined to the vertical and this accounts for the greater intensity of *a* as compared with *a′*.

See also Fig. 23.4 and legend.

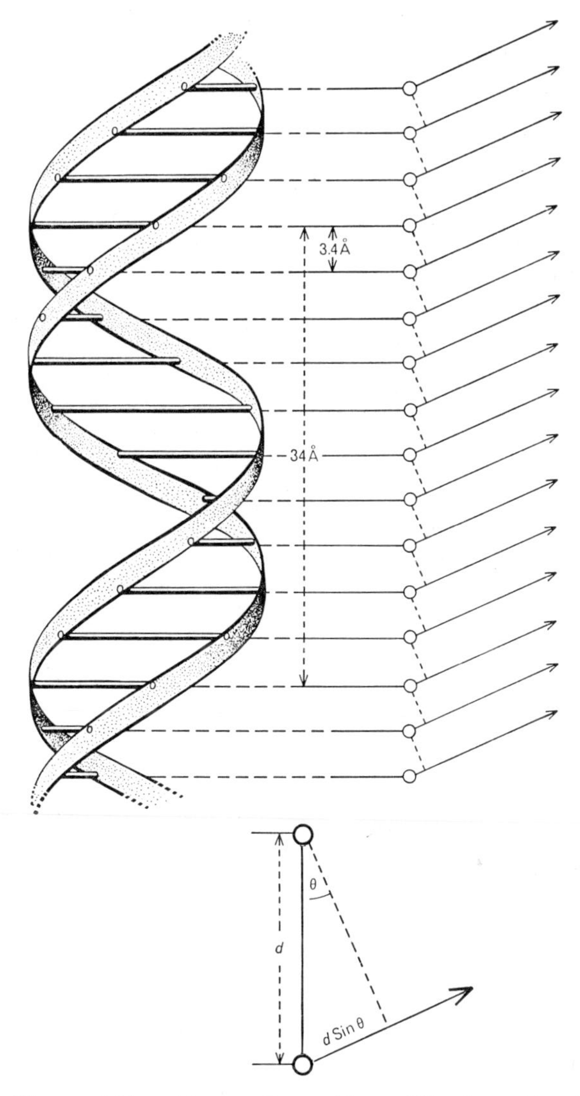

Fig. 23.4. The possibility of interpreting an X-ray diffraction pattern by simple inspection is very limited. The information which the pattern contains can only be extracted by measurement and computation.

In order to appreciate some of the more obvious features of the X-ray diffraction pattern of DNA (Fig. 23.3) imagine that the fibre may be represented by a diffraction grating made up of rods running at right angles to the axis of the fibre and spaced with their centres 3·4 Å apart. In general, a diffracted beam will be produced if

$$d \sin \theta = n\lambda$$

where n is a whole number and denotes the order (first, second, third, etc.) of the diffracted beam. The greater the spacing d, the smaller will be the angle θ.

situation of this kind is to extract the essential pieces of machinery, throw the rest away and try to reassemble the synthetic mechanism in a cell-free system *in vitro*. This was successfully accomplished for the synthetic mechanism of DNA by Kornberg. He embarked upon this

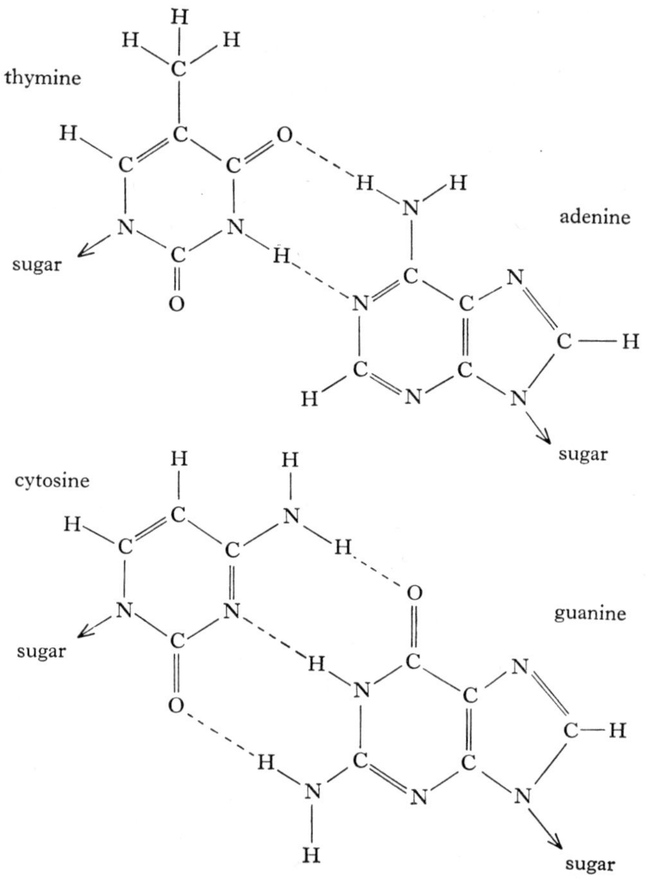

Fig. 23.5. Base pairing by hydrogen bonding between adenine and thymine (above) and between guanine and cytosine (below).

work on the assumptions: (1) that the synthesis of DNA is mediated by some enzyme or enzymes which must be universally present in cellular organisms and which it should be possible to extract and purify; (2) that it would be necessary to provide the enzyme system not only with raw materials in the form of nucleotides but also with a source of energy;

and (3) that it would be necessary to add some DNA as a 'primer' to provide the necessary information for the synthetic mechanism to work upon. The early experiments, first reported upon in 1957, were designed only to see if any sign of synthesis of DNA could be detected. A crude extract of the bacterium *Escherichia coli* was used as a source of the

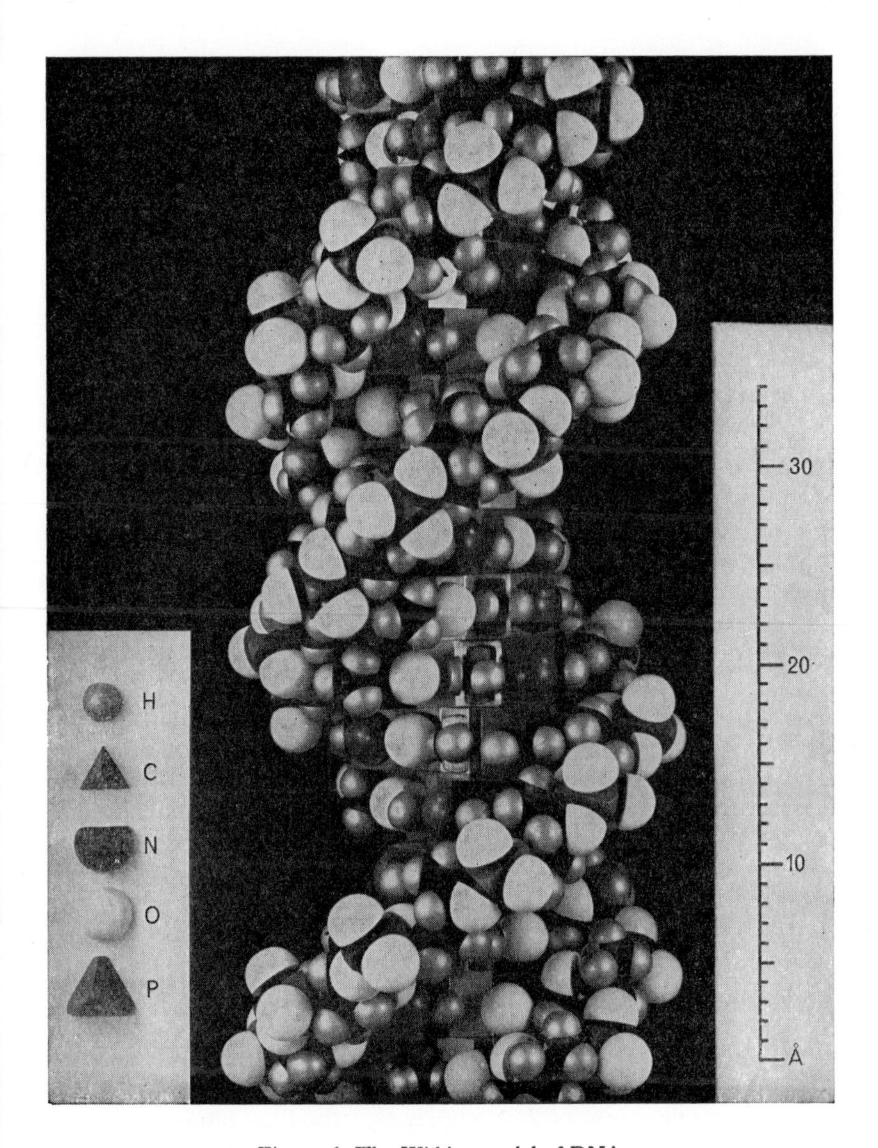

Fig. 23.6. The Wilkins model of DNA.

putative enzyme system. ATP was added as a source of energy and DNA from the same bacteria was added as primer. The nucleotides were heavily labelled with ^{14}C. After incubation the reacting mixture was separated into an acid-insoluble fraction (containing DNA) and an acid-soluble fraction (containing the free nucleotides) and the radioactivity of each fraction was measured. Only a minute amount of radioactivity was found in the acid-insoluble fraction, about 5 parts as compared with 100,000 in the other fraction. But it was a beginning.

At this stage the paramount necessity was for enzyme purification because crude extracts of cellular organisms contain a variety of enzymes which can break down DNA as fast as it is formed. Enzyme purification was, and continued to be, the main preoccupation; and although an improvement of several thousand fold was achieved, even the best preparations were still contaminated up to the time when this work was completed.

In later experiments ATP was dispensed with and instead the four nucleotides were supplied in the form of their triphosphates, in which (as in ATP) the two terminal phosphate groups are attached by energy-rich bonds. The six independently variable components of the synthetic system were: the purified enzyme preparation, primer DNA, and the four nucleotide triphosphates. With all six components present net synthesis of DNA, to an amount which exceeded the amount of primer by a factor of 20, was successfully demonstrated; if any one of the components was withheld no synthesis took place.

The technical difficulties having been sufficiently overcome, the rewards were now ready to be reaped. To denote the four bases of DNA, or the corresponding nucleotides, we will use their initial letters: A, G, C and T. As Chargaff showed, the ratio $(A+G)/(C+T)$ is always unity, irrespective of the source of the DNA. On the other hand, the ratio $(A+T)/(G+C)$ varies fairly widely and is characteristic of the source of the DNA. If our ideas about the importance of base pairing are correct we should expect to find that the DNA synthesised *in vitro* had the same $(A+T)/(G+C)$ ratio as that of the primer DNA. It is one of the great advantages of the cell-free system that an enzyme preparation from one organism can be used with primer DNA from another.

Using an enzyme preparation from *Escherichia coli* and primer DNA from three other organisms, Kornberg obtained the figures set out in Table 23.1. These results show clearly that the base ratios of the synthesised DNA conform to those of the primer DNA. The fact that the enzyme system comes from another organism makes no difference. We may then draw the conclusion that DNA carries all the information necessary for its own replication, and it is difficult to imagine how this can come about other than by the specific base pairing postulated by Watson and Crick.

Table 23.1

	$\frac{A+G}{C+T}$	$\frac{A+T}{G+C}$
Mycobacterium phlei		
Primer	1·01	0·49
Product	0·99	0·48
Escherichia coli		
Primer	0·98	0·97
Product	1·01	1·02
Calf thymus		
Primer	1·05	1·25
Product	1·02	1·29
Bacteriophage T_2		
Primer	0·98	1·92
Product	1·02	1·90

An essential feature of the replication process as envisaged by Watson and Crick is the separation of the two chains of the double helix, each chain regenerating a complementary partner. But it is not impossible that the original double helix remains intact and that a new double helix is built up alongside it; this would be called 'conservative replication'. The process whereby the chains separate and each regenerates a new partner is called 'semi-conservative replication'. How are we to decide whether replication is conservative or semi-conservative? This is a matter of some difficulty since it implies that we need to know not only the composition and structure of the molecule, but also its past history. Direct evidence for semi-conservative replication was provided by Meselson and Stahl in 1958, in an experiment of great elegance which merits a full description.

In principle the experiment is as follows. A culture of bacteria is allowed to grow upon a medium in which the nitrogen source contains exclusively the heavy isotope ^{15}N. This isotope becomes incorporated into the purine and pyrimidine bases of DNA, which is thereby of greater density than DNA containing the normal isotope ^{14}N. After a period of growth sufficient to ensure that virtually all the DNA is 'heavy' the medium is then changed for one containing only ^{14}N. Then, assuming that replication is semi-conservative, after the next division of the cells the DNA molecules will be composed of one original 'heavy' chain and one newly synthesised 'light' chain, and so will be of density intermediate between that of DNA in which both chains are 'heavy' and that of DNA in which both chains are 'light'. The DNA in which both chains are 'heavy' will disappear completely after the first division. At the second division DNA with both chains 'light' will appear for the first time, and will increase in amount at each subsequent division. The composite DNA, one chain 'heavy' and one chain 'light', will remain constant in total amount but will decrease in relative amount as the culture continues to grow and synthesise new 'light' DNA. On the other hand, if replication is conservative, the original DNA, both chains 'heavy', will persist; newly synthesised DNA will have both chains 'light' and no composite DNA will appear at any stage.

The technical problem is to devise a method of revealing small differences in the density of DNA. A time-honoured solution to the general problem of revealing density differences is to employ a density gradient. For example, if we wished to reveal a difference in density of two aqueous solutions we would prepare a density gradient by the partial mixing of two water-insoluble liquids of different density and we would allow a droplet of each solution to find its own level in the gradient. This simple approach is only practicable when insoluble particles or immiscible liquids are involved. When we are dealing with very small droplets or particles of molecular size their random movements under thermal agitation (Brownian movement) prevent them from settling to a definite level. It is a matter of the degree of thermal agitation in relation to density difference and to the strength of the gravitational field. By increasing the strength of the field we can create conditions in which the dispersive effect of thermal agitation is overcome.

When a solution of some heavy material such as caesium chloride (atomic weight of caesium, 132·8) is subjected to centrifugal forces of 100,000 *g* or over, the caesium chloride becomes concentrated in the 'lower' part of the tube containing the solution. The tendency of the caesium chloride to become concentrated at the bottom of the tube is opposed by thermal agitation which tends to produce uniformity in concentration. After the centrifuge has been running for some time an equilibrium is established between the concentrating effect of the centrifugal force and the dispersive effect of thermal agitation, with the production of a gradient in concentration of caesium chloride, increasing towards the bottom of the tube. This gradient in concentration is also a gradient in density. In this way we can produce and maintain a density gradient in an aqueous solution; and if we have also added some DNA to the contents of the tube this will settle to the level at which its buoyant density is equal to the density of the caesium chloride solution. The position of the DNA in the gradient can be ascertained by its absorption of ultraviolet light, purines and pyrimidines absorbing strongly in the region of 2600 Å. The principle of the system is shown in Fig. 23.7.

In the actual experiment the bacterium *Escherichia coli* was used, for which ammonium chloride serves as a source of nitrogen. The culture was grown for 14 generations on $^{15}NH_4Cl$, by which time 96·5% of the bacterial nitrogen was in the form of the heavy isotope. The nitrogen source was then abruptly changed from ^{15}N to ^{14}N by the addition of a great excess of $^{14}NH_4Cl$. Thereafter samples of the culture were taken at intervals, the DNA was extracted and added to centrifuge tubes containing a concentrated solution of caesium chloride. The tubes were run, in turn, in the centrifuge at 140,000 *g* for 20 hours and photographed at the end of this period.

One of the unavoidable imperfections of the experiment is that it is not possible to synchronise cell division in a culture of bacteria. As a consequence the composite DNA does not appear on the scene abruptly, but only gradually, as more and more cells undergo division after the change of medium. But the results, shown in Fig. 23.8, are unequivocal. The 'heavy' DNA disappears completely after sufficient time has been allowed for all the cells to divide at least once. A line corresponding to composite DNA appears over the period during which the first division is taking place and then gradually weakens in relation to the 'light' DNA which begins to appear with the second division and increases steadily thereafter.

The reader will no doubt have perceived that an alternative, though less probable, explanation of these results could be advanced. If the double helices of the 'heavy' DNA were to break transversely about the middle of the molecule and if new half-molecules composed of double helices of 'light' DNA were to be joined on to them, the same pictures would be obtained. But to this alternative a test can be applied. When a solution of DNA is boiled the two component chains of the double helix are separated, this relatively mild treatment being sufficient to break the weak hydrogen bonds which hold the base pairs together but not to break the co-valent bonds which hold the nucleotides together in a chain. When the solution containing composite DNA is boiled the

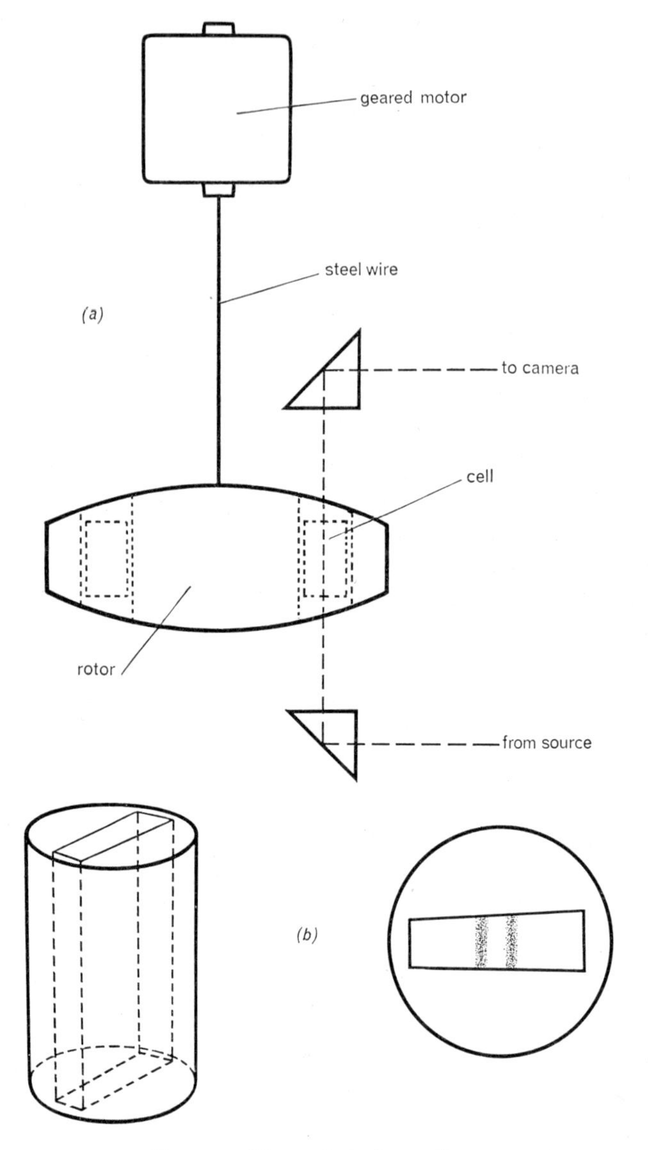

Fig. 23.7. The analytical centrifuge.

(*a*) The centrifuge is capable of rotating at speeds of 50,000 r.p.m. or over, and this raises difficult problems of design. The drive is provided by air or oil turbines or by a geared electric motor. The rotor which carries the cell is suspended from, and driven through, a flexible steel wire which accommodates any slight misbalance. The chamber enclosing the rotor is evacuated. (*b*) The cell is provided with a radial slot to contain the sample and with quartz windows through which the sample can be photographed with UV while the centrifuge is running.

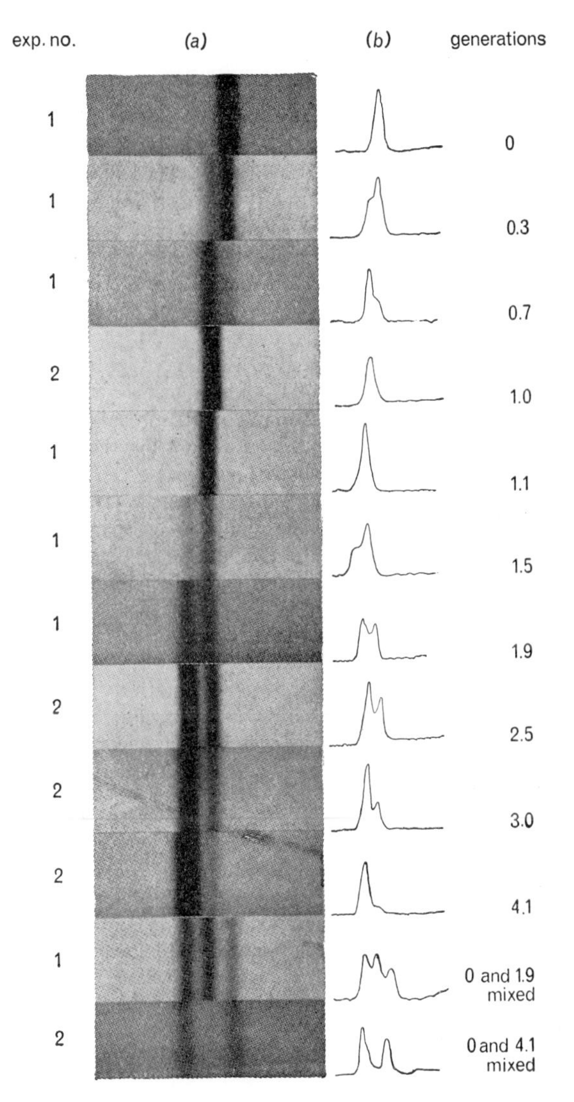

Fig. 23.8. The Meselson–Stahl experiment.

Under *a* are shown the UV photographs of the centrifuge cell, density increasing towards the right. Under *b* are plots of readings taken from the photographs with a densitometer.

Bacteria in culture do not divide synchronously, so by generations is meant average generation time. At the start of the experiment all the DNA is 'heavy'. After the first generation all the DNA is of intermediate density, as the Watson–Crick theory predicts; thereafter 'light' DNA increases in relative amount. Since there is no absolute scale of density against which the photographs can be read, the last two lines are necessary to establish the validity of the method.

corresponding line on the centrifuge picture disappears, and the lines of 'heavy' DNA and 'light' DNA appear instead. This result is not in accordance with the alternative explanation which would imply the persistence of DNA of intermediate density.

The Meselson–Stahl experiment demonstrates the semi-conservative replication of some duplex structure, not necessarily a double helix. But taken in conjunction with the X-ray data of Wilkins and the biochemical data of Chargaff and of Kornberg it scarcely leaves room for doubt; and we may therefore feel at liberty to accept in principle the idea advanced by Watson and Crick, namely that the replication of DNA is semi-conservative, the chains separating and each acquiring a complementary partner, with all the consequences for genetics which this implies.

By contrast with the brilliant success which has attended the investigation of DNA, progress with RNA has been slow. This is attributable primarily to the difficulty of obtaining crystalline preparations of this material. It has been suggested that the chains of RNA are branched, a possibility which arises from the presence of the hydroxyl group on carbon atom (2) of D-ribose, not present in 2-deoxy-D-ribose; this could afford a point of attachment for the phosphate group of an additional nucleotide, so that three nucleotide chains could converge upon a single sugar molecule. This may possibly be true of some kinds of RNA but fortunately it does not seem to be true of all. Preparations having a sufficient degree of crystalline order to give sharp X-ray diffraction patterns have been obtained by Wilkins, using a particular kind of RNA known as 'transfer' RNA. It will be more convenient, however, to defer consideration of the structure of 'transfer' RNA to the next chapter, in which its functions are to be discussed (p. 270).

24

THE SYNTHESIS OF PROTEIN

Before we go on to consider the mechanism of protein synthesis and what is known about it, it is perhaps necessary to be clear about the reasons why such very great importance is attached to the understanding of the synthesis of this particular class of substances.

The first reason arises out of logical difficulty. When we review what is known about the degradative and synthetic processes in organisms we find that almost invariably they are catalysed by enzymes. If protein synthesis follows the same lines as the synthesis of other classes of compound it too must require enzymes. All known enzymes are proteins. To synthesise protein we therefore need other proteins. Where does the first protein come from?

Secondly, proteins show much greater diversity than other classes of compound. Cellulose is a large and complex molecule, but it is nevertheless no more than a polymer, being built up of identical units—molecules of glucose—each joined to its neighbour by linkages of the same kind. In proteins, on the other hand, the units are amino acids, of which there are some 20, and we are not aware of any restrictions upon the order in which they can be joined together. This provides generously for the possibility of diversity; but in the only types of protein molecule which have as yet been fully analysed the order of amino acids is rigidly fixed and species-specific. To build up a protein having, say, 100 amino acids of 20 different kinds, joined together in a definite order, clearly requires very much more detailed information than is needed to build up a cellulose chain. This information, which must be passed on from the organism to its descendants, we believe to be carried in DNA. We must therefore contemplate that a very large part of the store of genetic information, certainly in micro-organisms, must be concerned with specification of the order of amino acids in protein molecules.

Thirdly, it appears that whereas carbohydrate polymers can be built up gradually, by successive addition of units, this does not seem to be true of proteins; analyses show the presence in organisms of proteins and of free amino acids, but peptides of intermediate size are conspicuously absent. Furthermore, if one necessary amino acid is with-

17-2

held from a growing organism no synthesis of protein takes place in its absence; and if we then supply the missing amino acid, at the same time withholding another, again no synthesis takes place. The necessary amino acids must all be available at the same time. This suggests that protein molecules are not built up piecemeal but are assembled in their completed form in a single operation.

There are various practical methods of establishing the occurrence of protein synthesis. We can determine the net protein synthesis by finding the amount of protein per organism before and after the experiment. More conveniently, we can supply the organism with radioactively labelled amino acids and observe the extent to which the label is incorporated into protein. But as we have seen, proteins are substances of great diversity, and to know that protein of some unspecified kind has been synthesised may not always be sufficiently informative; we may wish to know how much of some particular species of protein molecule has been synthesised. If so, we can make use of a specific antibody to precipitate the wanted protein from a mixture. Or, better still, if we choose to study the synthesis of an enzyme we can measure the amount synthesised in terms of its rate of action upon an appropriate substrate.

A connection between protein synthesis and RNA was noted independently by Caspersson in 1941 and by Brachet in 1942. Certain tissues, such as the vertebrate pancreas and the silk gland of the silkworm, are unquestionably the sites of protein synthesis and they contain large amounts of RNA; other tissues such as muscle and kidney, which have high metabolic activity but do not elaborate protein secretions, are relatively lacking in RNA. These original observations, made before the electron microscope came into general use, were soon followed up, and the cytoplasmic RNA was traced to the ribosomes of the endoplasmic reticulum.

If protein synthesis is dependent on RNA an obvious method of approach is to destroy the RNA by means of ribonuclease and to demonstrate the cessation of protein synthesis. This has been successfully accomplished on amoebae. After the exposure of the amoebae to ribonuclease the incorporation of labelled amino acids was reduced and the disappearance of RNA was confirmed by staining methods. When such amoebae were then treated with purified RNA from others of the same species their RNA content and power of incorporation were restored.

For any detailed study of the mechanism of protein synthesis the use of whole organisms has the obvious disadvantage that other metabolic activities are going on at the same time. What we need is some cell-free system which can be isolated and made to work *in vitro*, on which we can test the effects of various agents in which we may be interested. An early success in this direction was recorded by Gale and Folkes in 1954. They disrupted staphylococci by ultrasonic waves, and by centrifugation obtained a sediment consisting mainly of broken cell envelopes. The cell envelopes, provided with amino acids and ATP as a source of energy, were able to bring about the synthesis of the enzyme β-galactosidase, which attacks the disaccharide lactose. This synthetic ability was lost after treatment with ribonuclease and was restored by addition of RNA from a lactose-adapted strain of the same bacteria.

Evidence implicating ribosomes in protein synthesis continued to accumulate. Among many experiments in this field we may mention one carried out by Zamecnik and others in 1956 using the ribosomes of rat liver cells. The rats were each given an injection of ^{14}C-labelled leucine. At various times after the injection they were killed, their livers were cut out and with the greatest possible speed the ribosomes were separated as a sediment (A) by centrifugation. From the supernatant fluid (B) a precipitate of protein (C) was obtained. The amounts of radioactivity in each of these preparations at different times after the injection are shown in Fig. 24.1.

The first thing to be noticed is that the label appears very quickly in the ribosomes and then more slowly in protein. It is also to be noticed that the time course of radioactivity in A and C is very different according to whether the dose of labelled amino acid is large or small. After a small dose A soon reaches a peak and falls off, at which time C is reaching a steady level; after a large dose A quickly reaches a steady level while C continues to increase. Zamecnik's interpretation is that the amino acid is taken up in the ribosomes where it is incorporated into protein; the protein then leaves the ribosomes taking the label with it. Thus after a small dose incorporation is soon complete and the radioactivity of the ribosomes falls off, whereas after a large dose there is always excess of amino acid and the ribosomes are always saturated with it.

The problem of protein synthesis and its relation to nucleic acids is regarded by many people as the most important problem facing molecular biology today, and there is a widely approved working hypothesis which guides current investigations. This working hypothesis is as follows:

(i) the genetic information relating to the sequence of amino acids in protein is carried in the form of the sequence of nucleotides in DNA;

(ii) the nucleotide sequences of DNA in the nucleus are transcribed as nucleotide sequences of 'messenger' RNA† which then moves out from the nucleus into the ribosomes where it acts as a template;

(iii) the amino acids, each attached to a short length of 'transfer' RNA whose nucleotide sequence is specific for each amino acid, are lined up by base pairing between the 'messenger' RNA and the 'transfer' RNA;

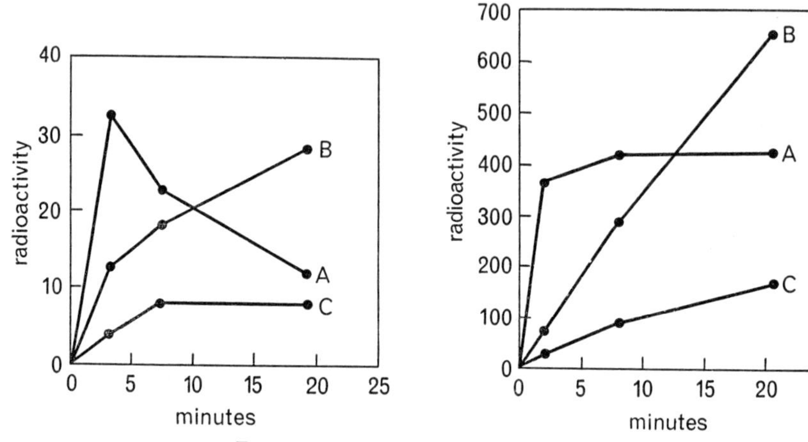

Fig. 24.1. Protein synthesis in ribosomes.

^{14}C-leucine was injected into rats. After some minutes the rats were killed and extracts of their tissues were prepared. The figures show the radioactivity in three fractions: A, precipitate of ribosomes; B, supernatant fluid; C, precipitate of protein from B.

For interpretation, see text.

(iv) the amino acids thus lined up are united by peptide links and the complete protein molecule is set free.

The evidence for the various steps in this hypothesis is certainly impressive in bulk. The number of original papers devoted to it is already in the region of a thousand and papers continue to appear at an ever-increasing rate. It is a story of a steady indirect approach, with much consolidation and filling in of detail, rather than a story of decisive experiments and dramatic advances. The present position has been built up in a large number of small steps, and for this reason it is difficult to present the evidence for it except in a very superficial way. We shall not

† Thymine in DNA being represented by uracil in RNA.

have time to go into experimental details but we will take a brief look at the nature of the evidence and note the points upon which it is satisfactory or unsatisfactory.

(i) The genetic information is carried in DNA in the form of nucleotide sequences

This proposition is commonly known as the 'coding hypothesis'. It can be examined by methods which have been known to cryptographers for many years and which are now assimilated to the general mathematical discipline of 'information theory'. Information theory cannot tell us whether the proposition is right or wrong. But if the formal problem is posed it can lay down certain limits within which a solution is possible.

The formal problem is this: there are 4 different nucleotides in DNA and some 20 different amino acids in protein; there are no known restrictions upon the order in which nucleotides may be arranged in DNA or amino acids in protein; what are the requirements of a code whereby the order of amino acids can be unambiguously specified by the order of nucleotides?

We do not need mathematicians to tell us that a code whereby a single nucleotide stands for a single amino acid is inadequate; with such a code we could only specify 4 amino acids and no more. If we say that two nucleotides are to stand for one amino acid we can specify $4^2 = 16$ amino acids—still not enough. We must therefore recognise that we need at least three nucleotides per amino acid, and this is our first conclusion.

Another question which we can try to answer is whether the code is overlapping or non-overlapping. The difference between overlapping and non-overlapping codes is explained in Fig. 24.2. Information theory tells us that even a partially overlapping code would impose restrictions upon the order in which amino acids could be arranged in protein. We do not find evidence of such restrictions in the proteins which have so far been analysed and we are therefore obliged to reject the idea of overlapping codes.

A further difficulty is that there are $4^3 = 64$ possible arrangements of nucleotides in triplets and only 20 amino acids; so either we must admit that more than one triplet can stand for one and the same amino acid, in which case the code is said to be degenerate, or we must postulate that of the 64 possible triplet configurations 20 stand for amino acids and make 'sense' while 44 do not stand for any amino acid and make 'nonsense'. There also arises the problem that if the message is a continuous sequence of letters some indication would have to be given of which groups of three letters were to be read as triplets. Such indication might be given by 'commas' (see Fig. 24.3). Alternatively, the whole sequence might be read off from a fixed point; this solution is at present in favour (see Chapter 25, p. 298). Another possible solution based itself on the

Nucleotide sequence	A	D	B	B	C	D	A	C	A
Overlapping code	A	D	B						
		D	B	B					
			B	B	C				
Partially overlapping code	A	D	B						
			B	B	C				
					C	D	A		
Non-overlapping code	A	D	B						
				B	C	D			
							A	C	A

Fig. 24.2. Overlapping codes.

In this figure, in order to preserve the pure formality of the treatment and to avoid the complication which base pairing would introduce, the 4 nucleotides are represented by the letters A, B, C, and D. The top line represents an imaginary nucleotide sequence. In the lower lines each set of three letters represents the triplet standing for an amino acid.

...ADBBCDACA...

could be read

...A,DBB,CDA,CA...

or

...AD,BBC,DAC,A...

or

...ADB,BCD,ACA...

Fig. 24.3. To illustrate the requirement for 'commas' to indicate which groups of three letters are to be read as triplets.

proposition that any two triplets which make sense can be placed end to end and that all triplets which overlap these make nonsense, as set out in Fig. 24.4. Now when this idea is followed up by mathematical methods it leads to the conclusion that 20, and no more than 20, 'sense' configurations are possible in a triplet code of 4 letters. Since there are, in fact, 20 amino acids in natural proteins this solution, a code without 'commas', seemed a particularly attractive one; but it has had to be discarded, because the code is now known to be degenerate.

These examples do not exhaust the complexities of the coding problem or the ingenuity of its devotees. They have been selected to illustrate the way in which information theory can be brought to bear upon the problem. Some critics have decried this approach on the ground that fanciful theorising will get us nowhere. But this criticism is ill-founded. The methods of information theory are entirely rigorous and far from encouraging fanciful theorising they serve the very useful purpose of setting bounds to speculation.

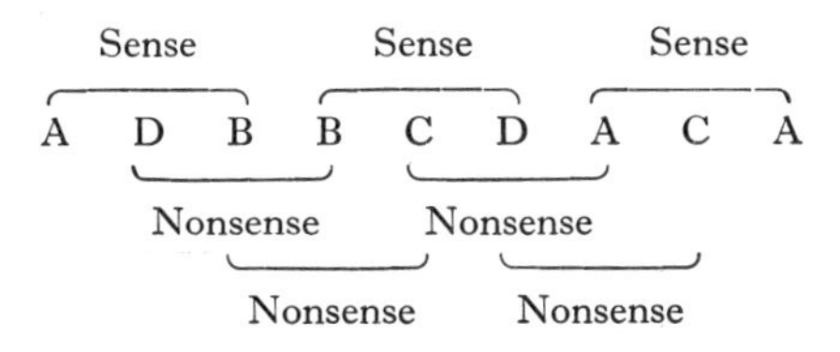

Fig. 24.4. To illustrate a possible solution to the coding problem, according to which there are 20 triplets each of which stands for one of the 20 amino acids and makes sense. Any sequence of 'sense' triplets can specify the primary structure of a protein. The remaining 44 triplets do not stand for any amino acids and make nonsense. The message can therefore be read in one way only and no 'commas' are necessary. Further explanation in the text.

The principal obstacle to progress is that we have no means of determining the sequence of nucleotides in nucleic acids as we have for determining the sequence of amino acids in protein. The best we have been able to do so far is to determine the 'nearest neighbour' relations of the nucleotides. The method of determining nearest neighbour relations is as follows.

The phosphate group of a naturally occurring free nucleotide is attached to carbon atom (5) of the sugar. We choose one particular nucleotide and label the phosphate group with ^{32}P. We then supply the labelled nucleotide to an actively growing micro-organism whereby it becomes incorporated into DNA. This incorporation takes place by the phosphate group of the labelled nucleotide becoming attached to carbon atom (3) of the nucleotide at the end of the chain—the nearest neighbour—see Fig. 24.5. After a period of growth we harvest the organisms and extract the DNA. By the use of appropriate enzymes which attack the bond between the phosphate groups and the carbon atoms (5) of the sugars we break down the DNA into nucleotides. The result of this is that the labelled phosphate group is transferred to the nearest neighbour. We then determine the proportion in which the label, attached to carbon atom (3), is distributed among all four nucleotides. We repeat this process with each of the other three nucleotides in the role of the original bearer of the label and so obtain a total of 16 values for nearest neighbour relations. The information is, of course,

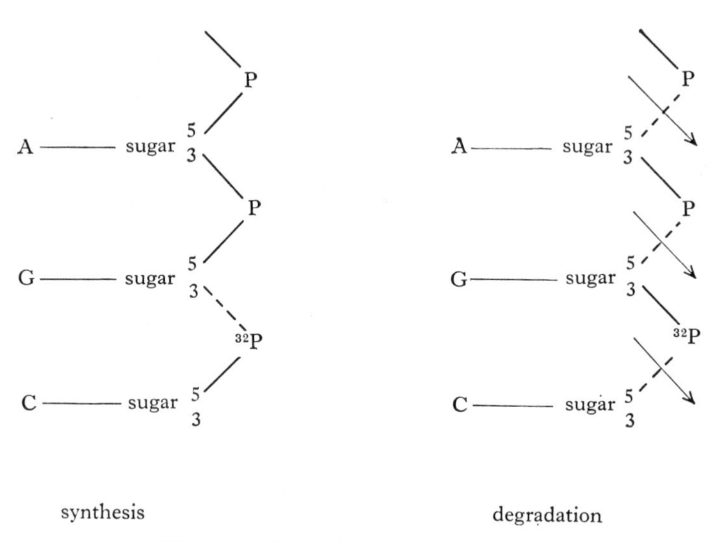

Fig. 24.5. The nearest neighbour method.

On the left a cytosine nucleotide labelled with ^{32}P is shown in the act of uniting with a guanine nucleotide at the end of the DNA chain. This union is effected between the labelled phosphate of the cytosine nucleotide and carbon atom (3) of the sugar of the guanine nucleotide. On the right the chain is shown being degraded under the action of an enzyme which attacks the bond between phosphate and carbon atom (5). As a result the labelled phosphate is now attached to the guanine nucleotide, the nearest neighbour.

statistical. It cannot give us the exact order of the nucleotides. It was hoped that some correlation might emerge between the nearest neighbour relations of nucleotides and the order of amino acids in proteins. But no correlation could be found.

(ii) The nucleotide sequences of DNA in the nucleus are transcribed as nucleotide sequences of 'messenger' RNA which then moves out from the nucleus into the ribosomes where it acts as a template

Before tracing the events which led to the discovery of 'messenger' RNA it may be well for us to note that most of the RNA content of cells belongs to one or other of two categories: (1) ribosomal RNA which appears to be the structural basis of the ribosomes and accounts for about 80% of the total, and (2) non-ribosomal (or soluble) RNA which accounts for about 20% of the total and is mostly 'transfer' RNA which will come up for consideration under subheading (iii). 'Messenger' RNA accounts for less than 5% of the total RNA and cannot be said to belong exclusively either to the ribosomal or to the non-ribosomal fraction since, although it is found mostly in association with ribosomes, it separates from them in the absence of magnesium.

It has long been generally agreed that protein is synthesised in association with ribosomes, and there was at first no good reason to suppose that the RNA which provided the structural basis of the ribosome was not also the template upon which protein was synthesised. But as time went on this view became increasingly difficult to maintain. It is estimated that a bacterial cell is capable of synthesising at least one thousand different species of protein molecule; if each ribosome has a built-in template and can only synthesise one protein species then for any one protein species the cell has on an average no more than 5–10 ribosomes available. But bacteria are known to switch their activities very rapidly from one protein species to another and a single enzyme can be produced at a rate which is as much as 7% of the total rate of protein synthesis. This would imply a prodigious rate of synthesis per ribosome. Another difficulty is that, on a statistical basis, the nucleotide base ratios of ribosomal RNA do not reflect the base ratios of homologous† DNA, as they should do if the information carried in DNA is copied in RNA.

To meet these difficulties Jacob and Monod put forward an alternative hypothesis. They suggested that the structural RNA of the ribosomes was not directly concerned in the transfer of information; instead, they proposed that the information was copied on a relatively short length of RNA, for which they proposed the name 'messenger' RNA, and that the 'messenger' RNA entered a ribosome where it played its part as a template and was broken down after the synthesis of one, or at most a few, protein molecules. According to this view the ribosome itself is an unspecialised machine which can make any sort of protein in accordance with the information supplied to it in the form of 'messenger' RNA. In putting forward this hypothesis, in 1961, Jacob and Monod pointed out that a candidate for the role of 'messenger' RNA was already in the field.

† i.e. DNA from the same species of organism.

Evidence for the existence of what later came to be known as 'messenger' RNA was gradually assembled in a series of investigations of the bacterium-bacteriophage relationship. The reproduction of phage within a bacterium affords particularly favourable opportunities for studying the dependence of protein synthesis upon nucleic acids. As described in Chapter 8, p. 126, the phage injects only DNA into its host, the host providing the apparatus whereby the phage builds up its own proteins. After infection of a bacterium by phage the synthesis of bacterial protein quickly declines and within a matter of minutes the synthesis of phage protein is in full swing.

In 1953 Hershey, in a study of the synthesis of phage DNA, noted in passing that there appeared to be a small fraction of the host RNA which had a high rate of incorporation of ^{32}P. Three years later Volkin and Astrachan confirmed and extended this observation. Having infected a culture of *Escherichia coli* with phage T_2 they added ^{32}P to the medium; any nucleic acid synthesised after that moment would therefore carry the label. Samples of the infected culture were taken at intervals, the RNA was extracted, broken down into mononucleotides and the amount of label in each of the four nucleotides was assayed. The interesting fact emerged that the base ratios of the labelled RNA were not the same as the base ratios of the total RNA (Table 24.1). Even more interesting was the similarity of the base ratios of the labelled RNA to the base ratios of the phage DNA.

Table 24.1

Nucleotide base	Total RNA	Labelled RNA	Phage DNA
Adenine	24	30	32·5
Uracil	22	30	32·6 (thymine)
Guanine	31	22	18·2
Cytosine	23	18	16·7

Thus in infected bacteria there appears to be a fraction of the total RNA which is rapidly synthesised and broken down and whose base ratios reflect those of the DNA which is, so to speak, giving directions to the system which is synthesising protein.

This was followed up by Brenner, Jacob and Meselson in 1961. Their experiments were of complicated design, involving multiple labelling, and to describe a single experiment, even in outline, would take too long. But in principle their approach was as follows. The bacteria were first grown upon a medium containing ^{15}N and ^{13}C.† Immediately after

† ^{13}C is a stable heavy isotope of carbon.

infection the culture was flooded with medium containing a large excess of ^{14}N and ^{12}C, and in addition either ^{32}P or ^{35}S. At the end of the experiment, which generally only lasted a few minutes, the RNA was extracted and separated into various fractions by centrifugation in a gradient of sucrose (p. 74). The use of stable isotopes made it possible to separate the 'heavy' components which had been synthesised before infection from the 'light' components which had been synthesised after infection. The purpose of adding ^{32}P was to label the rapidly exchanging RNA fraction which, being only 4% of the total RNA, could hardly otherwise be recognised. Correspondingly, ^{35}S was used to label newly synthesised protein.

The most important findings were that newly synthesised RNA, labelled with ^{32}P, and also newly synthesised protein, labelled with ^{35}S, were found at that position in the sucrose gradient which was occupied by 'heavy' ribosomes and not at the position which would have been occupied by 'light' ribosomes. It follows that the synthesis of phage protein does not have to wait upon the synthesis of new ribosomes but can be accomplished by the old ribosomes already present in the host before infection. The newly synthesised RNA which is found in the old ribosomes is presumably that fraction of rapid turn-over, previously identified by Volkin and Astrachan, and thus seems to meet the specification for 'messenger' RNA as laid down by Jacob and Monod.

Contemporaneously with the work of Brenner, Jacob and Meselson, Gros and five colleagues demonstrated the presence of a comparable 'messenger' RNA in uninfected bacteria, with a similar high rate of turn-over. But it is not to be supposed that a high rate of turn-over is always an essential attribute of 'messenger' RNA. In other systems, e.g. in the reticulocyte, to be discussed presently (p. 273), the ribosomes are capable of long-continued protein synthesis in circumstances which seem to preclude any renewal of 'messenger' RNA.†

Having established the relationship between 'messenger' RNA and ribosomes we turn now to the relationship between 'messenger' RNA and DNA. In accordance with the coding hypothesis we would expect that the transcription of the message would involve the separation of the two strands of the DNA double helix, following which a comple-

† It is also appropriate to mention at this point that recent work, in which more than usually gentle methods of disruption have been employed, suggests that the unit of synthetic activity of the reticulocyte is not the single ribosome but an aggregate of five ribosomes, and that it is 'messenger' RNA which holds the aggregate together.

mentary RNA strand would be synthesised alongside each DNA strand. This expectation is supported by a substantial body of evidence which can only be touched upon briefly.

(1) Following the general line of approach pioneered by Kornberg (p. 250) the enzymatic synthesis of RNA in the presence of a DNA primer has been achieved. The RNA thus synthesised has nucleotide ratios which reflect those of the primer DNA and the nearest neighbour relations are in conformity.

(2) As mentioned on p. 255, DNA can be denatured, with separation of the strands of the double helix, by boiling. RNA which has been enzymatically synthesised will then associate with the single-stranded primer DNA so as to form hybrid molecules which can be identified by centrifugation in a gradient of caesium chloride (p. 254). This approach has been extended to naturally occurring nucleic acids. 'Messenger' RNA from bacteria infected with phage T_2 has been successfully 'annealed' with denatured DNA from the same phage. It could not be 'annealed' with DNA from phage T_5 or from *E. coli* itself.

We may also recall in this context the observation mentioned on p. 92: namely, that there is a naturally occurring close association between RNA and DNA in the loops of the lampbrush chromosomes where, not impossibly, the transcription of the message takes place.

(iii) The amino acids, each attached to a short length of 'transfer' RNA whose nucleotide sequence is specific for that amino acid, are lined up by base pairing between the 'messenger' RNA and the 'transfer' RNA

The coding hypothesis, translated into practical terms, reveals the need for some kind of adaptor molecule by which an amino acid can be guided into its appropriate place. That is to say, if the information specifying a given amino acid is coded as a particular short sequence of unpaired nucleotides in a long chain of 'messenger' RNA (which is the template) we require the amino acid to be attached to a short nucleotide chain of complementary sequence which can find its proper place on the template chain by base pairing (Fig. 24.6). We must suppose that this adaptor molecule is composed of at least three nucleotides, and until we have some reason to do so we need not suppose that it has more than three nucleotides.

The attachment of amino acids to 'transfer' RNA was first observed by Hoagland, and by 1957 the original discovery had been confirmed and extended to cover all the 20 amino acids. For this attachment enzymes are needed, each specific for one amino acid, and ATP as a source of energy. But, inconveniently, the RNA molecules to which the amino acids are attached are not trinucleotides as the coding hypothesis would suggest, but are polynucleotides containing about 80 nucleotides in the chain. A polynucleotide of this size is difficult, though not impossible, to crystallise, and, as has been mentioned already (p. 258), X-ray diffraction patterns of 'transfer' RNA have been obtained by Wilkins. These patterns indicate a double helical structure, but for various reasons this cannot be interpreted in terms of two complementary chains as in DNA.

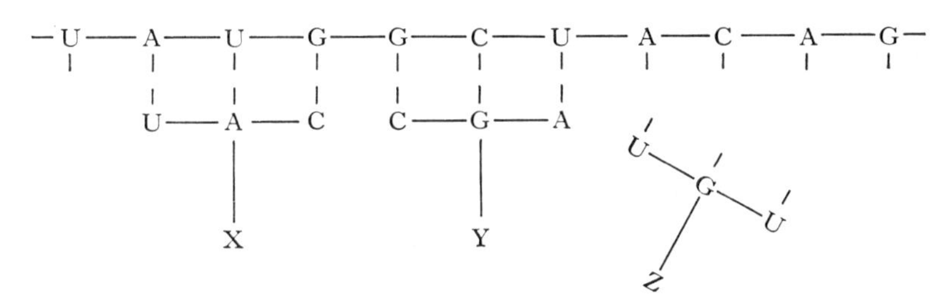

Fig. 24.6. This diagram indicates how in a hypothetical scheme the amino acids X, Y and Z, each attached to an adaptor triplet ('transfer' RNA), might become lined up on the template ('messenger' RNA) by base pairing.

The interpretation favoured by Wilkins is that the polynucleotide chain is bent back upon itself, with base pairing over part of the molecule, as illustrated in Fig. 24.7. Chemical evidence indicates that the nucleotide sequence at the end of the chain to which the amino acid is attached is –C—C—A, and this is the same for all species of 'transfer' RNA. In the region of the fold there is room for three or more unpaired nucleotides and one would like to believe that this is the region which becomes lined up on the template by base pairing. Evidence as yet available will not take us that far; but at least one can say that nothing emerges from the study of the structure of 'transfer' RNA which is incompatible with the role assigned to it by the coding hypothesis.

It should follow from the coding hypothesis that once an amino acid has become attached to a molecule of 'transfer' RNA the positions which it may occupy in a polypeptide chain are thereafter determined by the 'transfer' RNA and not by the amino acid. That is to say, if subsequent to its attachment to 'transfer' RNA an amino acid A is transformed into another amino acid B, then B will be found to occupy the positions in the polypeptide chain which are normally occupied by A. This has been put to the test by Chapeville and others and shown to be the case. A complete account of the experiment cannot be given here, but its essential features are as follows. The amino acid cysteine, attached to its appropriate 'transfer' RNA, was isolated. It was then

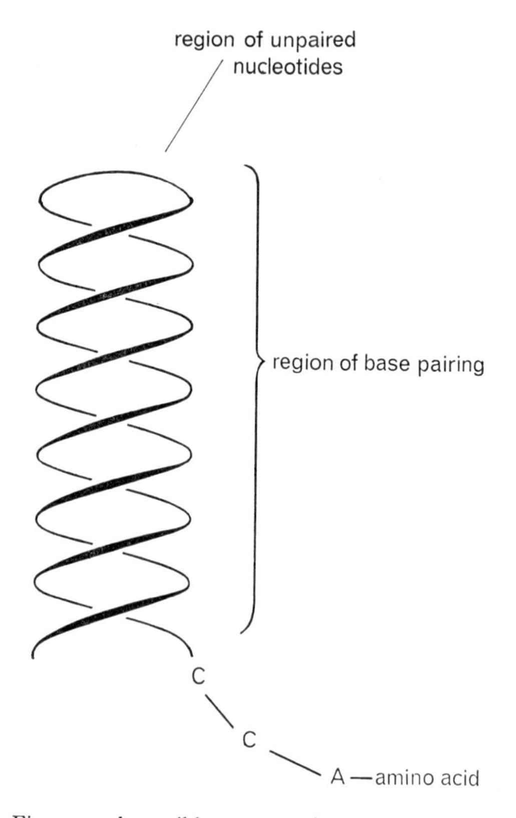

Fig. 24.7. A possible structure for 'transfer' RNA.

exposed to a catalyst known as Raney Nickel. This catalyst promotes the replacement of –SH groups by hydrogen atoms. In the presence of Raney Nickel cysteine was converted into alanine, without being detached from the 'transfer' RNA. The 'transfer' RNA, now combined with alanine, was added to a cell-free system capable of incorporating amino acids into polypeptide, whereupon it was found that the system incorporated the alanine as though it were cysteine.

(iv) The amino acids thus lined up are united by peptide bonds and the complete molecule is set free

This proposition scarcely merits independent status. If all the foregoing propositions are acceptable there is no great difficulty about this one. But it is useful to let the proposition stand, for it can serve as an introduction to yet another aspect of protein synthesis which, although not an essential feature of the hypothesis as a whole, has much intrinsic interest.

Earlier, we noted that protein molecules do not seem to be built up piecemeal but are assembled in their completed form in a single operation. In the light of what we already know this statement requires qualification. It is regarded as established that proteins are synthesised in ribosomes; what we really mean to say is that polypeptide chains are not released from ribosomes in an incomplete state. We do not have to believe that the mechanism of synthesis is like a stamping-out machine, all the links being established simultaneously.

It is possible that the amino acids find their way to their appropriate positions on the template in no particular order in time, the gaps between the early arrivals being filled in later. It is equally possible that they are assembled in orderly succession, starting from one end. We now have evidence that the second possibility is the correct one. This evidence, which has been provided by the work of Schweet and of Dintzis, comes from the synthesis of haemoglobin.

The mature red blood cell is without nucleus or ribosomes; it is charged with haemoglobin but no longer synthesises it. The synthesis of haemoglobin takes place in the immature cell, the reticulocyte, and can be followed in preparations of ribosomes from this cell; and haemoglobin is to all intents and purposes the only protein which is synthesised. Haemoglobin is made up of 4 component polypeptide chains, each comprising about 150 amino acids.

Now let us consider the situation in ribosomes which are in process of synthesising haemoglobin. According to our general working hypothesis some of the templates will be occupied by partially completed haemoglobin components. Suppose that we suddenly replace one of the amino acids—call it X—in the incubation mixture by labelled X; the as yet unoccupied X sites will thereafter be occupied by labelled X. If we can then break down the completed chain and identify the positions occupied by labelled X we can distinguish between the X sites which were already occupied before the change-over and the X sites which were occupied later. Reference should now be made to Fig. 24.8 which shows incomplete polypeptide chains assembled on templates at the moment when normal X is replaced by labelled X. If the amino acids are assembled in orderly succession starting from the left, as in Fig. 24.8*a*, then in the first completed chains to be set free the X sites towards the left will be occupied by normal X and the X sites towards

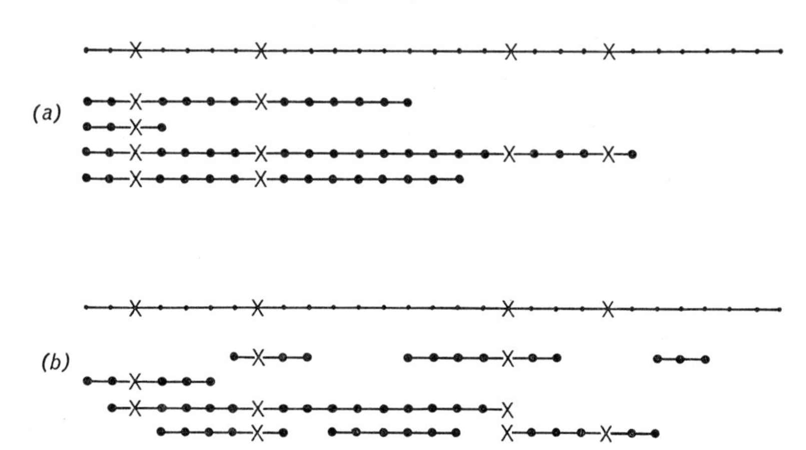

Fig. 24.8. A possible method of deciding whether amino acids are assembled in random order or in orderly succession.

In each figure the top line represents a polypeptide chain in which the positions occupied by some given amino acid, X, are indicated. The lower lines represent templates which are partly occupied by incomplete polypeptides in process of assembly.

At this point in time we supply the system with labelled X, replacing hitherto unlabelled X.

In *a* it is assumed that the amino acids are assembled in orderly succession from the left; when the completed chains are released from the template the label will be found in the X sites lying to the right. In *b* it is assumed that the amino acids are assembled in random order in time; when the completed chains are released the label will be distributed at random over all the X sites. If we can discover which sites are labelled in the completed polypeptide we can decide between these two alternatives.

the right will be occupied by labelled X. But if the amino acids are assembled in no particular order in time, as in Fig. 24.8*b*, then the distribution of the label will be random.

The foregoing account is to explain how, in principle, an experiment can be devised by which one can distinguish between two possible modes of assembly of amino acids into polypeptide chains. It is necessary to point out that the original experiments of Dintzis did not follow these lines exactly; the formidable difficulties of separating an amino acid into fractions sorted according to the positions which they had occupied in the polypeptide chain had yet to be overcome. What Dintzis did was to split the chain into some 30 oligopeptides and to follow the time course of the appearance of the label in these oligopeptides. Lack of precise information as to position in the chain made the results of the experiment less easy to interpret; but they were clearly in accordance with assembly in a definite order, starting from the *N*-terminal, and not in accordance with assembly in random order. In a later more refined experiment the actual positions of the labelled amino acid in the chain were established.

All the evidence so far cited, while it supports our working hypothesis in general terms, does not suggest any line of approach to the central problem of discovering the particular triplet of bases which stands for a given amino acid, as the hypothesis requires. But in fact a useful weapon with which to attack this problem is to hand.

As was mentioned in passing (p. 270), RNA can be enzymatically synthesised in cell-free systems. If the synthesis is brought about by the enzyme RNA polymerase in the presence of DNA the nucleotide ratios suggest that the synthesis of RNA is in some way directed by DNA. But synthesis of RNA can also be brought about in the absence of DNA by the enzyme polynucleotide phosphorylase, discovered by Ochoa, and in this case the nucleotide ratios of the synthetic RNA reflect the relative abundance of the various free mononucleotides from which it is built up. It is assumed that the nucleotides are joined together in random order.

It occurred to Ochoa and to Nirenberg, working independently, that synthetic RNA might be able to replace natural 'messenger' RNA in promoting the incorporation of amino acids into proteins in cell-free systems. It also seemed quite possible, on the coding hypothesis, that triplets all of the same base—AAA, GGG, CCC, UUU—might code for amino acids. If so, it should be possible to discover what these amino acids were; because although the various nucleotides might be assembled in random order out of a mixture, there could be no doubt about the composition of RNA which was synthesised in the presence of one single species of mononucleotide only.

The four polynucleotides—poly-A, poly-G, poly-C and poly-U—were synthesised and each in turn was used as a substitute for 'messenger' RNA in a preparation of ribosomes capable of incorporating amino acids into protein. Each of the 20 amino acids was tested in turn for incorporation, by the usual method of labelling with ^{14}C.

In this way it was first shown by Nirenberg, and almost immediately confirmed by Ochoa, that poly-U promoted the incorporation of the amino acid phenylalanine. Later it was found that the incorporation of proline was promoted by poly-C and that of lysine by poly-A, though not to such a marked extent as that of phenylalanine by poly-U.

For the incorporation of other amino acids more than one base has to be present, and with this requirement come difficulties. There being no method of establishing the order of nucleotides in nucleic acids, we must have recourse to the statistical approach. Given that a nucleotide co-polymer, say poly-UG, is built up of the nucleotides U and G in the proportions 3U:1G, and assuming that the nucleotides are assembled in random order, it is possible to calculate the relative frequencies with which the triplet groups 3U, 2U1G, 1U2G, 3G will occur in the polynucleotide. Using various synthetic polynucleotides with each labelled amino acid in turn we can discover the combinations of bases which result in maximum incorporation. This has been done for all 20 amino acids, and the results to date are set out in Fig. 24.9. No clear solution emerges as yet. Ambiguity arises partly as a consequence of the statistical approach, partly because the sequence of bases is unknown and partly, it seems, because the code is degenerate, more than one triplet standing for a single amino acid.

How far, then, have we got in elucidating the mechanism of protein synthesis, and what is the present state of the coding hypothesis?

Granting that genetic information is carried in DNA, the simplest alteration which we can envisage is the substitution of one nucleotide pair for another at some single position in the double helix. On the coding hypothesis this would alter the reading of a single triplet and might be expected to result in the replacement of one amino acid by another at the corresponding point on a polypeptide chain. An observation bearing upon this, and one to which very great significance is attached, was made by Ingram in 1956. He showed that the disease known as sickle-cell anaemia is associated with the replacement of glutamic acid by valine at one point in the haemoglobin molecule. Sickle-cell anaemia is a hereditary affliction, appearing in individuals who are homozygous for a recessive mutation. It is inherited in accordance with Mendelian principles. Ingram's observation establishes a connection spanning right across the field from the classical genetics of higher organisms to the recent discoveries of the mechanism of protein synthesis in cell-free systems, a connection of a kind which is entirely in accordance with expectation on the basis of the coding hypothesis.

We have seen that evidence is available at many places all along the line of the argument of the coding hypothesis, from DNA to protein, to support the various propositions which the hypothesis comprises. This evidence is impressive; but the massive support which is provided in

Amino acid	Combinations of three bases which promote incorporation			
Alanine	2C1G	CGU*		
Arginine	2C1G	2A1G	CGU*	
Asparagine	2A1C	2A1U		
Aspartic acid	AGU			
Cysteine	1G2U*	2G1U*		
Glutamic acid	2A1G	AGU*		
Glutamine	2A1C	2A1G	AGU*	
Glycine	2G1U	1A2G		
Histidine	1A2C			
Iso-leucine	1A2U	1U2A		
Leucine	1G2U	1C2U	1A2U	3U
Lysine	3A	2A1G*	2A1U*	
Methionine	AGU*			
Phenylalanine	3U			
Proline	3C*	2C1U*	1A2C*	2C1G*
Serine	1C2U	1U2C	CGU	
Threonine	1A2C	2A1C		
Tryptophane	2G1U			
Tyrosine	1A2U			
Valine	1G2U			

Fig. 24.9. Assuming a triplet code this figure indicates the bases which are likely to be concerned in specifying each of the 20 amino acids. Uncertainty still attaches to those which are marked with an asterisk. Only aspartic acid and methionine seem to require more than two bases.

The data in this figure represent no more than the progress made by the middle of 1963. Ideas may be expected to undergo extensive revision.

some places must not be allowed to distract our attention from other places where further support would be welcome. The central feature of the whole coding hypothesis is base pairing—the template idea—and the evidence for base pairing is almost entirely circumstantial. One of the strongest arguments for base pairing is that it enables us to combine the evidence of nucleotide regularities with the evidence of X-ray diffraction to provide a plausible structure for DNA. We may also quote in evidence that base pairing undoubtedly occurs in synthetic polynucleotides. But there is as yet no direct evidence that in DNA from natural sources adenine is paired with thymine and guanine with cytosine; there is no direct evidence of base pairing between 'messenger' RNA and 'transfer' RNA. Direct evidence continues to escape us because we have as yet no means of ascertaining the order of nucleotides in nucleic acid as we have for ascertaining the order of amino acids in protein.

Thus while we have direct evidence which unmistakably implicates both DNA and RNA in protein synthesis in the widest sense, we have only circumstantial evidence about the part which base pairing may play in the process. Much of what has been written about nucleic acids and protein synthesis, especially in the popular scientific press, fails to make this clear. 'This Disneyland of templates and pools and feed-backs' as Chargaff has called it 'will make us the laughing stock of later times.' In scientific circles it is freely admitted that the direct evidence for the coding hypothesis is at the best rather meagre. The circumstantial evidence, however, is so very strong that all the research workers in this field are firmly convinced that they are on the right lines. Progress is very rapid indeed, and it may well be that before these lines are in print a formal solution to the coding problem will have been found, meaning that a sequence of bases—and probably more than one sequence—can be assigned to each of the 20 amino acids. But this will not mean that the mechanism of protein synthesis has been completely elucidated.

There is no reason to be ashamed of speculative theories if these lead to experiments by which the theories can be tested. We may recall Krebs' opinion (p. 186) that the value of a hypothesis is to be assessed on the grounds of the discoveries to which it has given rise. On this basis the coding hypothesis, even in the unlikely event of its being proved to be wrong, has already justified itself many times over.

25

THE GENETICS OF MICRO-ORGANISMS

Since the time of the Second World War the most striking advances in genetics have come from the study of micro-organisms. No doubt the reason for previous neglect of micro-organisms is that differences in morphology between species are often impossible to establish. Different strains of pathogenic bacteria, otherwise indistinguishable, have long been recognised by their ability to infect one host species and not another, by their relative virulence and by the symptoms they produce. To the traditional geneticist, brought up on the easily recognisable differences between mutants of *Drosophila* and seeing the range of important problems still awaiting solution in this and other well studied types, micro-organisms must have seemed to be unattractive material. The advantages of a rapidly breeding stock were more than offset by the vagueness of the criteria by which genetic changes could be assessed. It was not until biochemists had shown that differences between strains could be defined in terms of requirement or non-requirement for a particular chemical compound that the possibilities of micro-organisms as genetical precision tools came to be recognised.

When we were considering the nutritional requirements of organisms in Chapter 2 we noted that some could be described as prototrophic, meaning that they were able to synthesise all the organic substances required for growth if they were supplied with the necessary elements in inorganic form, whereas others, spoken of as auxotrophic, were to a greater or lesser extent exacting in their requirements and had to be supplied with certain organic molecules which they were unable to synthesise for themselves. These differences provide the basis for biochemical genetics.

Genetical analysis is based upon measurement of recombination frequency, by which we mean the proportion of recombinants in the offspring of a cross between parents differing by two pairs of heritable characters. For characters within the same linkage group the closer the loci are together the lower will be the recombination frequency. In the case of organisms such as *Drosophila* the time and labour involved in examining the flies under the microscope sets a practical limit to the lowest recombination frequency which we can hope to detect. If we examine 10,000 offspring from one single type of cross and discover among them one recombinant we cannot assign any definite value to the recombination frequency; we can only say that it is likely to be less than 0·1 %, i.e. less than 1 in 10^3. In order to set a lower limit to the recombination frequency we would have to go on until we had counted, say, 10 recombinants, which might mean examining 100,000 flies—a formidable undertaking, but it has been done in more than one experiment. With micro-organisms, as we shall see presently, the use of selective media upon which only recombinants will grow reduces enormously the labour involved in their detection. In the investigations of bacteriophage, to be described later, the lower limit to the detection of recombination is in the region of 1 in 10^8. Thus the resolution (cf. optical resolution, Chapter 4) obtainable in genetical analysis of phage is many thousands of times greater than that obtainable with higher animals and plants.

1. FUNGI

The subject of biochemical genetics was effectively launched by the investigations of Beadle on the red bread mould *Neurospora*, and before going further it will be necessary to say something about this organism and about its mode of reproduction.

In its vegetative condition *Neurospora* grows as a mycelium of branched septate hyphae, each compartment containing several nuclei. It can reproduce asexually, but with this we need not concern ourselves here. Sexual reproduction involves the fusion of nuclei from different mycelia. No difference can be seen between the mycelia which take part in sexual reproduction and it is therefore impossible to describe one as male and the other as female; all we know is that mycelia belong to one or other of two types and that sexual reproduction will only take place between mycelia of different types. This condition is known as 'heterothallism'.

An important difference between *Neurospora* and higher organisms is the fact that the nuclei of the ordinary mycelium of *Neurospora* are haploid. In *Neurospora* the nucleus contains seven chromosomes. Because the nuclei are haploid sexual reproduction does not have to be preceded by meiosis; it can be effected by the fusion of any two nuclei from mycelia of opposite mating types. The zygote is of course diploid, but the diploid phase has only a brief existence. Meiosis follows at once, giving four haploid daughter nuclei, and each of these undergoes a further mitotic division to give eight nuclei in all. Each of the eight nuclei surrounds itself with a thick membrane and becomes a spore, and the eight spores are linearly arranged in a spore case known as an ascus. From the geneticist's point of view an important feature of this process is that the arrangement of the spores in the ascus faithfully represents the order in which division of the nuclei has taken place, as illustrated in Fig. 25.1. Since the ascospores can be dissected out of the ascus and

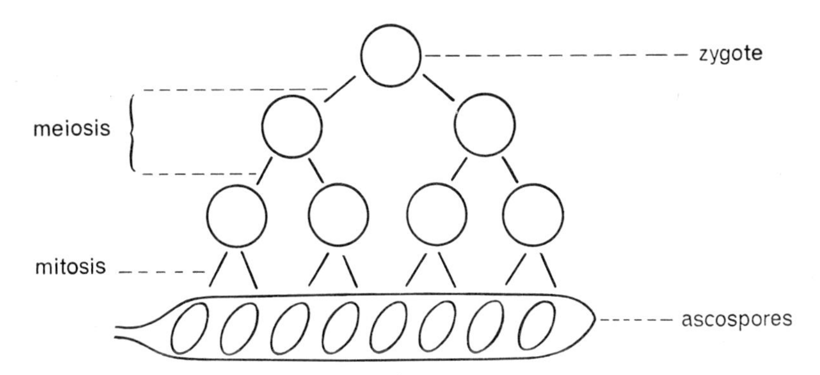

Fig. 25.1. The ordering of the ascospores in the ascus.

allowed to germinate in isolation, this means that the four products of any meiosis can be individually recovered; recourse to statistical analysis is thereby rendered unnecessary. Another advantage is that in a haploid organism the genetical situation is not complicated by dominance. Dominance (or recessiveness) is a feature of diploid organisms; a mutant character which might be recessive in the phenotype of a diploid will always be manifested in the phenotype of a haploid.

The wild type of *Neurospora* can maintain healthy growth and reproduction on an artificial medium of the following composition:

NaCl, $CaCl_2$, KH_2PO_4, $MgSO_4$
Traces of B, Mo, Fe, Cn, Mn, Zn
Ammonium tartrate and ammonium nitrate
Sucrose
Biotin (a vitamin)

This is known as the 'minimal medium' for *Neurospora*. Naturally occurring mutant forms which require other organic compounds in addition to the above are known, and a further range of mutant forms with a variety of additional requirements has been produced by exposure of the fungus to X-rays. The radiation is administered before sexual fusion. After the asci have formed, the ascospores are isolated and allowed to germinate on 'complete medium' which is conveniently prepared by adding yeast extract to minimal medium. When growth is well established portions of the mycelium are transferred to minimal medium to see if they are still able to flourish. If they do flourish it is concluded that no mutation affecting their nutritional requirements has occurred; if they do not flourish various other media are tested until eventually the substance which they require for growth is identified.

At this stage we may take a particular case for illustration. Fifteen mutants were isolated which required the amino acid arginine as an additional growth factor (designated 'arginine-*less*'). It was first necessary to establish the identity or otherwise of these mutants. If two mutant strains having the same nutritional requirement have arisen by mutation of the same gene, i.e. if they are alleles (p. 301) of the same gene, they can be bred together and will produce essentially only mutant offspring. If they have arisen by mutation of different genes recombination will be possible, giving the double mutant and the wild type. Non-identity can therefore be established by the appearance of mycelium able to grow on minimal medium; the absence of any such mycelium after extensive crossing of the mutant strains is taken to indicate that only one gene is involved. Of the 15 mutants originally isolated 7 proved to be genetically distinct.

It was known that the synthesis of arginine in higher organisms involves two intermediate compounds, ornithine and citrulline, and it was therefore natural to test these substances as possible substitutes for arginine. Of the 7 genetically distinct mutants one responded only to arginine, two responded to either arginine or citrulline, and four responded to arginine, citrulline or ornithine. There are good biochemical reasons for believing that the conversion of citrulline to arginine involves two steps and an intermediate compound. Ornithine is a precursor of citrulline but little is known about its origin in higher organisms. We may then interpret the results as in Fig. 25.2.

Another sequence of biochemical synthesis, known as the tryptophane cycle, runs in 6 steps from phenylalanine through tryptophane (both amino acids) to the vitamin nicotinic acid. Genes have been identified in *Neurospora* which specifically block 5 of the 6 steps.

Analysis of genetically controlled nutritional deficiencies has shown that in nearly every case the deficiency can be related to the requirement for a single organic compound. In *Neurospora* and in other fungi this relationship has been established in well over a thousand cases. On the basis of these findings, Beadle and Tatum put forward the 'one-gene-one-enzyme' hypothesis of the mechanism of gene action. According to this hypothesis the required substance is synthesised in the presence of an enzyme which is lacking in the mutant form. The evidence for the hypothesis is strong. In more than 50 cases the presence of the enzyme

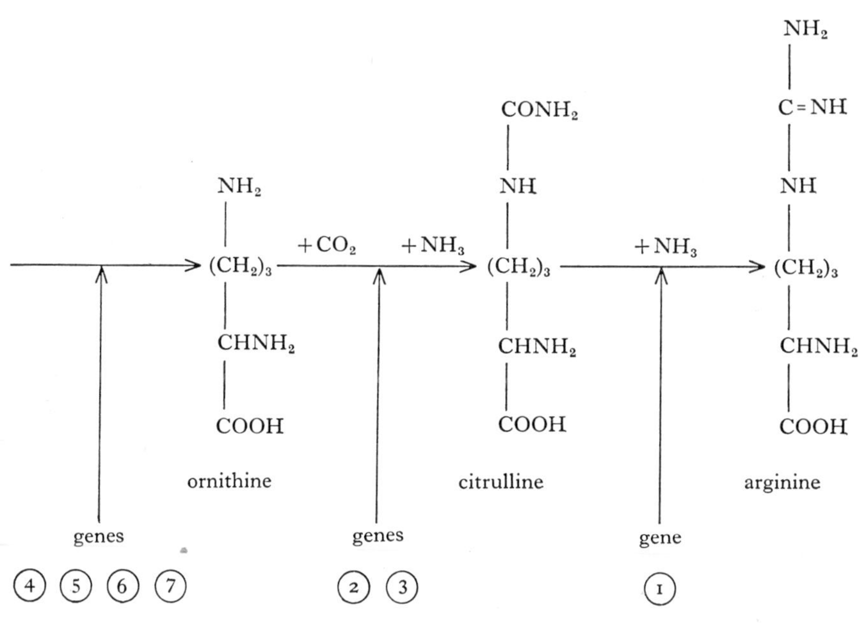

Fig. 25.2. The biosynthesis of arginine in *Neurospora*.
Mutation of the various genes blocks the synthesis at the points indicated.

in the wild type and its absence from the mutant has been confirmed by purely biochemical methods. But the evidence comes almost exclusively from micro-organisms.

2. BACTERIA

At the time of the Second World War it was by no means clear that bacteria 'had any genetics' in the sense that genetics was understood from the study of the higher organisms. As already mentioned, morphology does not always afford adequate criteria for the recognition of bacterial types, and nutritional requirements, which have proved so useful in the case of fungi, are in bacteria liable to extremely rapid adaptation. Bacteria which seemed at first to have a requirement for a particular organic compound often seemed to be able to learn to do without it. The existence in bacteria of something corresponding to the nucleus of higher organisms had only just been discovered and there was at that time no evidence that bacteria engaged in any form of sexual reproduction.

The nature and origin of adaptive variations were particularly puzzling. It was, for example, a matter of common knowledge that when bacteria were plated on agar with a bacteriophage to which they were susceptible most of them were destroyed, but a very small number survived to found colonies which were phage-resistant. There seemed to be two possible explanations. The first, which may be called the Lamarckian explanation, was that on exposure to phage a few of the bacteria acquired resistance (like a patient recovering from an infectious disease) and handed it on to their offspring. The second, which may be called the Darwinian (strictly, Neo-Darwinian) explanation, was that there were present in the culture a few bacteria which had already acquired resistance by spontaneous mutation, before exposure to the phage, and that these alone survived and produced resistant offspring.

The question remained open until 1943, when the fluctuation test was devised by Luria and Delbrück. The theoretical principle upon which this test is based is so simple, the technique of its application is so easy and its results are so unambiguous that it deserves our special attention. Luria and Delbrück argued as follows. If the second explanation —spontaneous mutation—is true, then mutation may occur at any time. Thus if we isolate a single phage-sensitive bacterium there is a small chance that a mutation will occur when it divides into two; one of its offspring will be phage-resistant and the other will be phage-sensitive. As growth proceeds, assuming that there is no difference in growth rate associated with the mutation, a bacterial population will be produced in which 50% of the bacteria will be phage-resistant. If the mutation occurs at some later division a correspondingly smaller proportion of the population will be phage-resistant; and if the period of growth is limited it is possible that no mutation will appear at all. Therefore if we set up a number of cultures each originating from a single bacterium of a phage-sensitive strain and allow them to grow for a limited period we will expect to find very great fluctuations between one culture and another in the proportion of phage-resistant bacteria which they contain. On the other hand, if the first explanation—acquired resistance after exposure to phage—is true, then there is a small chance that any bacterium, whatever its ancestry, will survive attack and will give rise to phage-resistant offspring. In this case there will be no difference between one culture and another; all will have the same proportion of phage-resistant bacteria. The reader will appreciate that what the fluctuation test will decide is whether conversion to phage-resistance can take place only after contact has been made with the phage or whether it can take place at any time before contact.

To carry out the test in practice a broth culture of *Escherichia coli* is started from a single bacterium of a phage-sensitive strain. This makes it certain that the population starts by being genetically homogeneous. From this stock culture several—say 10—subcultures are made. The technical difficulty of starting each subculture from a single bacterium can be avoided; all that is necessary is to ensure that the number of bacteria in the inoculum is such that the chance of a mutant being included in it is small. If we expect the proportion of phage-resistant bacteria in the stock culture to be of the order of 1 in 10^6 then it will be sufficient if each inoculum contains less than 1000 bacteria. After a short period of growth the cultures are tested by plating with phage to determine the proportion of phage-resistant bacteria. Figures obtained from an experiment of this type are reproduced in Table 25.1. The first column shows the number of phage-resistant colonies arising from 10 similar (i.e. containing approximately the same total number of bacteria) samples taken from one and the same culture. In this case a statistical test shows that the fluctuation between samples is no greater than we would expect from ordinary sampling error; but it is necessary to check this in order to be assured that no unforeseen variability is present under the conditions of the experiment. The second column shows the number of phage-resistant colonies arising from 10 similar samples taken one from each of the 10 subcultures. In this case the fluctuation between samples is obviously very much greater. We therefore conclude that conversion to phage-resistance can occur at any time and that it is presumably the result of spontaneous mutation.

The demonstration of mutation in bacteria was soon followed by the discovery of other features of inheritance already known in higher organisms. Genetic recombination in *Escherichia coli* was reported by Lederberg and Tatum in 1946. The principle of their method is the same as that which we have already met with in the case of *Neurospora*.

Two mutant strains, each requiring a different growth factor, are mixed and plated on minimal medium. If there is recombination we will obtain some double mutants (which like the bacteria of the parent strains will fail to grow) and some wild type which will grow and produce colonies.

Table 25.1. *Number of phage-resistant colonies produced*

Samples from same culture	Samples from different cultures
14	30
15	10
13	40
21	45
15	183
14	12
26	173
16	23
20	57
13	51

Mutants involving nutritional deficiencies arise spontaneously in bacteria and their appearance can be promoted by exposure to radiation. In order to select out a mutant strain which is nutritionally deficient use is made of the fact that many antibiotics will only destroy bacteria which are in process of active growth. The culture believed to contain mutants is inoculated into minimal medium to which antibiotic—penicillin, for example—has been added. Wild-type bacteria start to grow and are killed while the mutants fail to grow and survive. The bacteria are then washed free of antibiotic by suspending them in clean water and recovering them by centrifugation (repeated several times) and are then plated on complete medium. The deficient mutants grow into colonies from which they can be picked and tested further to discover the particular growth factors which they require.

A difficulty attending upon recombination experiments of this kind is that the wild type may appear not only as a result of recombination but also as a result of reverse mutation, from mutant to wild type. Lederberg and Tatum overcame this difficulty by using strains which were deficient for more than one growth factor; if the probability of reverse mutation is 1 in 10^7, the probability of two reverse mutations occurring in the same bacterium is 1 in 10^{14}, which is sufficiently small to be ignored. By successive mutagenic treatments with X-rays they produced and isolated two strains: (1) a strain which required as growth factors the vitamin biotin and the amino acids phenylalanine and cystine, which we may represent as $B^-Ph^-C^-$; and (2) a strain which required the vitamin thiamin (B_1) and the amino acids threonine and leucine, which we may represent as $B_1^-T^-L^-$. The wild type is then to be represented as $B^+Ph^+C^+B_1^+T^+L^+$. Strains (1) and (2), when plated separately on minimal medium at a density of 10^8 bacteria per plate, gave no colonies. The two strains were then plated together, 10^8 bacteria from each, on minimal medium and about 100 colonies were produced. As a further check, single bacteria were picked from these colonies and shown to be able to grow on minimal medium.

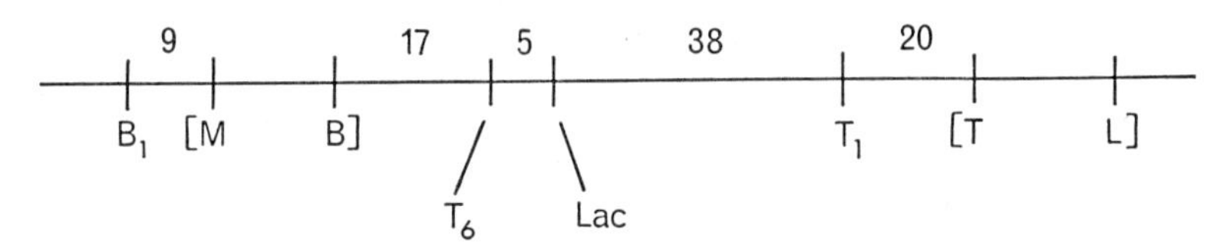

Fig. 25.3. The first genetic map of *Escherichia coli*, published by Lederberg in 1947. The symbols have the following meanings: requirement for threonine (T), leucine (L), methionine (M), biotin (B), thiamin (B_1), resistance to bacteriophages T_1 and T_6, ability to ferment lactose (Lac).

[MB] and [TL] are placed in brackets because they are close together and their actual distances apart were not determined. The figures above the line give the relative frequencies of recombination.

This demonstration of genetic recombination opened the way for genetical analysis of bacteria. The next step was of course to measure the frequency of recombination between different pairs of mutant characters, and a year later Lederberg was able to publish the first, very incomplete, genetic map of *Escherichia coli*. This map, shown in Fig. 25.3, has 8 loci. Altogether 15 mutants were examined in this work, and although 7 of these were not located on the map, it was clear that they belonged to the same linkage group and were not independently segregated. Lederberg therefore concluded that his evidence indicated a single bacterial 'chromosome'. Having observed no signs of dominance or recessiveness in the progeny of crosses, he further concluded that bacteria must be haploid. These conclusions still stand.

Now of course genetic recombination implies some sort of sexual process, and the fact of recombination, once established, reawakened interest in the possibility that bacteria might be capable of conjugation. Bacterial conjugation was demonstrated by Hayes and by Wollman and Jacob. Evidence was first obtained, about 1952, that recombination involved the transfer of genetic material from a donor strain, F^+, to a recipient strain, F^-, and not in the reverse direction. Such transfer was a relatively rare event and progress was slow until a mutant donor strain, *Hfr*, was isolated which showed much higher rates of recombination with the recipient strain F^-, by a factor of the order of 10^4. When *Hfr* and F^- were mixed, bacteria in conjugation could be seen with the electron microscope, a narrow tube apparently connecting the conjugants (Fig. 25.4).

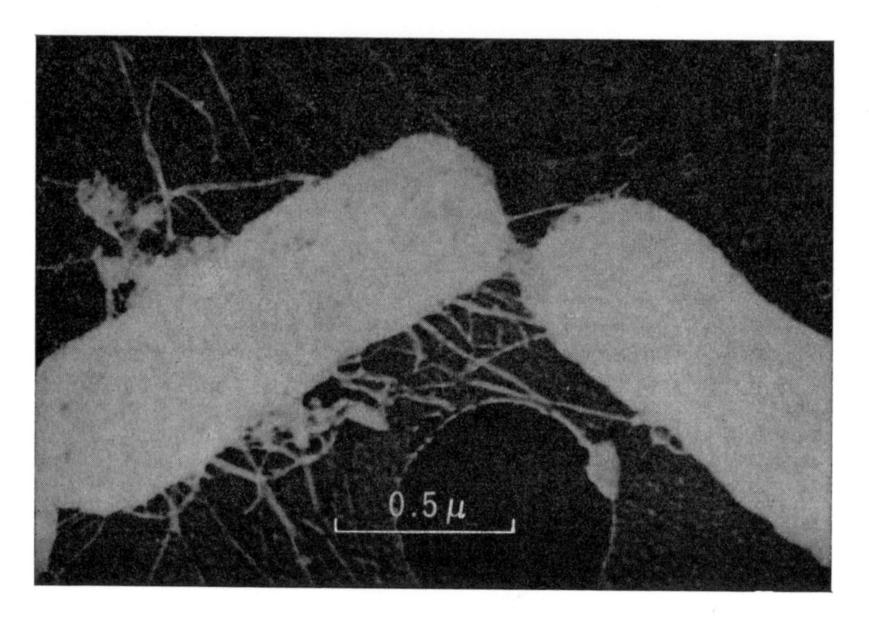

Fig. 25.4. Electronmicrograph (shadowed) of bacteria in conjugation, showing the connecting bridge.

The process of conjugation can be interrupted at any time before completion of the act by running the culture in a Waring blendor, which tears the conjugants apart. This makes it possible to study the time relations of the process of genetic transfer and some very interesting results are obtained.

To facilitate description of these experiments we will avoid naming the actual mutant characters and will merely symbolise them with the letters A, B, C, etc. We will suppose that these characters have already been subjected to genetic analysis and that their relative positions on the genetic map have already been established. In principle, an *Hfr* strain of constitution $A^+B^+C^+D^+E^+F^+S^s$ (where S^s denotes sensitivity to the antibiotic streptomycin) is mixed with an F^- strain of constitution $A^-B^-C^-D^-E^-F^-S^r$ (where S^r denotes resistance to streptomycin) and spread over a series of plates each deficient in one or other of the growth factors A, B, C, etc., and all containing streptomycin. The parental *Hfr* strain cannot reproduce in the presence of streptomycin, and bacteria

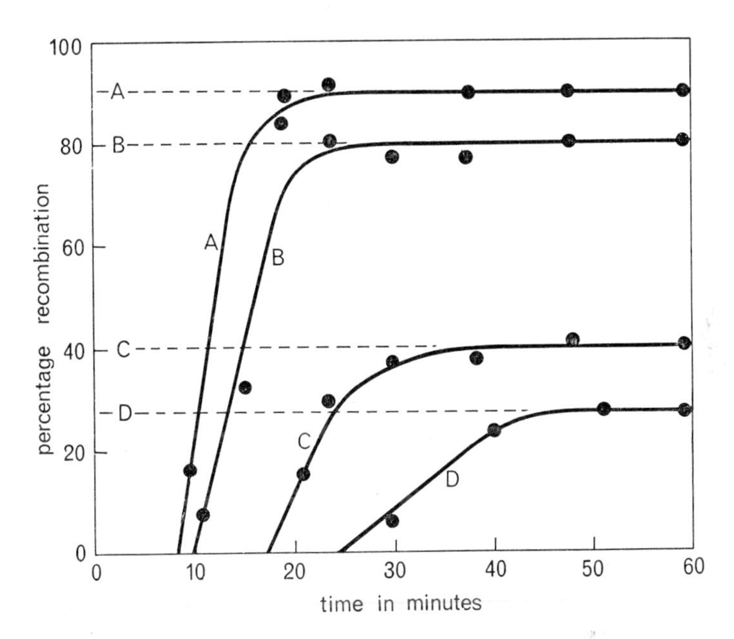

Fig. 25.5. Interrupted conjugation in *Escherichia coli.*

The dotted lines indicate the percentage recombination found in samples which are not run in the blendor but are allowed to stand for 60 minutes to allow conjugation to be completed. The full lines indicate the percentage recombination found in samples which are run in the blendor for the times indicated along the abscissa, reckoned from the time of mixing of the strains.

Further explanation in the text.

of the F^- strain will only be able to grow if they have received the appropriate genes A^+, B^+, C^+, etc., from the *Hfr* strain by conjugation.

When crosses of this kind are made, without any interruption of the process of conjugation, it is found that there is considerable variation between one character and another in frequency of recombination. Thus 90% of the bacteria tested may show recombination for A, 80% for B, 40% for C, 30% for D and so on. When conjugation is interrupted at various times after the strains have been mixed it is found that if the period of conjugation is less than 10 minutes no recombination occurs. As the period of conjugation is increased, recombinants appear in increasing numbers; A is the first to enter into recombination, followed by B, C, D, etc. in regular order. The results of an experiment of this type are shown in Fig. 25.5.

What these results tell us is first, that the genes determining the characters under study enter the bacterium in a definite order in time, which is the same (with certain qualifications) as their order on the genetic map; and second, that genetic transfer is ordinarily incomplete, the later arrivals being less likely to be transferred at all. The results may be interpreted on the basis that a portion of the 'chromosome' of the donor bacterium is transferred to the recipient in conjugation, that one of the two ends always enters first, and that the conjugants are liable to separate spontaneously before transfer is complete.

Table 25.2

Hfr strains	Order in which genes enter recipient						
1	A	B	C	D	E	F	G
2	C	B	A	G	F	E	D
3	E	D	C	B	A	G	F

There is not just one *Hfr* strain of *Escherichia coli*, but several. When these strains are tested by the method just described it is again found that the genes enter the recipient in a definite order, but the order is not the same for all *Hfr* strains. None the less, the order is never without some relationship to the order upon the genetic map. Table 25.2 makes this clear.

A further step in interpretation now becomes possible. The bacterial 'chromosome' can be represented as a circle (Fig. 25.6). The character which we know as *Hfr* consists in the breaking of the circle at some point which is different for the different *Hfr* strains. In this process some kind of polarity is conferred upon the 'chromosome' so that one end and not the other always enters first.

Electronmicrographs recently obtained have shown that deoxyribonucleoprotein extracted from bacteria by the most 'gentle' methods is in the form of threads about 1 mm in length, and that these threads are often in the form of closed loops.

3. BACTERIOPHAGE

Our knowledge of the genetics of viruses relates almost exclusively to bacteriophage; inheritance has been little studied in other types of virus.

Escherichia coli is subject to attack by 7 types of phage, distinguishable by their plaque morphology, serological reactions and a variety of other factors. Within each of these types there are different strains. The strains can be recognised by differences in host range, i.e. a given strain of phage

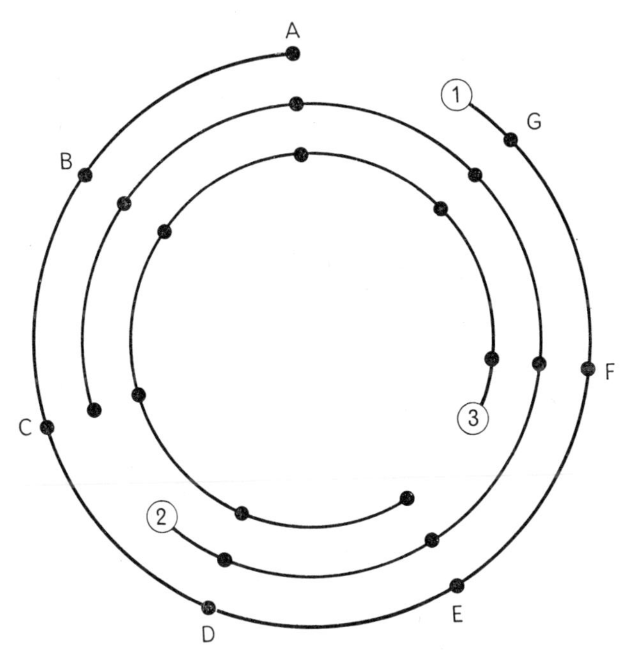

Fig. 25.6. Representation of the 'chromosomes' of bacteria of *Hfr* strains (1), (2) and (3) of Table 25.2. The numbered end of the 'chromosome' is the last to enter the recipient.

will attack some strains of bacteria but not others. They can also be recognised by differences in plaque morphology; some produce large plaques, others produce minute plaques, others produce fuzzy plaques and so on (Fig. 25.7).

The occurrence of spontaneous changes in phage strains has been recognised for over 30 years, but it was not until 1946 that any systematic study of phage genetics was undertaken. When sensitive bacteria

 19-2

are exposed to a mixture of two different types of phage, such as types T_1 and T_2, there is often mutual exclusion—only one type of phage multiplies in any one bacterium. But when bacteria are exposed to two strains of the same type both strains can infect one and the same bacterium. When this happens we can observe recombination of the char-

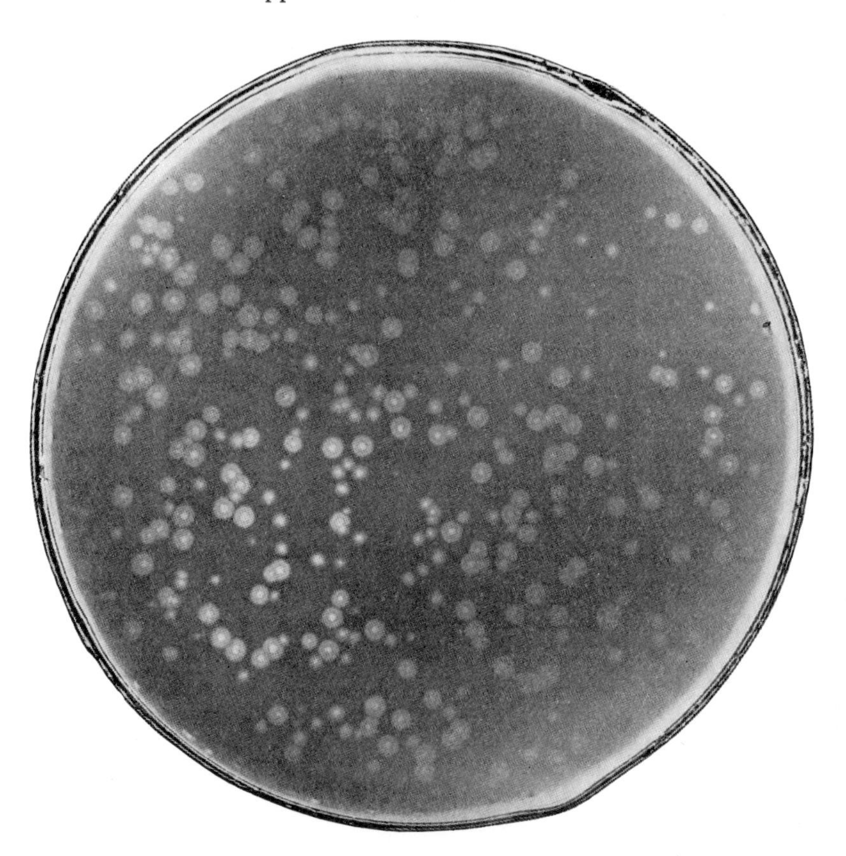

Fig. 25.7. Variations in plaque morphology of bacteriophage strains.
Two strains of bacteriophage are present on this plate: (1) the wild type, which produces small fuzzy plaques, and (2) a mutant, which produces larger and more clearly defined plaques.

acters by which the strains differ. Recombination, as we have seen, implies some form of sexual reproduction. Since only the DNA content of the phage—the phage 'chromosome'—enters the host one can only suppose that sexual reproduction in phage must involve the association of 'chromosomes' from the different strains, akin to the process of synapsis (Chapter 6, p. 86) in higher organisms, with subsequent crossing-over. But this is hypothetical; not even the electron microscope has revealed anything about the mating habits of phage.

First in 1955 and later in 1957 Benzer published the results of a genetical analysis of certain mutant types of phage. As already mentioned (p. 279) the possibility of high resolution depends upon 'screening' techniques whereby mutants and recombinants can be recognised and selected, and in order to understand the tactics of the breeding experiments one must understand how they are related to the screening techniques.

The wild type of phage T_2, which can be obtained from mammalian faeces, produces relatively small plaques when plated with *E. coli*. Certain naturally occurring mutants can be found which produce larger plaques, as a result of more rapid lysis of the host bacteria. These mutants are known collectively as the '*r*' mutants. The *r* mutants may be referred to one or other of three categories, *rI*, *rII* and *rIII*, by differences in plaque morphology when they are plated with three different strains of *E. coli*: strain S, strain B and strain K. These relationships are set out in Table 25.3. From this table it can be seen that mutants of class *rII* are particularly advantageous for genetical study for the reason that they do not produce plaques on strain K. Strain K can therefore be used as a 'screen' to show up wild-type phage which arises by recombination of *rII* mutants, since such recombination will give the double mutant and the wild type and the latter alone will produce plaques on strain K. As far as the screening technique *per se* is concerned the possibility of resolution is virtually infinite. The practical limit to resolution is set by the rate of reverse mutation from mutant to wild type.

Table 25.3. *Plaque morphology of strains of phage* T_2

Phage strain	Bacterial host strains		
	B	S	K
Wild type	Wild	Wild	Wild
Mutant, *rI*	*r*	*r*	*r*
Mutant, *rII*	*r*	Wild	No plaques†
Mutant, *rIII*	*r*	Wild	Wild

† The bacteria are infected and killed, but do not undergo lysis.

The procedure for selecting and crossing *rII* mutants is as follows. Wild-type phage is plated on strain B. Many such plates are made and are examined for large '*r*' plaques. When such a plaque is detected it is picked and allowed to multiply on strain S. This stock is then tested by plating on strain K. If it produces plaques it is assumed to belong to class *rI* or class *rIII* and is discarded. If it produces no plaques, or only very few, it is assumed to be an *rII* mutant and the stock is set aside for further investigation.

In order to make a cross between two *rII* mutants a broth culture of B is infected with a mixture of the mutants, each at the density of about 3 phage particles per bacterium. Any unattached phage particles are removed by centrifugation. After the bacteria have been lysed the free phage particles released will include the original mutants and, if there has been any recombination, the double mutant and the wild type. The phage suspension from B is then plated on B and on K. All four of these categories will be capable of forming plaques on B, but only one of the recombinants, the wild type, will be capable of forming plaques on K. The percentage of recombination will then be given by

$$\%\ \text{recombination} = \frac{2 \times \text{number of plaques on K}}{\text{number of plaques on B}} \times 100.$$

Benzer first chose 8 different *rII* mutants and carried out most of the possible crosses. From the percentages of recombination he was able to construct a genetic map, which is set out in Fig. 25.8. Examination of the percentages of recombination will show that additivity is not strictly preserved in that a long distance tends to be less than the sum of the component shorter distances.† But the serial order of the loci is unambiguously established.

A further classification of these loci can be established by tests in which a mixture of two *rII* mutants is plated directly upon K (i.e. without having been previously crossed on B, as in the procedure outlined above). When this is done one or other of two quite distinct results is obtained. Either (i) there are no plaques—or at most very few, or (ii) there is extensive lysis of the bacteria all over the plate. This is interpreted as distinguishing complementary from non-complementary pairs of mutants. That is to say, if by analogy with the mutants of other micro-organisms we assume a mutant to be deficient in some way, then in a mixed infection it is possible that one mutant can supply the factor which the other

† The technique of the three-point cross (p. 228) cannot be used here because there are only two phenotypic differences—plaques on B and plaques on K—which we can observe.

lacks, so that at least one and possibly both can grow in a situation where neither could grow alone; such mutants are said to be complementary. The interesting feature of the situation as revealed in phage is that the linear order of the loci can be divided into two segments, *A* and *B*, such that every mutant in segment *A* is complementary towards every mutant

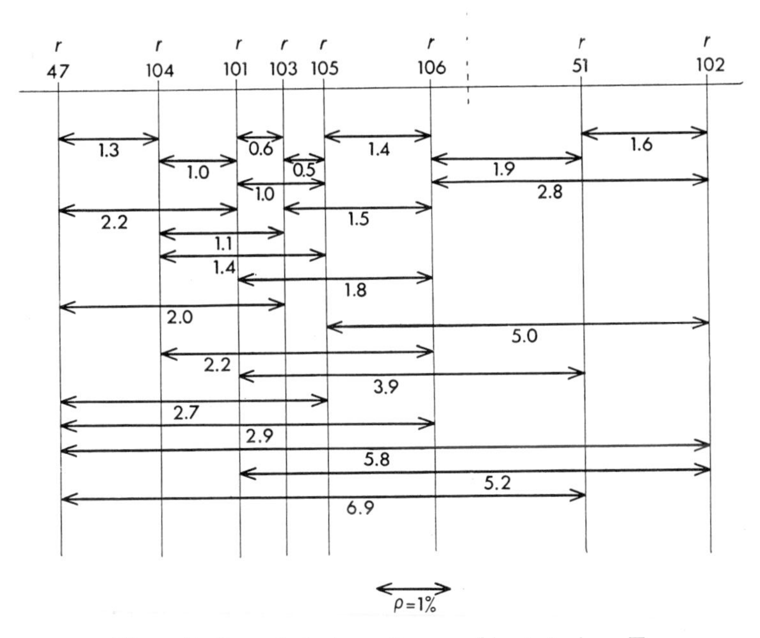

Fig. 25.8. Benzer's first genetic map of bacteriophage T_2.
The identifying numbers of the mutants are set out along the top. Below are given the percentages of recombination for the various crosses as indicated. The broken vertical line shows the demarcation between segment *A* (left) and segment *B* (right).

in segment *B* and is non-complementary towards every other mutant in segment A—and conversely.

The nature of the complementation phenomenon will be more clearly understood if reference is made to Fig. 25.9. Each of the rectangles is supposed to represent a bacterium of strain K into which two phage 'chromosomes' have been injected. A short vertical line upon the 'chromosome' represents a mutant locus, and the vertical dotted line represents the line of demarcation between segments *A* and *B*. By 'active' is meant that the combination will produce extensive lysis of strain K and by 'inactive' is meant the opposite. When a mixed infection is made with an *rII* mutant and a wild type, as in 1, 2 and 3, the absence of mutation from both *A* and *B* segments of the wild-type 'chromosome' ensures the production of whatever is necessary to cause lysis. In 5, 6 and 7 we have mixed infection with two different *rII* mutants. Where both mutant loci lie in the same segment, that segment is defective in both 'chromosomes' and whatever it is that the segment

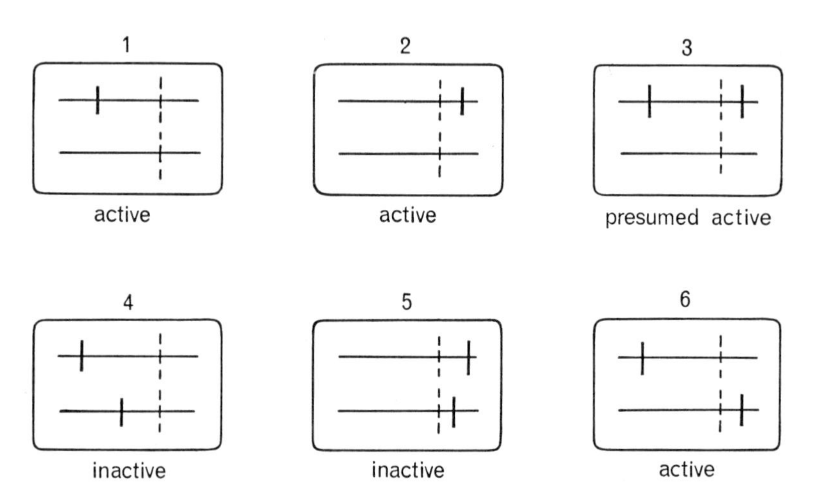

Fig. 25.9. Each diagram represents a diploid heterozygote as simulated by mixed infection of a bacterium of strain K with two types of phage containing the mutations indicated by short vertical lines. Active means extensive lysis of the mixedly infected bacteria; inactive means very little lysis. The vertical dotted line represents the dividing point in the *rII* region, separating segment *A* from segment *B*. Further explanation in the text.

should provide, this fails to be provided and there is no lysis. But when the two mutations lie in different segments, as in 6, there is a normal *A* segment on one 'chromosome' and a normal *B* segment on the other, the one complements the other and lysis of the bacterium follows. A segment of a phage 'chromosome' which can be defined by this test of complementarity is called by Benzer a 'cistron'.†

Having established the applicability of general genetical principles to bacteriophage Benzer then brought the full power of the method to bear upon the study of one single group of mutant loci occupying a short region of the genetic map. In this experiment 931 *r* mutants were isolated and of these 145 belonged to the chosen microcluster within the

† This word 'cistron' is coined from the *cis-trans* test for 'pseudo-alleles' in diploid organisms (p. 306).

rII region. The 145 mutants on further investigation could be referred to 11 separate 'species'.† Within the microcluster the smallest recombination frequency observed was 0·02 %, and this was observed for only one pair of 'species'.

Now since the resolution of the method is such that we could expect to detect recombination as low as 0·0001 % it seems that a lower limit of recombination frequency has been reached. It is therefore of some interest to inquire whether this limit has any special significance in relation to the role of DNA as the repository of genetic information. The calculations which follow are only very approximate because of considerable areas of uncertainty, some of which will be touched upon.

It is not particularly difficult to arrive at a figure for the amount of DNA contained in a single phage particle; reckoned as nucleotides, the total DNA in a phage particle corresponds to 4×10^5 nucleotides. Benzer had reasons (which cannot be gone into here) for believing that only about 40 % of this DNA was genetically important. On this basis he reckoned that the genetic information was carried in $1{\cdot}6 \times 10^5$ nucleotides, or 8×10^4 nucleotide pairs.

The total length of the genetic map is not accurately known. As explained on p. 232, we cannot ascertain the total length by crossing mutants whose loci are far apart, since the possibility of multiple crossing-over is so great that the recombination frequency tends towards a maximum value of 50 %, which is what we find in the case of unlinked, independently segregating characters. We can only estimate the total length by the summing of short distances, and when this is done for all the known mutants of phage T_2 we arrive at a map length of 800 units (1 map unit corresponds to 1 % recombination) and this is the best value available.

Putting these figures together, we observe 800 % recombination over a length of 80,000 nucleotide pairs, from which by simple proportion we obtain 0·01 % recombination per nucleotide pair. 'That is to say, if two mutants, having mutations one nucleotide pair apart, are crossed, the proportion of recombinants in the progeny should be 0·01 %.' It is very difficult to imagine any change in DNA which could be of smaller scale than the distance between adjacent pairs of nucleotides. Bearing in mind, then, the very approximate nature of the calculation just made,

† The word 'species' is here used not with its conventional taxonomic connotation but to denote a group of mutants which when crossed among themselves do not yield any wild-type recombinants.

and that it may well be in error by a factor of 10, it does nevertheless seem possible that in these investigations of recombination in phage we have in fact come up against the limits set by the chemical nature of the material basis of inheritance.

How these discoveries affect our conception of the nature of the gene will be taken up again in the concluding chapter of this book. But before we leave the subject of phage genetics we must take note of a recent investigation by Crick and others in which a very significant advance in the coding problem has been made.

When we were discussing this problem we noted in passing (p. 264) that a possible solution to the 'comma' difficulty is that the message is read off from a fixed point. If this is true then the addition or deletion of a base pair at any point on the message would wholly alter the reading beyond that point. We know that deletion can occur and we may entertain as a working hypothesis the possibility that addition may also occur.

Suppose, then, that a mutation occurs which results in the addition of a base pair into one of the *rII* cistrons not far from the end at which the message originates. This mutation, which we will designate as +, will alter the rest of the message beyond the mutant locus; the function of the cistron will be abolished and the mutants will produce *r*-type plaques. Now suppose that close to the locus of the first mutation there occurs a second mutation which involves the deletion of a base pair; we will designate this second mutation as −. Apart from the short stretch between the two mutant loci the message will now read as it did originally and it is possible that the function of the cistron will be restored (Fig. 25.10). If we observe that wild-type plaques are now produced we might be tempted to interpret the observations as mutation followed by reverse mutation; but since the second mutation occurs at a different locus it is not a true reverse mutation but rather a suppressor of the first mutation; and in many cases it is noticeable that reversion by suppression produces strains whose plaque morphology is not precisely identical with that of the wild type, as might be expected if a small part of the message makes nonsense.

Now consider the possibility of the suppressor (−) mutation occurring by itself without the original (+) mutation. Again, the message will be altered beyond the mutant locus and the function of the cistron will be lost, with the appearance of the *r*-type plaques. The function can then be restored if another mutation, this time +, occurs near the suppressor locus. This new mutation is then the suppressor of a suppressor. Extending this line of thought we can contemplate a suppressor of a suppressor of a suppressor, which must then be −, and so on. We have of course no means of knowing whether the original mutation (assumed to involve a single base pair) is an addition or a deletion, but if we arbitrarily designate it as + then the suppressor must be −, the suppressor of the suppressor must be +, and so on.

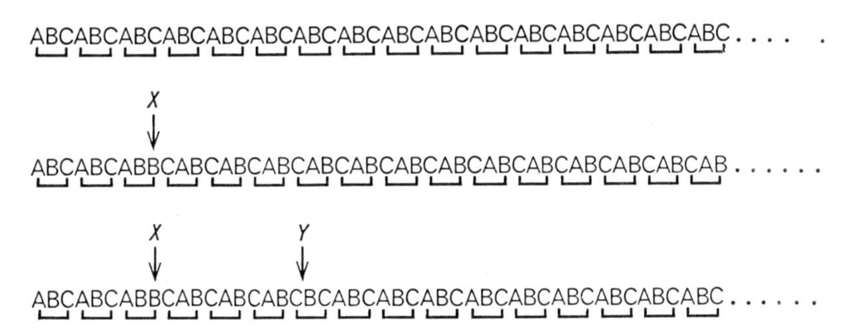

Fig. 25.10. To illustrate a possible explanation of suppressor mutation. For simplicity we suppose that the message is ABCABCABC...... and is read off in triplets from the left (line 1). A mutation occurs at *X* involving the addition of a base pair (line 2). As a consequence the message to the right of *X* reads differently. A further (suppressor) mutation then occurs at *Y*, some little distance to the right of *X*, involving the deletion of a base pair (line 3). As a consequence the message to the right of *Y* is restored to its original form.

By suitable crossing and recombination we can obtain any two mutants in the same piece of genetic material, and if the hypothesis is correct any combination of + with − should give a double mutant producing wild-type (or pseudo-wild-type) plaques; any combination of + with + or − with − should give *r*-type plaques. This was tested in 4 cases of + with + and in 10 cases of − with −, and in all 14 cases the double mutant produced *r*-type plaques, as theory predicted.

We have already seen (p. 263) that the minimal coding ratio is likely to be 3 base pairs for one amino acid. On this basis we should expect that if we produce triple mutants, either + + + or − − −, the effect should be to restore the sense of the message on the far side of the last mutant locus and thereby to restore the function of the cistron. This was tested in 6 cases of triple + mutants and in all cases the plaques were wild-type or pseudo-wild-type.

This simplified account has deliberately avoided mention of certain complications, but these do not affect the principle of the method. The experimental results provide evidence on two points relating to the coding hypothesis. First, they suggest that the coding ratio is 3 (or some multiple of 3). Secondly, they are compatible with the idea that the message is read off from a fixed point. This is of special interest when considered in relation to the experiments described on p. 273 which indicate that amino acids are assembled into protein likewise starting from a fixed point.

26

THE EVOLUTION OF THE CONCEPT OF THE GENE

Neither Mendel himself, nor most of the geneticists who followed up his lead during the early years of the twentieth century, were greatly preoccupied with the problem of how the heritable characters were represented within the germ cells. That they were determined by agencies associated with the chromosomes became the generally accepted view from 1902 onwards and in the absence of stronger evidence this was about as far as a working hypothesis could be legitimately projected. Nevertheless we cannot but believe that these investigators, although they were not able to come to grips with the problem, were well aware of its existence. In fact it became necessary to have some word to denote the genetic determinants of hereditary characters, and the word 'gene', introduced by Johanssen in 1911, soon came into general use.

The earliest conception of the mode of gene action and the nature of mutation may be described in retrospect as the 'presence-or-absence' theory. Since nearly all mutants were found to be recessive to the wild type it seemed reasonable to suppose that the dominant manifestation in the phenotype was due to the presence of the gene and the recessive to its absence. This view was shaken by the discovery of a few mutations which were dominant to the wild type and was finally discredited by the discovery of multiple alleles. The idea of multiple alleles is forced upon us by the observation that more than two differentiating characters can be associated with a single locus. The mutant *white E* in *Drosophila* is one of a series of mutants having a graded series of eye colours lying between white eye and the red eye of the wild type. These intermediate mutants of which there are 11—they have such names as *cherry*, *apricot*, etc.—are quantitatively different as regards depth of pigmentation but are qualitatively the same in affecting eye colour. When first discovered and investigated at low resolution they did not show any recombination with *white* or among themselves. This was interpreted to mean that they all occupied the same locus. We must therefore admit that mutation can involve something less drastic than the complete loss of a gene and we must recognise that a gene can exist in several different forms—alleles—each with a recognisable effect on the phenotype.

Another misconception which can be discerned in retrospect may be called the 'mosaic' theory. It was at one time supposed that each gene had special and exclusive charge of a certain feature of the phenotype—eye colour, for example—and that the final integrated form of the phenotype was the result of harmonious co-operation between a multitude of semi-independent agencies. This simple view soon collapsed in the face of gathering evidence. There are some 30 genes in *Drosophila* all affecting eye colour and all at different loci. Furthermore, a gene may affect more than one part of the phenotype. Generally there is a primary or most conspicuous effect, from which the gene is named, but careful examination usually reveals changes in other parts of the body as well. The mutation *stubble*, for example, is so named because the primary effect of the mutation is to shorten the bristles; but it also shortens the legs, changes the shape of the wings and alters the number of branches on the antenna.

Whatever mutation we choose to study we generally find that its manifestations are in some degree affected by the presence of other mutations. Indeed some genes have been identified whose only recognisable effect is that they alter the effects of mutation in other genes, from barely perceptible modification to complete suppression.

Another observation which we must take into account is that the phenotypic expression of mutation in a gene can be affected by the position which that gene occupies in the chromosome complement. To understand how this 'position effect' was discovered it is necessary to know that gross chromosomal aberrations can occur, sometimes spontaneously, sometimes as a result of exposure to X-rays. Very often these aberrations consist in the breaking-off of a length of one chromosome and its attachment to another. The genes pertaining to the translocated length are then to be found in a different linkage group, as can be demonstrated by breeding experiments. The effects of mutation in these genes are still recognisable for what they were in the original position, yet at the same time unmistakable minor differences are discernible.

All these facts as they accumulated necessitated a reappraisal of ideas upon the nature of genes and their mode of action. The concept of the gene as an autonomous agent was replaced by the concept of the 'gene complex', according to which the action of any one gene was held to be conditioned by the environment of the other genes in which it found itself.

Into this situation of increasing difficulty and complexity the one-gene-one-enzyme hypothesis—or in its more recent form the one-gene-one-polypeptide hypothesis—came as a gust of fresh air. Although this theory is still somewhat precariously poised upon evidence which relates only to micro-organisms it recommends itself not only as being in line with modern biochemical thought but also as restoring to genetics some of its original simplicity which is smothered in the idea of the gene complex. According to this view what we see in the adult mutant phenotype of *Drosophila* is something which results from a basic biochemical disturbance. This disturbance is immensely distant in origin from the effects to which it gives rise, the final product being reached through an inconceivably complex sequence of growth processes in which the ultimate effect of some particular biochemical abnormality is almost unpredictable. Genetics thus neatly divests itself of these unpromising aspects of the subject and relegates them to morphogenesis.

How big is a gene? This is a question which has exercised the minds of geneticists for decades. It poses a problem which can be approached in various ways.

One of the most promising lines would appear to be through the giant salivary chromosomes of *Drosophila*. These have been described in Chapter 6, but we have yet to discuss the information which they provide for cytogenetics. First, it was possible, by careful comparative studies of chromosome abnormalities in relation to phenotypic characters, to confirm that the 4 linkage groups had been correctly assigned to the 4 pairs of chromosomes. Secondly, by extension of the same method, it was possible to establish the positions of the genes on the chromosomes. In this way *cytological* maps were worked out and could be compared directly with the *linkage* maps which had been arrived at from breeding experiments. It was a matter of great satisfaction to find that the two methods were in complete agreement as regards the order of the genes. The distances between the genes, as determined by the two methods, were fairly widely discrepant (Fig. 26.1), but this occasioned no concern; for in constructing the linkage maps it had been assumed, in the absence of any evidence to the contrary, that crossing-over occurred with equal facility at all points on the chromosomes. This assumption must now be discarded and the positions of the genes as given on the cytological map must be accepted as correct.

Another problem which seemed capable of solution was the question of whether the bands on the giant chromosomes could be identified with genes. This was taken up by Muller and Prokofyeva in 1935. They made a very careful examination of a short region at the end of the *X*-chromosome and were able to show that a chromosomal abnormality involving the deletion of the second band from the end also involved the deletion of 4 genes. There being more than one gene present within the length occupied by a single band, the attractive idea that the bands were co-extensive with the genes had to be given up. As a by-product of this

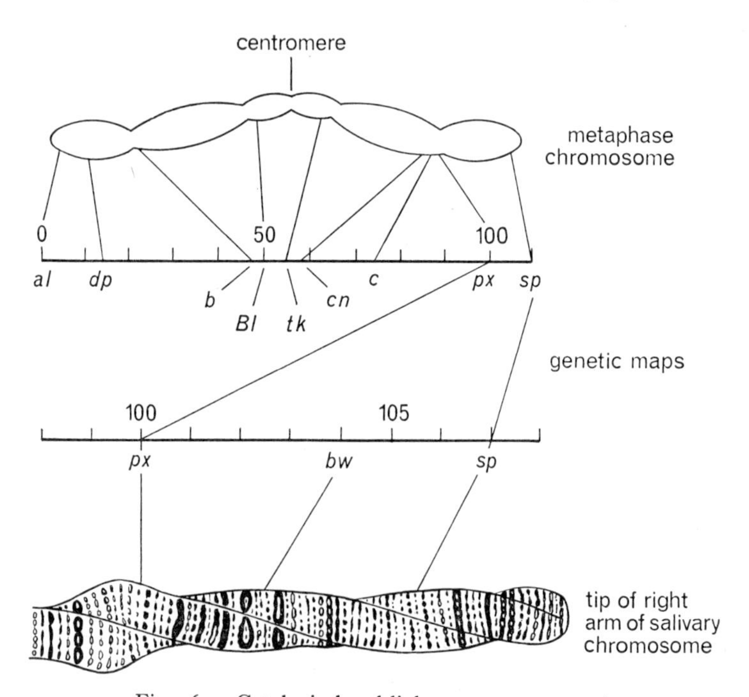

Fig. 26.1. Cytological and linkage maps compared.

investigation, however, Muller and Prokofyeva were able to make an estimate of the size of a gene. The band in question, measured with a UV microscope, was 0·5 μ (5000 Å) in length. It contains at least 4 genes. Therefore the average length of chromosome per gene cannot exceed 1250 Å in this case.

Another method of estimating the size of the gene makes use of the effect of X-rays or other ionising radiation in causing mutation. The data relating X-ray dose to number of mutations produced are interpreted in terms of 'target theory'. According to this theory it is assumed

(i) that there is a sensitive volume within which a single ionisation will cause a mutation, (ii) that the sensitive volume is spherical, and (iii) that all mutations so produced are detected. The calculations, which are too complex to be gone into here, lead to a value of 40–90 Å for the diameter of the sensitive volume.

For comparison with these estimates which are based upon experiment we can make a calculation which takes the one-gene-one-polypeptide theory as its basis. An average figure for the molecular weight of a single polypeptide chain from a protein molecule may be taken as 50,000. Such a polypeptide would be likely to comprise some 300 amino acids. According to the coding hypothesis we would need 3×300, say 1000, nucleotide pairs of DNA to specify the order of these amino acids. The distance between adjacent nucleotide pairs being 3·4 Å, our gene is then about 3500 Å long. Alternatively, if we suppose that this length of DNA is rolled up into a ball, the diameter of the ball would be about 110 Å.

Estimates of the size of the gene thus range from 1250 to 3500 Å in terms of length and from 40 to 110 Å as the diameter of a sphere. In view of the bold assumptions and drastic approximations upon which the estimates are based it is perhaps surprising that they agree to within an order of magnitude. But before we attempt to draw any conclusions from these estimates we would do well to re-examine the whole structure of genetic theory in order that we may get a clearer view of what it is we are trying to measure.

The idea of a unit runs through the whole of classical genetics. Mendel spoke of 'elements', to which the name 'gene' was later given. Although it does not seem to have been formally so stated, the gene was always conceived of as being particulate, and for some 50 years geneticists vaguely pictured the gene as some sort of corpuscle and the chromosome as an array of such corpuscles, like beads on a string. Yet in the theory of the gene as stated by Morgan nothing is said about the size or the nature of the gene, or about the changes which occur in mutation or about the way in which the gene expresses itself in the phenotype. The only attribute of Morgan's gene is its position. In effect Morgan defined the gene as the unit of recombination: if recombination occurs between two characters two genes are involved; if there is no recombination the characters are determined by alleles of the same gene. Nothing in the evidence or in its interpretation requires us to think of the gene as a particle. On the other hand, the qualitative and quantitative differences between the effects of different genes make it impossible to conceive of the gene as a point, having position but no magnitude. The gene may be defined in other ways. It may be defined as the unit of mutation—the smallest part which when changed is reproduced in the changed form. It may also be defined as the unit of physiological activity, as, for example, in the one-gene-one-enzyme hypothesis, for which the test is complementarity.

For the vast majority of genetical situations no confusion arises between these three definitions of the gene; the limits set by one definition do not transcend the limits set by the others. But when we have to deal with very closely linked genes difficulties may arise, even in higher organisms. There is a mutant form of *Drosophila*, known as *lozenge E*, in which the pigmentation of the eye is affected, and various alleles of the gene are known. M. M. and K. C. Green, studying this mutation in 1949, discovered that there was a very small, but quite definite, percentage of recombination between the alleles. Let *a* and *b* be two alleles of the *lozenge* gene. Now if, following Morgan, the gene is to be defined as the unit of recombination, then according to classical genetics *a* and *b* are not alleles of the same gene but, instead, there are two *lozenge* genes, *lozenge a* and *lozenge b*, close together. Recombination between *a* and *b* —a rare event—provides us with two different genotypes:

$$\frac{ab}{++}$$

the *cis*-form, and

$$\frac{a+}{+b}$$

the *trans*-form. According to classical genetics, since *a* and *b* are recessive, both these heterozygotes should be wild-type in appearance. In fact, the *cis*-form is wild and the *trans*-form is mutant. Thus by the test of recombination there are two genes, whereas by the test of complementarity there is one gene. Alleles which can be separated by crossing-over but which are non-complementary in the *trans*-form are called 'pseudo-alleles'.

Thus the classical theory of the particulate gene, according to which all definitions conform to the same limits on the genetic map, runs into difficulties even in the genetics of higher organisms. In the genetics of phage it fails extensively. Benzer has found it expedient to introduce a completely new set of terms,† operationally defined strictly in accordance with what is actually observable in practice.

'The *recon*: the smallest element of the genetic map which is interchangeable but not divisible, as shown by the test of recombination.

The *muton*: the smallest element of the genetic map which when altered gives rise to a mutant form.

The *cistron*: a map segment corresponding to a function which is unitary as shown by the test of complementarity.'

The new facts which have emerged, particularly from phage genetics, cannot fail to strengthen the case of those who have argued for a revision of the classical concept of the gene. Goldschmidt drew attention to its weaknesses as long ago as 1938. He maintained that the idea of the gene as a corpuscle had no foundation and was simply an extension of the idea of a mutant locus. A great many mutations, recognisable by their effects upon the phenotype, have been traced to rearrangement within the chromosomes, i.e. to position effect. Goldschmidt at one time suggested that all mutations could be regarded as position effects, brought about by rearrangement within the chromosomes on a greater or lesser scale.

Having thus seen the tidy concept of the corpuscular gene effectively undermined we may now appropriately turn to other directions in which it has become necessary to give further extension to the gene concept.

It has long been known that certain strains of bacteria are 'lysogenic'. By this is meant that when bacteria of a lysogenic strain are mixed with certain other bacteria, the latter undergo lysis. When this happens large numbers of free phage particles appear in the culture. The lysogenic bacteria are carriers of the phage, in much the same way as some people are carriers of typhoid bacteria without showing any symptoms of the disease. 'Virulent' phage always destroys the bacteria which it infects. 'Temperate' phage may destroy the bacteria, by multiplying at their expense and causing their lysis, in a type of life cycle which we may therefore designate as 'lytic'; or it may be carried for an indefinite time, without destruction of the host, in a type of life cycle which we call 'lysogenic' (Fig. 26.2). We know that it is capable of reproducing itself within the lysogenic bacteria, because their ability to infect others is not diminished over many generations of growth; but it does not multiply to the extent of causing their death and lysis. Exceptionally, however,

† Of these terms, only the cistron seems likely to be generally adopted.

just this happens; and the occasional carrier which succumbs and liberates a crop of phage particles is the means of infecting a susceptible strain.

Temperate phage can be passed from one bacterium to another in the process of conjugation. This is not in itself surprising. What is surprising is that the phage is transmitted in association with the bacterial 'chromosome'. By the method of interrupted conjugation (p. 288) it can be shown that the phage has a definite place in the order in which the genes are

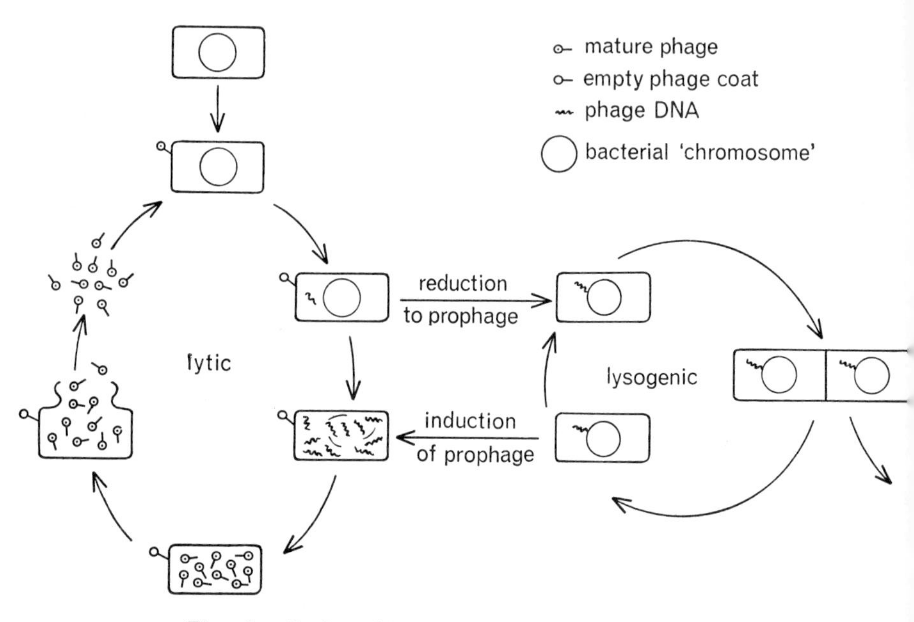

Fig. 26.2. Lytic and lysogenic life cycles of bacteriophage.

transmitted. This is confirmed by breeding experiments in which it can be shown that the phage can undergo orderly recombination with other genes. We cannot escape the conclusion that the phage is attached to the bacterial 'chromosome' at a definite position. In this attached state we speak of it as 'prophage'. All the indications are that prophage replicates itself at each division of the host bacterium just as though it were a bacterial gene. By various treatments, notably by exposure to UV light, the prophage can be 'induced', whereupon it multiplies rapidly and the host bacterium is destroyed with the liberation of free phage particles. We cannot but suppose that induction involves the detachment of the prophage which is then free to reproduce at a rapid rate and eventually kills its host by lysis. The relation between the lytic and the lysogenic types of life cycle is shown in Fig. 26.2.

Thus it appears that the DNA injected into a bacterium by a phage can attach itself to the bacterial 'chromosome'. When so attached it undergoes orderly replication and recombination with other genes. In much the same way one or more bacterial genes can become attached to the phage 'chromosome'. This is a feature of the phenomenon of transduction', first described by Zinder and Lederberg in 1952.

The aim of their research was to demonstrate, if possible, the occurrence of genetic recombination in the bacterium *Salmonella typhimurium*. Experiments planned and conducted along the same lines as those of Lederberg and Tatum (p. 285) gave anomalous results which seemed more likely to be due to transformation by free DNA (p. 235) than to the orderly recombination of a sexual process. To test this possibility they constructed a U-tube having the two limbs separated by a bacterial filter sealed in across the bottom of the U. Suspensions of two different strains of bacteria were placed in the U-tube, one in each arm. By this arrangement it was possible to keep the two strains out of direct contact, but to allow the medium, with any DNA in solution, to be flushed through from one side to the other. It seemed that transformation still took place; and this would probably have been accepted as the true explanation had it not been that it continued in the presence of DNase which would have destroyed any free DNA. Examination of the medium revealed the presence of phage particles, and when their origin had been ascertained another explanation was seen to be possible. The strain of bacteria which acted as recipient in this 'cross' was found to be harbouring a prophage. Occasional induction of this prophage liberated a small crop of phage particles and these were small enough to pass through the filter. The other strain, the donor, was susceptible to this phage which multiplied rapidly at its expense. The explanation favoured by Zinder and Lederberg was that some of the phage particles incorporated a part of the host DNA and, passing back through the filter, injected it into the bacteria of the recipient strain. Further study of this and other cases of transduction has shown that the bacterial genes which are transduced by phage are those which occupy positions close to the prophage on the bacterial 'chromosome'; and there is also evidence, which it would take too long to describe, that part of the phage 'chromosome' can sometimes be exchanged for part of the bacterial 'chromosome', giving a phage which is active in transduction but defective in other ways.

A phage is therefore capable of existing in one or other of three forms: as free phage, as vegetative phage and as prophage. As free phage it is metabolically inactive and enclosed in a coat, like a bacterial spore. As vegetative phage it is an autonomous unit growing rapidly within the host, replicating and directing the synthesis of the special proteins which form the coat and ancillary apparatus of the free phage. As prophage it is attached to the bacterial 'chromosome' at a definite position and is functionally integrated with it; it replicates in step with the replication of the bacterial 'chromosome'; it can recombine with bacterial genes by crossing-over; it can affect the character of the bacterium in so far as its presence protects the bacterium from lytic infection by other phages of the same type. In fact, as prophage it has all the essential attributes of the classical gene. Yet in response to mild treatment it can leave its position on the bacterial 'chromosome'; and having done so it can reproduce independently of the host and can go on to enjoy an independent free existence. If we are to admit that prophage is a gene, then clearly our ideas about what genes can do must come under drastic revision.

There are other self-replicating bodies in bacteria which also have the faculty of being able to pass between the autonomous and the integrated states. The sex factor, *F*, is one of these. This factor is present in the autonomous state in bacteria of donor mating strains which have been designated F^+ (p. 288). It reproduces independently, though not at an excessive rate, and it can be transmitted independently of the 'chromosome' during conjugation, whereupon it confers the F^+ character upon the previously F^- recipient. In the integrated state it attaches itself to the 'chromosome' at a definite position, and this attachment appears to result in the opening of the 'chromosome' loop at that position. This attachment and subsequent opening of the loop is in fact the essential feature of the *Hfr* mutation; and the sex factor is to be found at that end of the 'chromosome' which is the last to enter the recipient during conjugation.

Prophage, sex factor and other semi-autonomous self-replicating genetic elements in bacteria have been called 'episomes' by Jacob and Wollman. They are to be regarded as non-essential features since many bacteria seem to be without them. They can be acquired from other bacteria by direct infection or by conjugation or, conceivably, by transduction or transformation. Once acquired they may exist either in the autonomous state or in the integrated state in which they are attached to the bacterial 'chromosome' at a definite position. It seems that 'attached' is the right word for this; such evidence as there is does not indicate that episomes are incorporated into the bacterial 'chromosome' by intercalation. These relationships, as proposed by Jacob and Wollman, are summarised in Fig. 26.3.

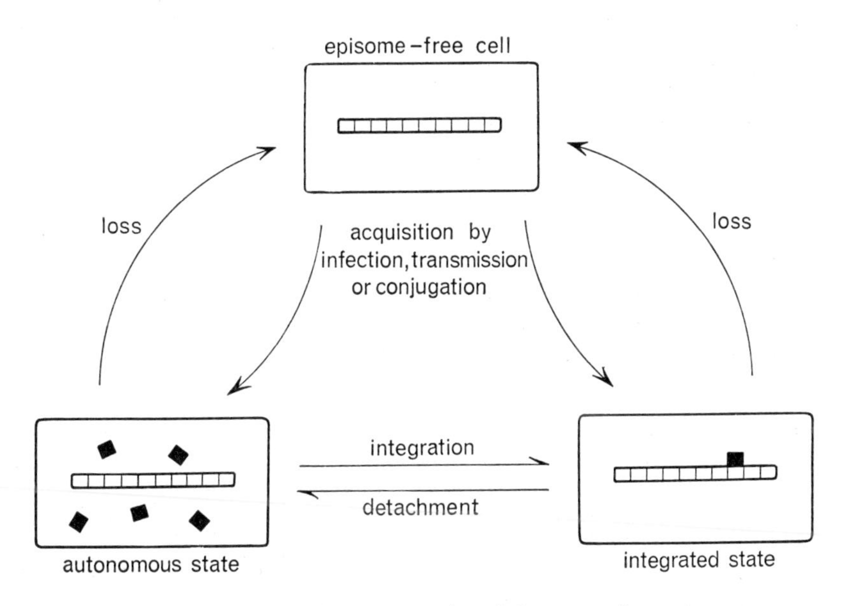

Fig. 26.3. Diagrammatic representation of the states of an episome.

What, then, are we to understand by the word 'gene'?

If the present situation seems more confused than it was 30 years ago this is not because the classical experiments have been found to be unsound or their interpretations incorrect. The experiments of Mendel and of Morgan remain the twin pillars of genetics as they always have been. It was natural and proper that the results of the early experiments in genetics should be assimilated to a theory in which the gene was at once the unit of recombination, the unit of mutation and the unit of genetic function. This was in accordance with the principle of William of Occam, which forbids us to adopt a complicated explanation where a simpler explanation is adequate. The idea of a unit, coupled with the demonstration that the units were arranged in linear order, could hardly fail to grow into the 'string of beads' which allows us to form an agreeably simple mental picture of the chromosome. This simple picture will no longer do. We have to take away the beads and leave only a string with certain properties which cannot be defined except operationally.

Although the classical view of the gene is no longer tenable, this does not mean that the word has lost its usefulness. In the great majority of genetical situations and in the applied aspects of the subject—'everyday genetics'—we can continue to think of genes as we have done in the past. Only when we carry genetic analysis to high degrees of resolution or try to relate the results of breeding experiments to cytological or biochemical phenomena do we find the classical gene unsatisfactory.

The nearest thing to a classical gene that we are left with is the cistron. This is a segment of the genetic map, operationally defined by the test of complementarity; it cannot be identified as yet with anything which can be seen in the microscope or prepared in a test-tube. On the basis of the coding hypothesis and the one-gene-one-polypeptide hypothesis we may picture the cistron as a length of DNA comprising anything from one hundred to several thousand nucleotide pairs. What we have yet to decide is whether or not there is discontinuity between adjacent cistrons. There must be some indication of where the message begins and ends, corresponding to the two ends of the polypeptide chain. These might be indicated by the intercalation of 'nonsense' sequences of nucleotides, in which case there would be discontinuity between cistrons and we would be on our way back to the string of beads. On the other hand, it may turn out that cistrons overlap one another and that some other device is used to indicate the beginning and end of the message.

The inadequacy of the classical gene was never at any time a serious bar to progress in genetics and for this reason the clearing up of earlier misconceptions is unlikely of itself to herald any spectacular new advances. More important are two other ways in which our outlook has been widened. First, we need no longer suppose that all the genetic material is at all times fully integrated into chromosomes. The recognition of the chromosome-episome relationship seems likely to clarify some of the mystery surrounding cytoplasmic inheritance and the so-called 'plasma-genes' of higher organisms. Secondly, we need no longer think of sexual reproduction as being the only way in which genetic information can be recombined. The phenomena of transformation and transduction, so far known only from micro-organisms, may now be looked for in higher organisms as well; or, if they cannot be found in nature we may attempt to bring them about by artificial means with reasonable hope of success. This seems likely to open up a new field of investigation, the genetics of somatic cells, which many people believe to be the direction in which we are most likely to discover the causes of cancer.

APPENDIX

1. AMINO ACIDS

The following amino acids are found in protein. Standard abbreviations in brackets.

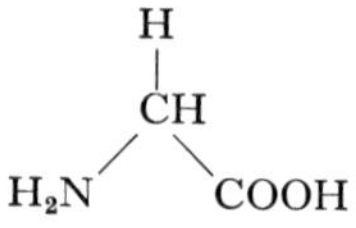

H
CH
H_2N COOH
glycine (gly)

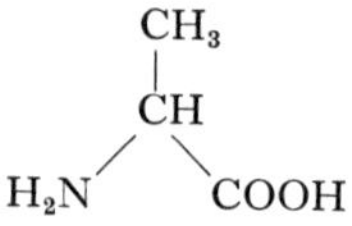

CH_3
CH
H_2N COOH
alanine (ala)

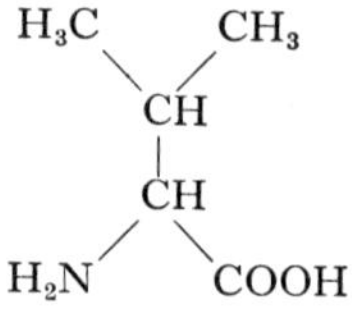

H_3C CH_3
CH
CH
H_2N COOH
valine (val)

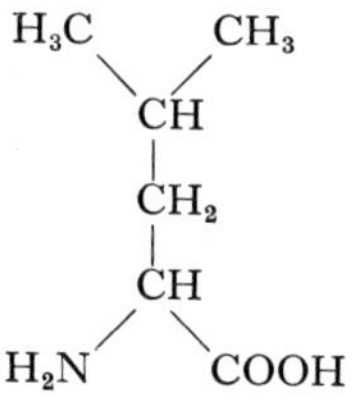

H_3C CH_3
CH
CH_2
CH
H_2N COOH
leucine (leu)

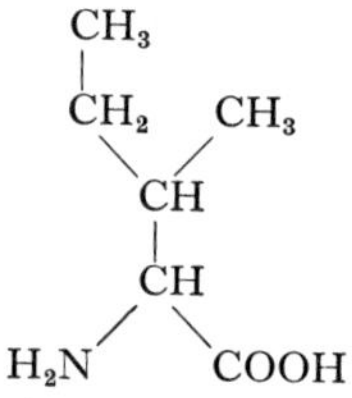

CH_3
CH_2 CH_3
CH
CH
H_2N COOH
iso-leucine (ileu)

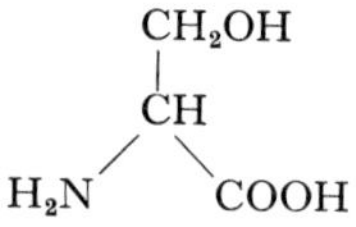

CH_2OH
CH
H_2N COOH
serine (ser)

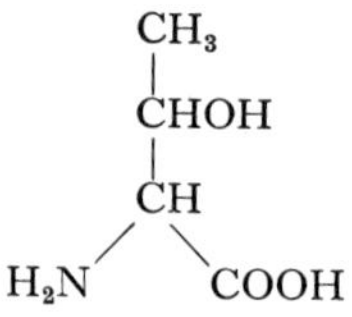

CH_3
CHOH
CH
H_2N COOH
threonine (thr)

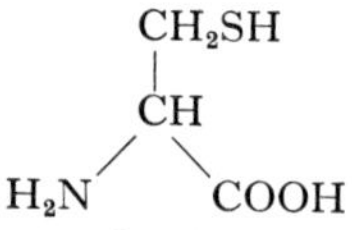

CH_2SH
CH
H_2N COOH
cysteine (cysSH)

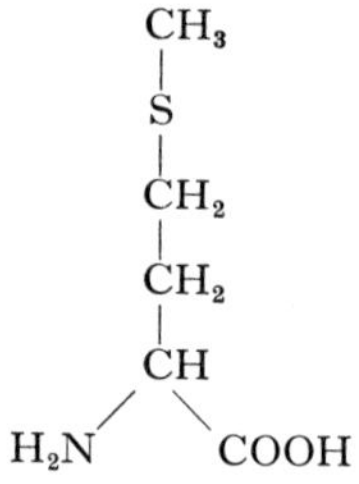

CH_3
S
CH_2
CH_2
CH
H_2N COOH
methionine (met)

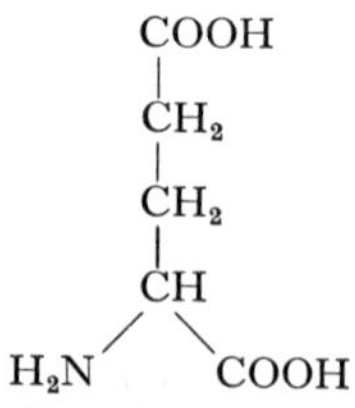

COOH
CH_2
CH_2
CH
H_2N COOH
glutamic acid (glu)

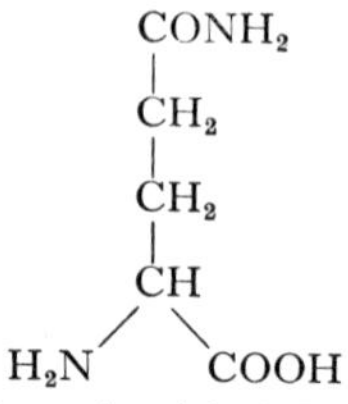

$CONH_2$
CH_2
CH_2
CH
H_2N COOH
glutamine (glu-NH_2)

APPENDIX

COOH
CH_2
CH
H_2N COOH
aspartic acid (asp)

$CONH_2$
CH_2
CH
H_2N COOH
asparagine (asp-NH_2)

CH_2NH_2
CH_2
CH_2
CH_2
CH
H_2N COOH
lysine (lys)

HN NH_2
C
NH
CH_2
CH_2
CH_2
CH
H_2N COOH
arginine (arg)

H
H C H
C C
C C
H C H
CH_2
CH
H_2N COOH
phenylalanine (phe)

OH
H C H
C C
C C
H C H
CH_2
CH
H_2N COOH
tyrosine (tyr)

H NH_2
H C CH_2—CH
C C——C COOH
C C C
H C N H
H H
tryptophane (try)

H CH_2—CH NH_2
C═══C COOH
N N
C H
H
histidine (his)

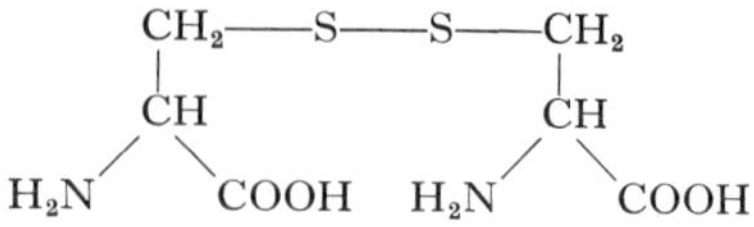

cystine (cysS)

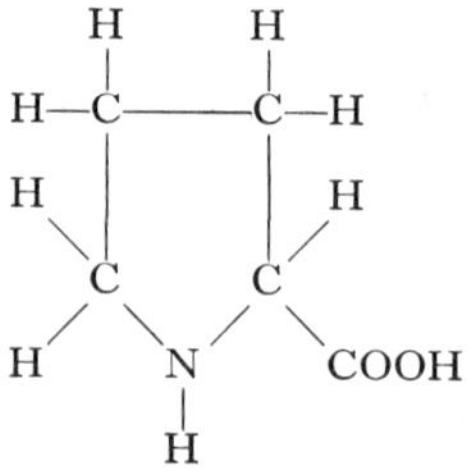

proline (pro)

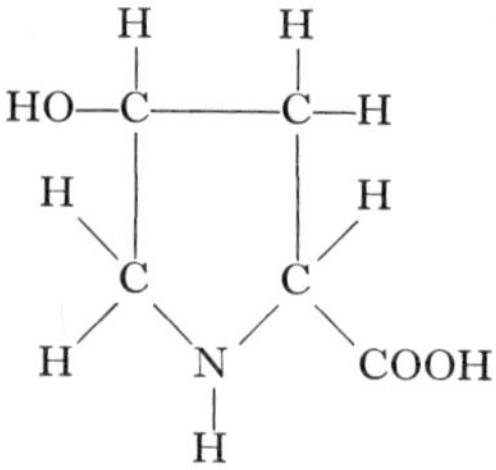

hydroxyproline (hypro)

2. SUGARS

(i) Monosaccharides

Glucose and most other monosaccharides were originally thought to be open-chain compounds. From consideration of their optical properties Emil Fischer and others showed that they are cyclic compounds. In glucose the atoms are arranged in a ring which incorporates 5 carbon atoms and 1 oxygen atom.

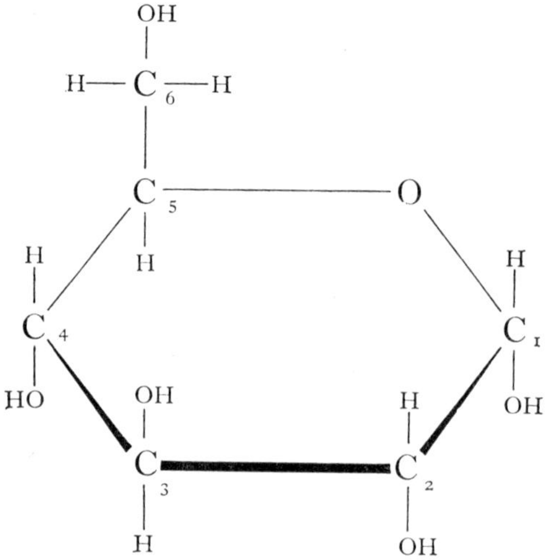

α-D-glucopyranose

This ring is also found in the heterocyclic compound pyrane, and a sugar having this configuration is called a pyranose.

Conventionally, the ring is represented as lying in a horizontal plane, the heavy lines connecting carbon atoms (1), (2), (3), (4) indicating that this side of the ring is nearer to the observer. The hydrogen atoms, the hydroxyl groups and the terminal $-CH_2OH$ group can then be shown as lying above or below the plane of the ring.

H H H OH H O
HO—C—C—C—C—C—C
H OH OH H OH H

CH_2OH
C OH
H H
C CHO
OH H
HO C C
H OH

CH_2OH CH_2OH
C O C O
H H H OH
H C H C
C C
OH H OH H
HO C C OH HO C C H
H OH H OH

α-form (36%) β-form (64%)

There are 32 possible isomers of this structure. In the D-series the $-CH_2OH$ group lies above the ring; in the L-series (not found in nature) it lies below. Permutations of the hydroxyls at carbon atoms (2), (3), (4) give 8 possible D-isomers, of which three (glucose, galactose and mannose) are found in nature. If the hydroxyl on carbon atom (1) is on the same side of the ring as carbon atom (6) the sugar is said to have the β-configuration; if these are on opposite sides it is said to have the α-configuration. The α- and β-forms are in equilibrium with the open-chain form.

In other monosaccharides, e.g. ribose, the ring may contain 4 carbon atoms only, as in the heterocyclic compound furane; the sugar is then called a furanose.

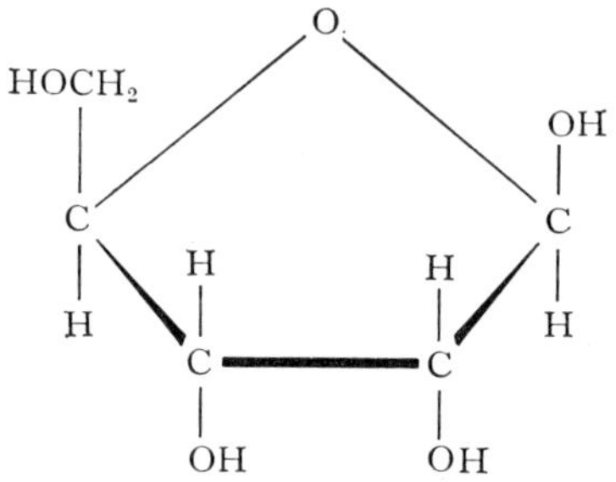

```
                       O
        HOCH2       /     \
          |        /        \         OH
          |       /           \        |
          C                      C
          | \    H        H    / |
          |  \   |        |   /  |
          H   \  |        |  /   H
                C ━━━━━━━ C
                |         |
                OH        OH
```

β-D-ribofuranose

Ribulose, unlike the aforementioned sugars, does not exist in the heterocyclic form, but only in the open-chain form.

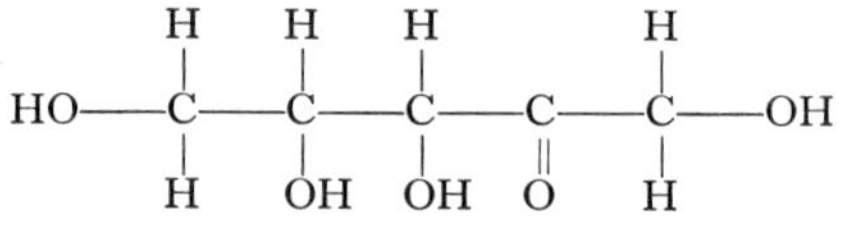

```
        H    H    H         H
        |    |    |         |
  HO----C----C----C----C----C----OH
        |    |    |    ||   |
        H    OH   OH   O    H
```

ribulose

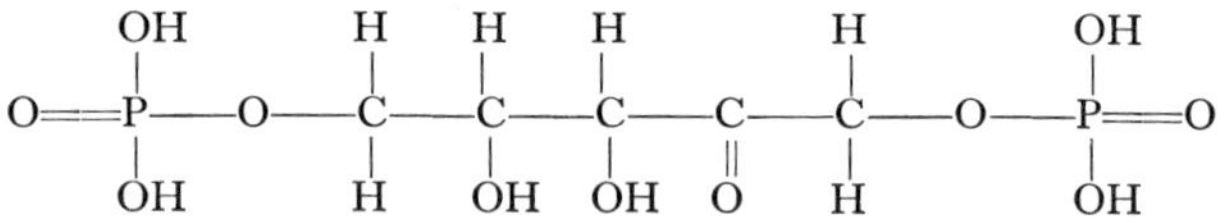

```
     OH         H    H    H         H         OH
     |          |    |    |         |         |
  O==P----O-----C----C----C----C----C----O----P==O
     |          |    |    |    ||   |         |
     OH         H    OH   OH   O    H         OH
```

ribulose 1,5-diphosphate

3. NUCLEOTIDES

As explained on p. 241 nucleic acids are built up of units known as mononucleotides. The mononucleotides obtained from nucleic acids have the following general form:

purine or pyrimidine—pentose—phosphate.

The term 'nucleotide', originally used to describe these degradation products of nucleic acids, has now been extended to include compounds in which the nitrogenous base is not necessarily a purine or a pyrimidine.

The term 'nucleoside' is used to denote the compound of a nitrogenous base with a pentose sugar, lacking the phosphate group.

(i) Purines

The purines found in nucleic acids are adenine and guanine.

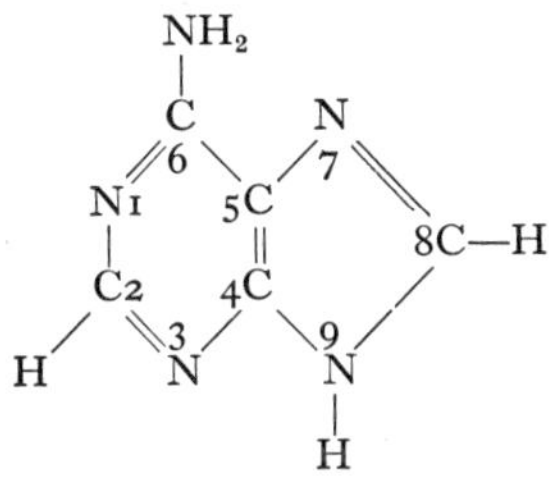

adenine

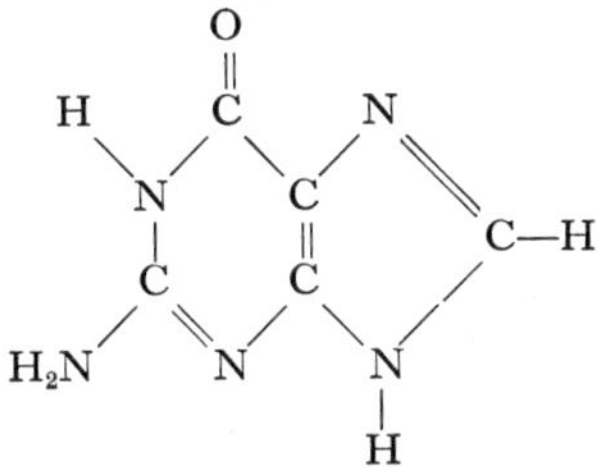

guanine

(ii) Pyrimidines

The pyrimidines found in nucleic acids are thymine, cytosine and uracil.

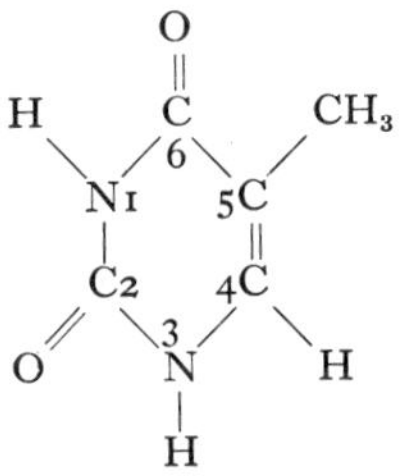

thymine

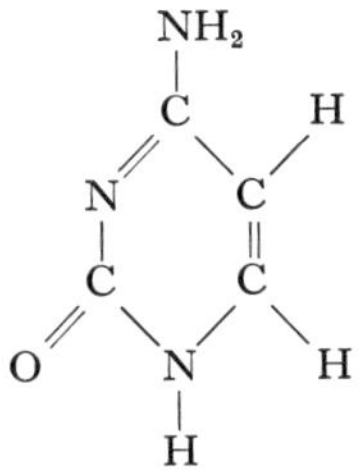

cytosine

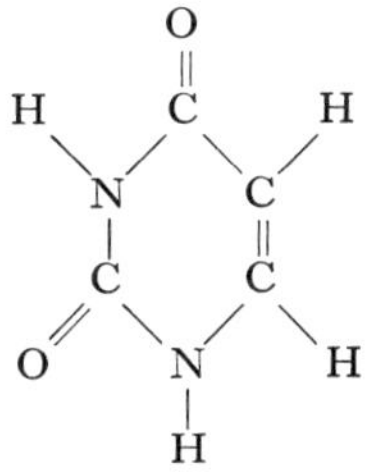

uracil

(iii) Purine nucleosides, e.g. adenosine

These are formed by condensation of a purine with a pentose which may be either β-D-ribose or β-2-deoxy-D-ribose. The nitrogen atom at position 9 in the purine is attached to carbon atom (1) of the pentose.

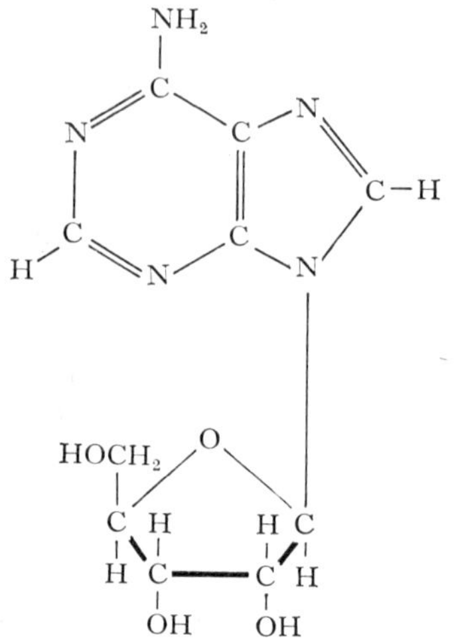

adenosine

The other purine nucleoside is known as guanosine.

(iv) Purine nucleotides, e.g. adenylic acid

In the purine nucleotides obtained from nucleic acids the phosphate group may be found attached either to carbon atom (3) or to carbon atom (5) of the pentose, depending upon the manner in which the nucleic acid is hydrolysed.

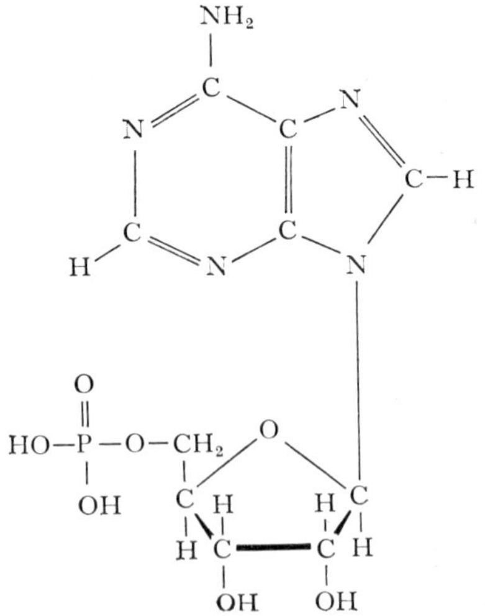

Adenylic acid, also known as adenosine monophosphate, AMP

The other purine nucleotide is known as guanylic acid.

Under the heading of purine nucleotides should be included adenosine triphosphate, ATP.

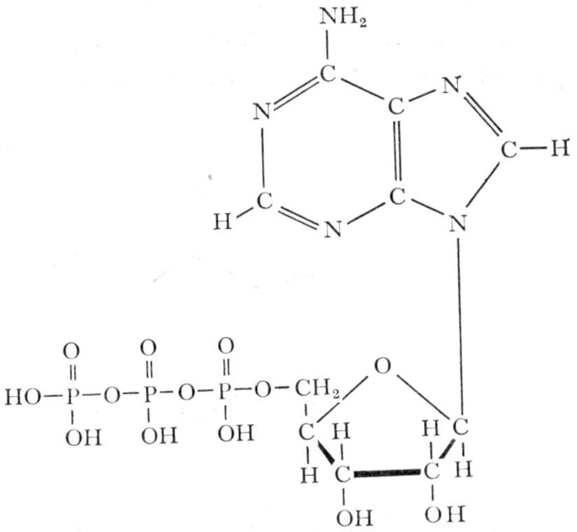

adenosine triphosphate, ATP

(v) **Pyrimidine nucleosides**, e.g. thymidine

These are formed by condensation of a pyrimidine with a pentose, the nitrogen atom at position 3 in the pyrimidine being attached to carbon atom (1) of the pentose. The other pyrimidine nucleosides are known as cytidine and uridine. In the pyrimidine nucleosides obtained from nucleic acids the pentose is ribose in uridine and deoxyribose in thymidine; either pentose may be found in cytidine.

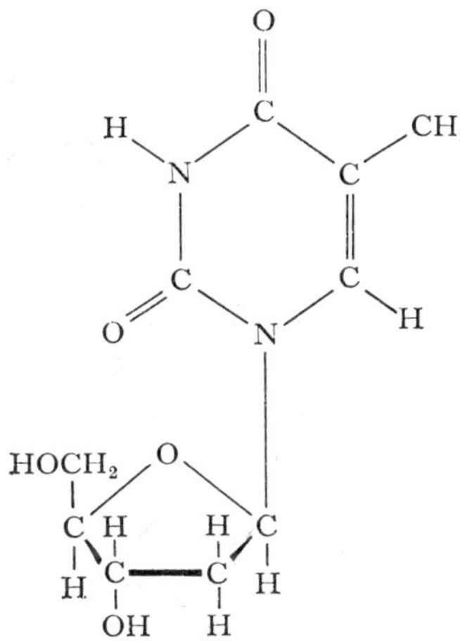

thymidine

(vi) Pyrimidine nucleotides, e.g. thymidylic acid

As in the case of the purine nucleotides, the phosphate group may be attached either to carbon atom (3) or to carbon atom (5) of the pentose.

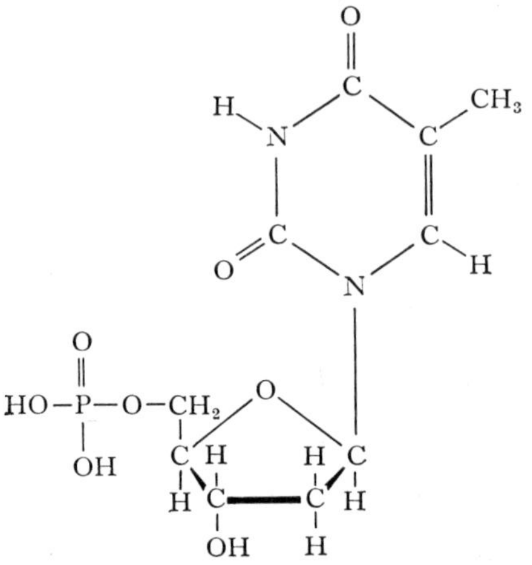

O
C
H N C CH_3
O C C H
N
O
HO—P—O—CH_2 O
OH C H H C
H C C H
OH H

thymidylic acid

The other pyrimidine nucleotides are known as cytidylic acid and uridylic acid.

(vii) Pyridine nucleotides

The nitrogenous base pyridine has the formula:

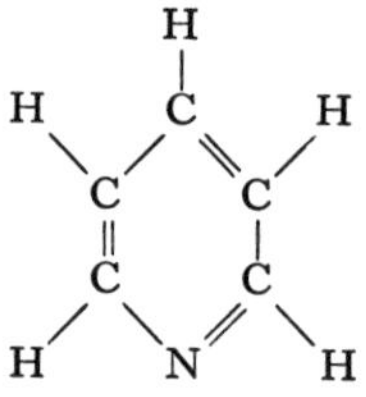

H
H C H
C C
C C
H N H

pyridine

The form of pyridine which occurs in hydrogen carriers is nicotinamide.

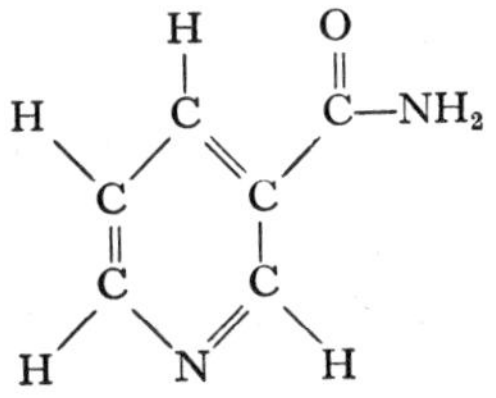

nicotinamide

The mononucleotide of nicotinamide is not found free in nature. It is found combined with adenylic acid as a dinucleotide.

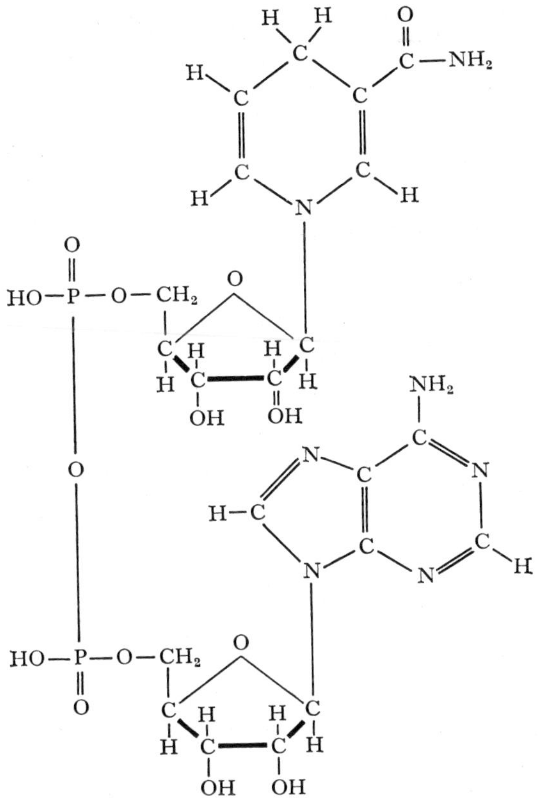

Reduced nicotine-adenine dinucleotide, $NADH_2$. Also known as diphosphopyridine nucleotide, $DPNH_2$, or as co-enzyme I

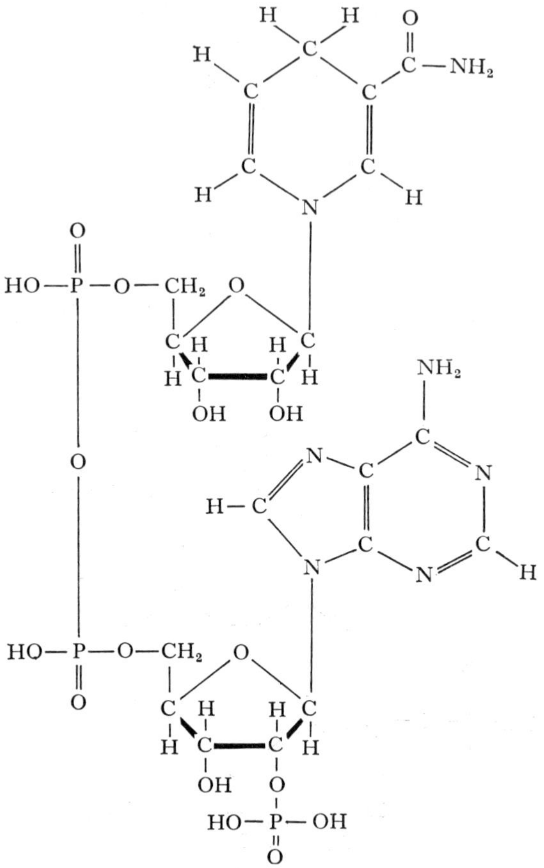

Reduced nicotine-adenine dinucleotide phosphate, $NADPH_2$. Also known as triphosphopyridine nucleotide, $TPNH_2$, or as co-enzyme II

Pyridine nucleotides act as hydrogen carriers in a rather complicated way. Hitherto they have been represented in the reduced form. When oxidised, the nicotinamide group gives up a hydrogen atom and an electron while a hydrogen ion comes from the nearest phosphate group.

Some authorities write the oxidised form as DPN^+ and the reduced form as DPNH.

reduced form

oxidised form

(viii) Flavin nucleotides

The nitrogenous base here is of the alloxazine type.

alloxazine

The form of alloxazine which occurs in hydrogen carriers is 6,7-dimethyl *iso*-alloxazine.

6,7-dimethyl *iso*-alloxazine

It is condensed not with a pentose sugar but with a 5-carbon alcohol, ribitol.

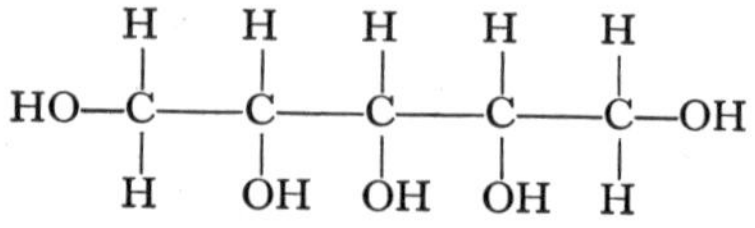

```
         H    H    H    H    H
         |    |    |    |    |
    HO—C————C————C————C————C—OH
         |    |    |    |    |
         H    OH   OH   OH   H
```

ribitol

The nucleoside, known as riboflavin, is a yellow compound largely responsible for the colour of egg-yolk.

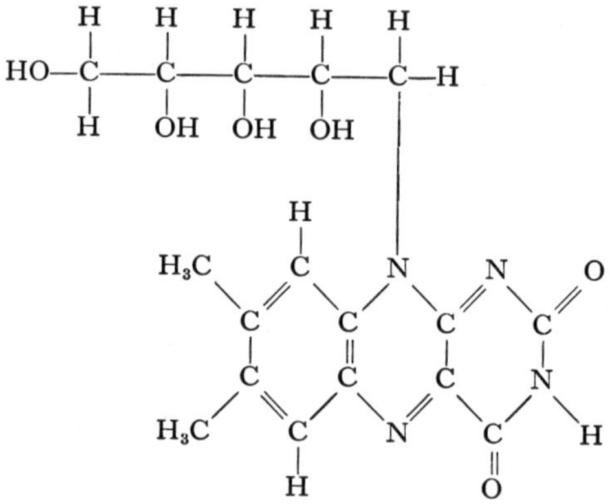

```
        H    H    H    H    H
        |    |    |    |    |
   HO—C————C————C————C————C—H
        |    |    |    |    |
        H    OH   OH   OH   |
                              |
                 H            |
                 |            |
      H3C        C            N          N         O
          \    //  \        /   \      // \       //
            C         C             C        C
            |         ||            |        |
            C         C             C        N
          /   \\    /   \         //  \    /   \
      H3C         C           N          C       H
                  |                      ||
                  H                      O
```

riboflavin

The nucleotide, flavin monophosphate, FMN, is a hydrogen carrier normally found in combination with a protein.

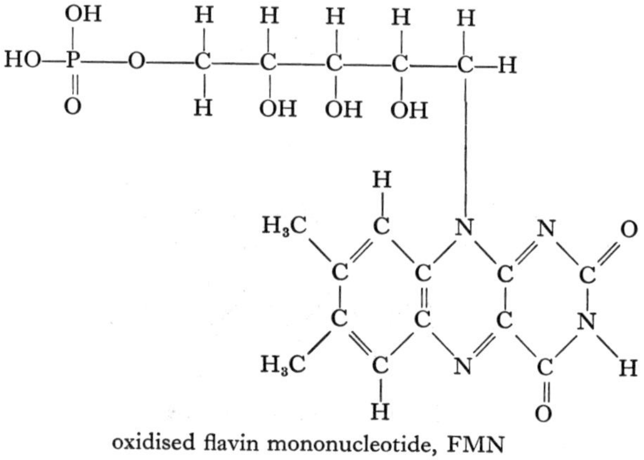

```
       OH         H    H    H    H    H
       |          |    |    |    |    |
   HO—P————O————C————C————C————C————C—H
       ||         |    |    |    |    |
       O          H    OH   OH   OH   |
                                        |
                            H           |
                            |           |
                 H3C        C           N          N         O
                     \    //  \       /   \      // \       //
                       C         C            C        C
                       |         ||           |        |
                       C         C            C        N
                     /   \\    /   \        //  \    /   \
                 H3C         C          N          C       H
                             |                     ||
                             H                     O
```

oxidised flavin mononucleotide, FMN

Flavin monophosphate forms a dinucleotide with adenylic acid, known as flavin-adenine dinucleotide, FAD. This is also a hydrogen carrier, normally found in combination with a protein.

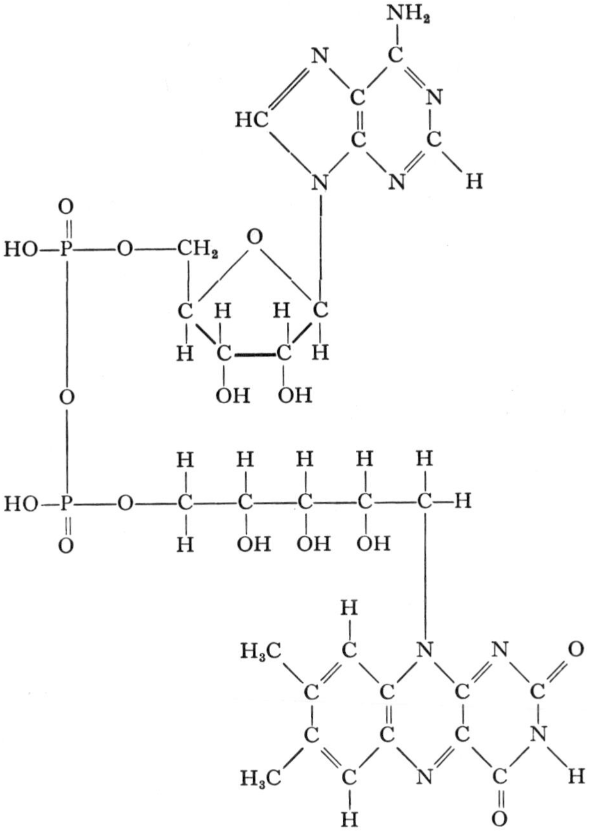

oxidised flavin-adenine dinucleotide, FAD

Flavin nucleotides act as hydrogen carriers by taking up two hydrogen atoms in the alloxazine rings.

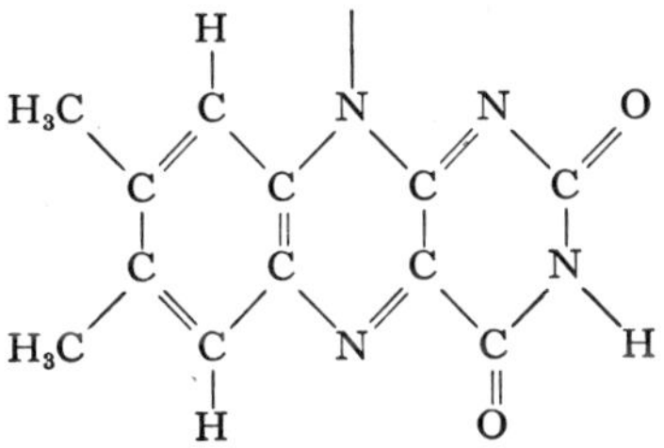

oxidised form

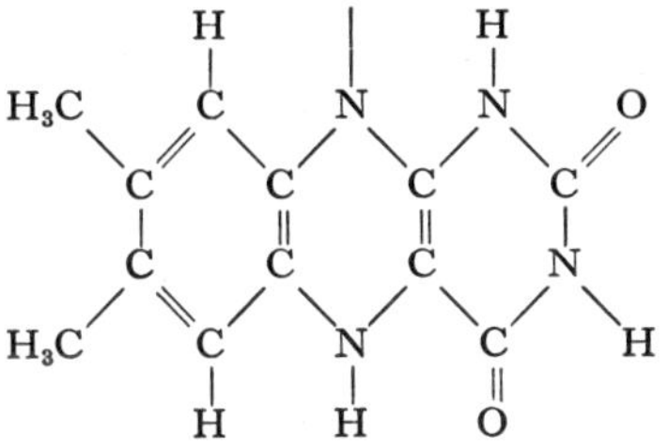

reduced form

(ix) Co-enzyme A

This is a compound of adenosine and the non-cyclic molecule pantetheine, united by two phosphate groups.

```
                    CH3
        O      H     |   H  O  H  H  H  O  H  H  H
        ||     |     |   |  ||  |  |  |  ||  |  |  |
HO — P — O — C — C — C — C — N — C — C — C — N — C — C — SH
        |      |     |   |         |  |         |  |
        |      H     |   OH        H  H         H  H
        |           CH3
        |                                 NH2
        |                                  |
        |                                  C
        O                            N  /    \\
        |                           // \\  C      N
        |                      H — C      ||      |
        |                           \\    C      C — H
        |                             N /   \\ N //
        |                             |
HO — P — O — CH2      O               |
        ||          |   /    \\           |
        O          C  H    H  C
                   |   \\ |    | / |
                   H    C ——— C   H
                        |     |
                        O     OH
                        |
                   HO — P — OH
                        ||
                        O
```

co-enzyme A

The biochemically important reactions take place at the sulphur atom.

$$R\text{—}CH_3\text{—}SH \;+\; CH_3\text{—}CO\text{—}COOH \;+\; DPN \longrightarrow R\text{—}CH_2\text{—}S\text{—}CO\text{—}CH_3 \;+\; CO_2 + DPNH_2$$

co-enzyme A   pyruvic acid   acetyl co-enzyme A

4. PORPHYRINS

The fundamental configuration is porphin, four pyrrole rings united by —CH= units.

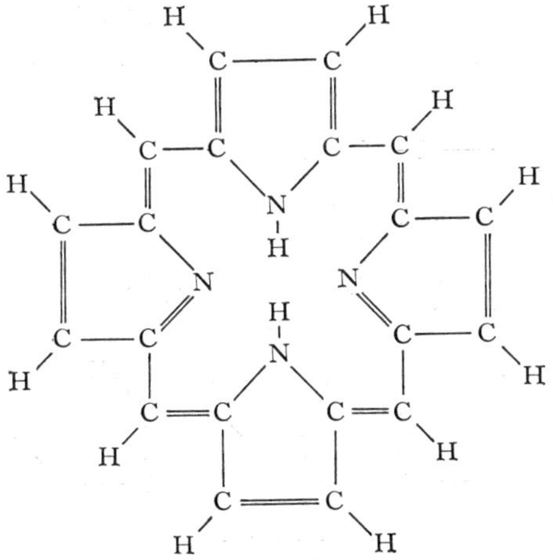

porphin

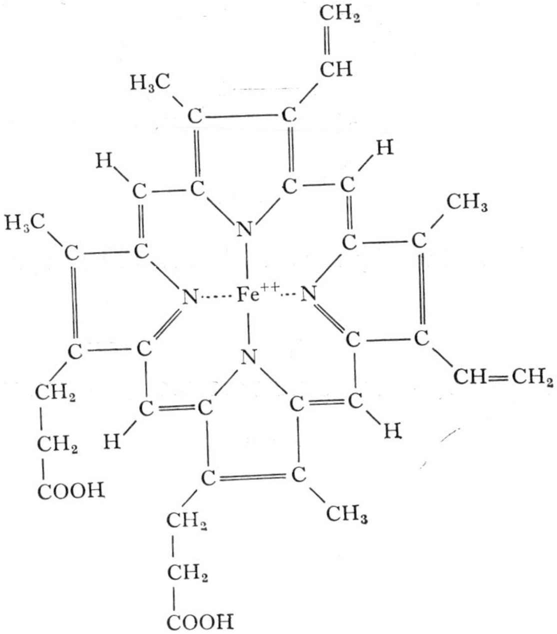

haem

In porphyrins various side-chains are attached to the porphin framework, and the centre of the framework is often occupied by a metal atom which is bonded to the four nitrogen atoms in a very special manner, too complicated to be described here.

Among the iron-porphyrins the most important is haem.

Chlorophyll is also a porphyrin in which the metal is magnesium.

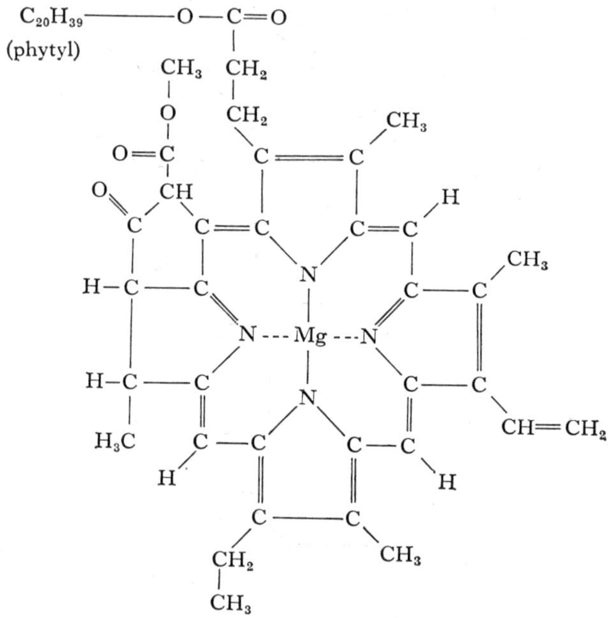

```
C20H39————O—C=O
(phytyl)        |
          CH3  CH2
           |    |
           O   CH2        CH3
           |     \        /
       O=C       C=====C
           |       |     |          H
   O      CH       |     |         /
    \\   /  \      |     |        /
      C      C==C         C==C
      |      |    \     /     |          CH3
      |      |      N         |          /
  H—C————C        |         C————C
      |        \\     |      //        ||
      |          N---Mg---N            ||
      |         /     |      \         ||
  H—C————C        |         C————C
      |      ||       N         ||       \
    H3C      C——C   /   \     C——C      CH=CH2
            /     ||       ||     \
           H      ||       ||      H
                  C————C
                 /         \
              CH2           CH3
               |
              CH3
```

chlorophyll *a*

A long-chain alcohol, phytol, is attached to the porphyrin.

$$\begin{array}{c}
\qquad CH_3 \qquad\qquad\qquad\qquad\qquad\quad CH_3 \qquad\qquad\qquad\qquad\qquad\qquad CH_3 \\
CH_3.\overset{|}{CH}.CH_2.(CH_2.CH_2.\overset{|}{CH}.CH_2)_2.CH_2.CH_2.\overset{|}{C} = CH.CH_2OH
\end{array}$$

phytol

SUGGESTIONS FOR FURTHER READING

I have been taken to task for having failed to provide a list of references with the first printing of this book.

The book is addressed to school leavers who are going on to the University. I find it depressing to contemplate the ever increasing load of factual knowledge which the first-year undergraduate is asked to carry. I do not wish even to seem to suggest that it should be added to. Have we not all seen that flattened look on the face of the student who has brightly asked an innocent question and now holds in his hand the list of references which an over-zealous instructor has advised him to consult? One should hesitate, I think, before encouraging first-year undergraduates to spend time in reading advanced texts or original papers. Only too often these cannot be understood without further background reading, and much time may be taken up in this way without any adequate return in interest or enlightenment.

Nevertheless I do agree that I ought to indicate how the reader who wishes to know more can best proceed to the next stage, bearing in mind that he has not all the time in the world. I have therefore assembled a short list of books, most of which are of less than 250 pages, and I have added some notes about their aims and content.

ANFINSEN, C. B. *The Molecular Basis of Evolution*. John Wiley, 1959. 228 pp. Price 14*s*. (paperback). Although it is now getting on in years this is still a runner. It is mainly about proteins and nucleic acids, together with some genetics and a little evolutionary theory. It was written not for the benefit of any particular class of student but for the author's pleasure—and it is a pleasure to read. The references are to reviews and to original papers.

BALDWIN, E. H. F. *Dynamic Aspects of Biochemistry*. 4th ed. Cambridge University Press, 1963. 519 pp. Price 42*s*. This is a text-book written for the first-year undergraduate embarking upon a course in biochemistry. It is now in its fourth edition, a testimony to its excellence, and I note with approval that the author has apparently resisted suggestions that he load it down with references to the latest papers.

SUGGESTIONS FOR FURTHER READING

FINCHAM, J. R. S. *Microbial and Molecular Genetics.* In the series *Modern Biology*, ed. J. E. Webb, English Universities Press, 1965. 149 pp. Price 15*s.* This assumes no previous knowledge of genetics. It covers much the same ground as Part III of the present book, but in more detail and with the inclusion of more recent work. It gives references to more advanced books and to original papers.

HAGGIS, G. H., MICHIE, D., MUIR, A. R., ROBERTS, K. B. and WALKER, P. M. B. *Introduction to Molecular Biology.* Longmans, 1964. 401 pp. Price 40*s.* This is addressed to second-year or third-year undergraduates. It assumes no knowledge of biology but a knowledge of physics and chemistry equivalent to British 'A' level. It covers the ground of Parts I and III of the present book, but in more detail and at a hotter pace. The references are to advanced books and original papers.

HARRISON, K. *A Guide-book to Biochemistry.* 2nd edition. Cambridge University Press, 1965. 150 pp. Price 12*s.* 6*d.* (paperback). This is an introduction to the subject, eminently suitable for those who are about to enter a university. One of its merits is that it schematises some of the complex metabolic pathways in forms that can readily be taken in at a glance and can be the more easily remembered. There are no references.

LOEWY, A. G. and SIEKEWITZ, P. *Cell Structure and Function.* In the Modern Biology Series by Holt, Rinehart and Winston, 1963. 228 pp. Price 20*s.* This is intended for beginners and assumes little previous knowledge, but is written at a more detailed level than the present book. The emphasis is on biochemistry. The references are to advanced books and reviews.

MAHAN, B. H. *Elementary Chemical Thermodynamics.* In the General Chemistry Monograph Series, ed. Russell Johnson. W. A. Benjamin, 1964. 152 pp. Price 19*s.* (paperback). Probably as simple a introduction to chemical thermodynamics as has ever appeared in book form. The reader is referred to a small number of suitable text books of more advanced character.

PERUTZ, M. F. *Proteins and Nucleic Acids.* Elsevier, 1962. 211 pp. Price 50*s.* Based on the Weizmann Memorial Lectures, 1961, this is written for the sophisticated newcomer to the field of molecular biology rather than for the first-year undergraduate. The emphasis is upon molecular structure. The references are mainly to original papers.

INDEX

Italic figures refer to the Appendix

INDEX

INDEX